T0338184

DURABILITY DESIGN OF CONCRETE STRUCTURES

DURABILITY DESIGN OF CONCRETE STRUCTURES

PHENOMENA, MODELING, AND PRACTICE

Kefei Li
Tsinghua University, China

WILEY

Registered Office
John Wiley & Sons, Singapore Pte. Ltd., 1 Fusionopolis Walk, #07-01 Solaris South Tower, Singapore 138628.

For details of our global editorial offices, for customer services and for information about how to apply for permission to reuse the copyright material in this book please see our website at www.wiley.com.

Library of Congress Cataloging-in-Publication Data

Names: Li, Kefei, 1972– author.
Title: Durability design of concrete structures : phenomena, modeling, and practice/Kefei Li.
Description: Solaris South Tower, Singapore : John Wiley & Sons, Inc., [2016] |
    Includes bibliographical references and index.
Identifiers: LCCN 2016014283 | ISBN 9781118910092 (cloth) | ISBN 9781118910122(epub)
Subjects: LCSH: Concrete construction. | Concrete–Deterioration.
Classification: LCC TA681.5 .L525 2016 | DDC 624.1/834–dc23
LC record available at https://lccn.loc.gov/2016014283

Set in 10/12pt Times by SPi Global, Pondicherry, India
Printed and bound in Singapore by Markono Print Media Pte Ltd

10 9 8 7 6 5 4 3 2 1

Contents

Preface   ix
Acknowledgments   xv

Part I   DETERIORATION OF CONCRETE MATERIALS   1

1   Carbonation and Induced Steel Corrosion   3
   1.1   Phenomena and Observations   3
   1.2   Carbonation of Concrete   7
         1.2.1   Mechanisms   7
         1.2.2   Influential Factors   9
         1.2.3   Models   12
   1.3   Steel Corrosion by Carbonation   18
         1.3.1   Mechanism   18
         1.3.2   Influential Factors   21
         1.3.3   Models   22
   1.4   Basis for Design   25
         1.4.1   Structural Consequence   25
         1.4.2   Design Considerations   27

2   Chloride Ingress and Induced Steel Corrosion   29
   2.1   Phenomena and Observations   29
   2.2   Chloride Ingress   32
         2.2.1   Mechanism   32
         2.2.2   Influential Factors   34
         2.2.3   Models   42
   2.3   Steel Corrosion by Chloride Ingress   46
         2.3.1   Mechanisms   46
         2.3.2   Influential Factors   49
         2.3.3   Models   51

2.4 Basis for Design   53
 2.4.1 *Structural Consequence*   53
 2.4.2 *Design Considerations*   54

3 Freeze–Thaw Damage   **56**
3.1 Phenomena and Observations   56
3.2 Mechanisms and Influential Factors   57
 3.2.1 *Mechanisms*   57
 3.2.2 *Influential Factors*   64
3.3 Modeling for Engineering Use   71
 3.3.1 *Model FT-1: Critical Saturation Model*   71
 3.3.2 *Model FT-2: Crystallization Stress Model*   72
3.4 Basis for Design   76
 3.4.1 *Structural Consequence*   76
 3.4.2 *Design Considerations*   76

4 Leaching   **78**
4.1 Phenomena and Observations   78
4.2 Mechanisms and Influential Factors   80
 4.2.1 *Mechanisms*   80
 4.2.2 *Influential Factors*   83
4.3 Modeling for Engineering Use   85
 4.3.1 *Model L-1: CH Dissolution Model*   85
 4.3.2 *Model L-2: CH + C-S-H Leaching Model*   87
 4.3.3 *Further Analysis of Surface Conditions*   92
4.4 Basis for Design   94
 4.4.1 *Structural Consequence*   94
 4.4.2 *Design Considerations*   94

5 Salt Crystallization   **96**
5.1 Phenomena and Observations   96
5.2 Mechanisms and Influential Factors   99
 5.2.1 *Mechanisms*   99
 5.2.2 *Influential Factors*   102
5.3 Modeling for Engineering Use   106
 5.3.1 *Model CT-1: Critical Supersaturation Model*   106
 5.3.2 *Model CT-2: Crystallization Stress Model*   107
5.4 Basis for Design   110

Part II FROM MATERIALS TO STRUCTURES   **113**

6 Deterioration in Structural Contexts   **115**
6.1 Loading and Cracking   115
 6.1.1 *Mechanical Loading*   116
 6.1.2 *Effect of Cracks: Single Crack*   118
 6.1.3 *Effect of Cracks: Multi-cracks*   128

6.2 Multi-fields Problems 130
    6.2.1 Thermal Field 132
    6.2.2 Moisture Field 135
    6.2.3 Multi-field Problems 141
6.3 Drying–Wetting Actions 144
    6.3.1 Basis for Drying–Wetting Actions 144
    6.3.2 Drying–Wetting Depth 147
    6.3.3 Moisture Transport under Drying–Wetting Actions 151

Part III DURABILITY DESIGN OF CONCRETE STRUCTURES 155

7 Durability Design: Approaches and Methods 157
7.1 Fundamentals 157
    7.1.1 Performance Deterioration 158
    7.1.2 Durability Limit States 158
    7.1.3 Service Life 161
7.2 Approaches and Methods 164
    7.2.1 Objectives 164
    7.2.2 Global Approaches 166
    7.2.3 Model-based Methods 168
7.3 Life Cycle Consideration 170
    7.3.1 Fundamentals for Life-cycle Engineering 171
    7.3.2 Life-cycle Cost Analysis 171
    7.3.3 Maintenance Design 173

8 Durability Design: Properties and Indicators 183
8.1 Basic Properties for Durability 183
    8.1.1 Chemical Properties 184
    8.1.2 Microstructure and Related Properties 185
    8.1.3 Transport Properties 187
    8.1.4 Mechanical Properties 194
    8.1.5 Fundamental Relationships 195
8.2 Characterization of Durability-related Properties 198
    8.2.1 Characterization of Chemical
    and Microstructural Properties 198
    8.2.2 Characterization of Transport and Mechanical Properties 200
    8.2.3 Durability Performance Tests 201
8.3 Durability Indicators for Design 205
    8.3.1 Nature of Durability Indicators 205
    8.3.2 Durability Indicators for Deterioration 206
    8.3.3 Durability Indicators: State of the Art 208

9 Durability Design: Applications 210
9.1 Sea Link Project for 120 Years 210
    9.1.1 Project Introduction 210
    9.1.2 Durability Design: The Philosophy 213

	9.1.3	Model-based Design for Chloride Ingress	214
	9.1.4	Quality Control for Design	219
9.2	High-Integrity Container for 300 Years		220
	9.2.1	High-Integrity Container and Near-Surface Disposal	220
	9.2.2	Design Context	223
	9.2.3	Design Models for Control Processes	226
	9.2.4	Model-based Design for 300 Years	227
9.3	Further Considerations for Long Service Life Design		232

10 Codes for Durability Design   234

10.1	Codes and Standards: State of the Art		234
	10.1.1	Eurocode	234
	10.1.2	ACI Code	235
	10.1.3	JSCE Code	237
	10.1.4	China Codes	239
10.2	GB/T 50476: Design Basis		240
	10.2.1	Environmental Classification	241
	10.2.2	Design Lives and Durability Limit States	242
	10.2.3	Durability Prescriptions	243
10.3	GB/T 50476: Requirements for Durability		244
	10.3.1	Atmospheric Environment	245
	10.3.2	Freeze–Thaw Environment	246
	10.3.3	Marine and Deicing Salts Environments	249
	10.3.4	Sulfate Environment	253
	10.3.5	Post-tensioned Prestressed Structures	255

References   259
Index   270

Preface

Durability of Concrete Structures: State of the Art

Durability is a term related to both performance and time, reflecting the degree to which a structure/infrastructure meets its intended functions for a given duration of time. This description applies to all types of structure and infrastructures in civil engineering. Actually, during the service life a structure displays time-dependent behaviors by ageing of the structural materials. The ageing processes can be intrinsic to the structural materials or induced by the interactions between the service conditions and the structural materials. This picture holds for all structures and their constitutive materials. In fact, concrete structures have transient behaviors due to some well-known time-dependent properties of structural concrete, such as shrinkage and creep. Take creep, for example. Engineers had been challenged by this evolving property as early as the 1900s, the very beginning of concrete structures coming into use. During the following years the lack of consideration of creep, surely due to lack of knowledge, had caused some serious accidents in structural engineering; for example, the collapse of the Koror–Babelthuap Bridge, Palau, in 1996. The past century has witnessed considerable research efforts dedicated to this subject, and the colossal creep models and established databases. The awareness of the deterioration of concrete properties by environmental actions comes much later. In the early 1990s, field investigations from various sources showed that concrete structures, massively constructed during 1950s and 1960s, were in very poor condition. The cost of the maintenance works due to deterioration by environmental actions was reported to be reaching an alarming level and generated heavy financial burdens on the structure owners. This situation makes durability a worldwide concern for decision-makers, structural designers, and material suppliers. Accordingly, the past three decades witnessed enormous efforts dedicated to intensive research on deterioration processes of structural elements and concretes, and the durability specifications for concrete structures at the design level.

Today, the term "durability" is somewhat standardized in a technical sense. The standard ISO-13823 provided the definition as the "capability of a structure or any component to satisfy, with planned maintenance, the design performance requirements over a specified period of time under the influence of the environmental actions, or as a result of a self-ageing process"; the ACI Concrete Terminology gives the definition as the "ability of a material to resist weathering action, chemical attack, abrasion, or any other process of deterioration." Evidently, the former definition is more adapted to structural engineering, while the latter is more oriented to concrete materials. However, one can notice that both definitions exclude the most evident time-dependent properties of structural concrete: shrinkage and creep. This is doubtless due to the fact that the recent engineering concern, as well as the corresponding efforts, mainly focuses on the environmental actions, the reason why the term "environmental actions" is explicitly expressed in both definitions. In this book, this established terminology is also followed, though shrinkage and creep remain the most important transient properties of structural concretes. The awareness of structural durability leads to two important changes in structural design. First, the design changes from a "static" mode to an "evolving" mode and the evolution of certain structural and materials properties must be taken into account through appropriate approaches. The design service life, or design working life, becomes an independent design parameter and target for the design procedure. Hence, the design changes from a loading-based procedure to a service life-based one. Second, durability awareness enables the life-cycle concept in structural design and management. Modern civil engineering is a highly multidisciplinary domain, connected with more fundamental social stakes, such as sustainability and ecological impacts. The life-cycle concept introduces into the structural design procedure, besides the structural requirements, the requirements of structural demolition, reuse, material recycling, and other ecological considerations.

Durability Design: Multilevel Procedure and Challenge

Performing a design for durability is by no means a trivial task. First of all, durability design is by nature a multilevel problem: durability design has different meanings for the whole structure, structural elements, and structural materials. For a structure as a whole, the durability design aims, for a given service life, given environmental actions and given budgetary constraints, to ensure the most rational structural element assemblage and global layout so that the transient performance can always be maintained to an expected level. Furthermore, a rational partition of initial investment in the construction phase and subsequent investment on maintenance works is also expected. For structural elements, the durability design, following the design strategy at the whole structure level, is to fulfill the design service life through more specified requirements, such as bearing capacity, section details, concrete cover thickness, and material properties. On this level, the durability design focuses on the technical requirements, and less on the budgetary factors. Structural design transfers also on this level the technical requirements on durability onto the material level through specified material properties. Then comes the material design part. On this level, material engineers should design the concrete mixture appropriately, both in order to satisfy the specified material properties transferred from the structural design, and to ensure good workability of the concrete mixture for in-place operation. Good workmanship is crucial to achieving durability of concrete structures in construction, since concretes need in-place operation and curing to grow into a structural materials.

This multilevel design process necessitates good communication between the structural design part and the material design part. This procedure is a performance-based one and also an ideal one. Although easy to understand, this performance-based procedure relies heavily on the available knowledge of the deterioration processes. Thus far, the state-of-the-art of the knowledge on durability is unfortunately far from homogeneous. For the processes such as concrete carbonation and chloride ingress, the available knowledge can provide models and support quantitative requirements for design of given service life and environmental actions. This is by no means the case for other processes, like salt attack and pore crystallization. As a result, only empirical and qualitative requirements can be formulated for the material design against these processes. Actually, this empirical format of durability design existed long before the performance-based format was established, and is still used in design codes such as ACI-318 code and Eurocode2.

The second challenge comes from the concrete material itself. Modern concretes change…, and quite radically. Owing to the importance of CO_2 emission from cement clinker production, modern concretes incorporate more and more secondary cementitious materials into the binder, including fly ash, ground granulated blast-furnace slag, and lime powder. The alkali-activated binder even contains no Portland cement clinkers. Also, from ecological considerations, recycled and artificial aggregates are incorporated into concrete to replace natural aggregates. Concretes made from these composites can have quite different properties and behaviors compared with ordinary Portland cement (OPC) concretes. Historically, it is the OPC concretes that have undergone intensive research and own more return of experiences from existing structures. The technical requirements, quantitative or qualitative, are based heavily on the accumulation of such data. With the dearth of systematic data and experiences with these new concrete composites, extrapolation from the available knowledge to the appropriate specification of these new composites for durability design seems highly challenging. The knowledge on the deterioration process in time scale constitutes the last challenge. To establish a reliable deterioration law, one needs to have a reliable model formulated from correct mechanisms and validated by real-scale tests or in-place structural investigations. Here, the term "scale" refers to time. Normally, the deterioration process under natural environmental actions is extremely slow; for example, it takes normally 15–30 years to obtain meaningful chloride ingress results for specimens stored in marine exposure stations. However, most deterioration research is conducted in the laboratory under artificial environmental actions to accelerate the process to obtain measurable data within an acceptable time scale. Since the similarity between these accelerated tests and the natural deteriorations is rather low, how to extrapolate the laboratory observations to natural processes remains always a tricky problem.

Given all the realistic aspects related to the multilevel design process, we can find ourselves in a dilemma, between the need to formulate requirements and specification for a given service life and the constant lack of sufficient data and experiences to support them due to the use of new concrete composites, new exposure conditions, or new service conditions. So, one can imagine that the durability design following a mixed performance–empirical format will always be a design option, a rather realistic one.

Modeling of Durability Processes: Common Basis

The deterioration of concrete materials under environmental actions is fundamental knowledge for durability design. The very reason that concrete can deteriorate stems from the facts that concrete is a porous material and through the pore network the material can have mass and energy exchange with the external environment or within its internal components. The term "deterioration" encompasses actually all the intrinsic or action-induced exchange processes. Accordingly, deterioration has a multi-physical nature. Mastering such processes is normally difficult, if not impossible, due to the multi-nature of concrete.

First, concrete is a heterogeneous composite with hardened cement paste as matrix and aggregates and fillers as inclusions. The heterogeneity depends on the size distribution of aggregates and fillers, and also on the properties under investigation. In particular, many durability processes occur on the concrete surface, where the boundary effect of aggregates cannot be neglected. Second, concrete is a porous medium, with pore structure playing a primary role in the related transport processes. The pore size of concrete covers more than six orders of magnitude: the typical interlayer distance within calcium silicate hydrates (C-S-H) is about 2 nm, the intergranular gap among C-S-H bundles is about 5 nm, the size of capillary pores ranges between 10 nm and 1 μm, the thickness of the interfacial transition zone (ITZ) is on the order of 10 μm, the entrained air bubble is about 100 μm, and the residual air voids in concrete can reach 1–2 mm. Pores and voids with different sizes have very different contributions to the physical properties of concrete. To complicate the scenario further, concrete pores can be saturated partially with pore solution, normally a highly alkaline electrolytic solution in equilibrium with the surrounding hydrates and minerals, and partially with pore gas phases. Thus, all masses, ions or gas molecules, transport through the pore structure of this complexity.

Third, concretes have only partial multiscaling property. The hardened cement paste can be regarded as a multiscale assemblage: C-S-H with its inherent porosity as the basic scale (~1 nm), different packing patterns of C-S-H bundles as the second scale (~10 nm), then hydrates cells containing C-S-H bundles and other hydrates as the third scale (~10 μm). The hardened cement paste was reported to have good multiscale property for the elastic properties. However, this property breaks down from cement paste to mortar because the coarse pores and weak mechanical properties of the ITZ perturb this multiscaling property.

Given all these aspects, a correct description of deterioration of concrete materials is far from an easy task, even with the physical and chemical mechanisms clarified. Fortunately, we talk about modeling rather than reality here. Modeling is basically an approximation of what is really going on. The good news is that, for a specified purpose, the facts and the mechanisms can be simplified, but not too much to lose its capacity to predict. For engineering use, this point is crucial and the key to depicting sophisticated phenomena. The last several decades witnessed some powerful tools for concrete modeling, such as poromechanics, micromechanics, and multi-physical transport theory. In Part 1 of this book the deterioration mechanisms will be presented in the individual chapters treating different processes, and the relevant models are introduced for engineering use. For a book aiming at durability design, no attempt is taken to establish a general theoretical framework for all deterioration processes. Instead, regardless of the theoretical basis, the following three principles are used for modeling the durability-related processes.

Principle 1. Correct estimation of the length scale. Concrete is a heterogeneous material, but the material components distribute randomly in space, allowing for the definition of a representative elementary volume (REV). The size of the REV scales to such an extent that concrete can be regarded as a homogeneous medium. As a rule of thumb, the REV size of concrete is usually estimated to five times the maximum aggregate size for transport properties. If a process occurs on a length scale smaller than the REV size, it can only be described in a heterogeneous context. The mass transport across a concrete surface with a length scale smaller than the REV size belongs to this case.

Principle 2. Correct estimation of the time scale. Normally the time scale of deterioration processes in durability design is on the order of the service life. Thus, all transient phenomena with a characteristic time much shorter than this time scale can be simplified as instantaneous, a powerful tool to propose simplified models. Take the transport process, for example. As ions transport in the pores of concrete, the pore-wall hydrates adsorb the ions. The adsorption has its own characteristic time; for example, 2–3 days for chloride ions. Compared with the target time scale (>10 years), all adsorptions can be regarded as instantaneous events. Accordingly, transport models need not include a transient adsorption.

Principle 3. Correct estimation of modeling error. The principal role of a model is to predict, and the capacity of a model is judged by its prediction accuracy. The raw materials of concrete, except binder materials, are all natural materials. Their properties and compositions have quite significant variation, and so do the properties of concrete. If the model takes into account detailed physical/chemical mechanisms, then the dispersion of model prediction must be estimated considering the randomness of related properties. As the dispersion is important, the model had better include statistical characteristics of the properties. This aspect is quite important for structural-level specification to leave a sufficient safety margin for design.

Besides the above three principles, a physically meaningful model is always a better choice than otherwise (e.g., purely empirical models). Physically meaningful models have explicit mechanisms behind them, whereas empirical models provide merely a fit of experimental data. As the environmental conditions or the concrete composition change, empirical models are in a situation where they need to extrapolate their prediction to other cases not included in the model fitting data, and thus are less probable to remain predictive. On the last point, validation is crucial for the engineering use of models. Normally, laboratory research tends to build a model only on a few sets of experiments, which is far from sufficient for engineering use. In some cases the model is even validated by the same set of data from which the model is made. Indeed, a big contrast exists between the huge number of models proposed in the literature and the few reliable ones really used in durability design work. The underlying reason is that most literature models have not been validated by enough in-site data. Reliable model validation should best be performed on real structures with long-term monitoring data.

This Book: Structure and Audience

This book attempts to treat durability design, addressing simultaneously the material and structural disciplines in civil engineering. The book follows a basic logic line from concrete materials to structural design, and the content is accordingly divided into three parts. Part 1 is dedicated to the deterioration of concrete materials under different environmental actions, including carbonation, chloride ingress, freeze–thaw, leaching, and salt crystallization, treated respectively in Chapters 1 to 5. In each chapter, the deterioration process is presented following a phenomena–mechanism–modeling logic line. Part 2, consisting of a single chapter, Chapter 6, treats the subject of concrete deterioration in a structural context, including the effect of mechanical loading, the impact of cracks, the multi-field problems, and drying–wetting actions. Part 3 elaborates the topics of durability design for concrete structures in four chapters, Chapters 7–10, treating respectively the global method and approach, durability indicators, design applications, and the codes and standards.

The intended audience of this book includes structural design engineers and civil engineering students in their postgraduate programs or later part of their bachelor programs. This book can be used as a reference for durability design work or as a textbook for graduate courses on durability of concrete structures. Part of this book can be assigned to undergraduate students as extensive reading materials for standard courses on structural design and construction materials. Structural engineers can also benefit from this book by learning the deterioration processes and the related models for engineering use, and researchers of concrete materials can benefit from this book by discovering how knowledge on the material level is transferred to the structural level.

Acknowledgments

When I bravely proposed to write this book, I had no idea how deep is the water I was stepping in. This book cannot see its day without the support of people behind. The content of this book comes from the work of my team on the durability of concrete materials and structures in the past 10 years in Tsinghua University, China. The contributions of my former PhD students and postdoc fellow, Chunqiu Li, Chunsheng Zhou, Qiang Zeng, Xiaoyun Pang, Le Li, and Linhu Yang, are deeply acknowledged. I am indebted to Professor Olivier Coussy for the spirit of modeling for engineering use: always keeping equilibrium between scientifically correct and practically usable, which I hope to have achieved in this book. Part of this book has been used as the teaching materials in the postgraduate course "Durability and Assessment of Engineering Structures" in Tsinghua University. This course has been taught for 10 years, and the feedback from each class has substantially enriched the related content. A part of Chapter 10 is extracted from the drafting works of the China design code GB/T 50476-2008 and its revision work in 2014–2015. Thus, I would like to thank the drafting team for the collective work and support for this book, especially Professor Zhaoyuan Chen and Professor Huizhen Lian, initiators of durability standards in China. Moreover, my colleagues are gratefully acknowledged for providing the photographs to enhance the illustration of this book, especially Dr Tingyu Hao. Finally, I express my deep gratitude to my wife and my son for their support during the writing of this book, and they may include me in the regular family events from now on.

Part One

Deterioration of Concrete Materials

1

Carbonation and Induced Steel Corrosion

This chapter treats the first important durability process of concrete materials and structures: the carbonation and the induced corrosion of steel bars in concrete. The carbonation of concrete originates from the reaction between the alkaline pore solution of concrete and the carbon dioxide (CO_2) gas migrating into the pores. The carbonation does not compromise the material properties but decreases the alkalinity of the pore solution, which has an adverse effect on the electrochemical stability of steel bars in concrete. The risk of steel corrosion can be substantially enhanced in a carbonated concrete. This chapter begins with the phenomena of concrete carbonation and its effect on the long-term durability of concrete materials and structures. Then the detailed mechanisms are presented, according to the state of the art of knowledge, for concrete carbonation and the induced steel corrosion, together with a comprehensive analysis on the main influential factors for these processes. On the basis of the available knowledge, the modeling aspect is brought forth through mechanism-based and empirical models for engineering use. Since the valid scope and the uncertainty are two fundamental aspects for model application, the critical analysis is given to the models presented and their application. Some basis for durability design against the carbonation and the induced corrosion is given at the end.

1.1 Phenomena and Observations

As concrete is exposed to the atmosphere, the CO_2 present in the atmosphere can migrate into the material through the pore structure and react with the cement hydrates such as portlandite ($Ca(OH)_2$ or CH) and the calcium silicate hydrates (C-S-H). These reactions are termed the "carbonation" of concrete materials. The direct consequence of carbonation is the consumption of CH, eventually C-S-H, and the decrease in pH value of the pore solution. Under a less alkaline environment, the electrochemical stability of the embedded steel bars in concrete can be destroyed, the steel can be depassivated and the electrochemical process of corrosion can

Figure 1.1 Rebar corrosion of RC slabs in a fine-art gallery of age 38 years by concrete carbonation. The building was constructed in 1962 with C25 concrete for the RC beams. The concrete binder is ordinary Portland cement (OPC), and the building was exposed to a rather dry environment with average temperature of 11 °C and average humidity of 57%. *Source:* courtesy of Tianshen Zhang.

occur. As the corrosion develops to a significant extent, the reaction products from corrosion accumulate at the concrete–steel interface and can fracture the concrete cover. Since all concrete structures are built and used in the envelope of the atmosphere, carbonation is a basic and fundamental process for the long-term durability of concrete elements. Note that the detrimental aspect of carbonation resides mainly in the corrosion risk for the embedded steel bars and the carbonation itself is not found detrimental to concrete materials. Relevant studies show that the products of the carbonation reaction can notably reduce the porosity of hardened concrete, and the carbonation can also be used as a pretreatment technique for the recycled coarse aggregates to reduce the water sorptivity (Thiery *et al.*, 2013).

Concrete carbonation has been well investigated with regard to both the reaction mechanism and the alteration of properties of material and structural elements. The mechanism of carbonation is to be detailed later, while the most severe deterioration of concrete elements by carbonation is usually due to the less compacted concrete, insufficient protection of steel bars from the concrete cover, and the favorable moisture conditions for steel corrosion. For residential buildings, most reinforced concrete (RC) elements, like slabs and walls, usually bear surface lining or protection layers; thus, relatively few severe deterioration cases are reported for these elements by carbonation until a very late stage of service life. Nevertheless, the RC elements exposed directly to atmospheric precipitation, such as roofs, can show very advanced deterioration due to carbonation. Figure 1.1 illustrates an advanced state of carbonation-induced corrosion of reinforced steel bars of RC slabs in a fine-art gallery of age 38 years.

Compared with residential buildings, certain industrial buildings contain more aggressive environments for the concrete elements; for example, the concrete roofs of steel process

(a)   (b)

Figure 1.2   Advanced deterioration of concrete elements in a steel process workshop constructed in 1980, aged 24 years at inspection. The RC elements in the steel workshop were exposed to high concentration of corrosive gas rich in chloride and sulfate, high temperature, and humidity (a). The local degradation was manifested through advanced corrosion of steel bars, concrete spalling, and leachates from concrete cracks (b). *Source:* courtesy of Tingyu Hao.

workshops are exposed to high temperature, high humidity and other corrosive gases in addition to CO_2. These corrosive gases can enhance the consumption of CH in pore solution; thus, the deterioration rate of these elements is faster than those exposed to the normal atmosphere. Such an example is presented in Figure 1.2 for a steel process workshop of age 24 years.

Actually, bridge structures on highways, railways, or in urban areas are usually more affected by the deterioration of concrete carbonation due to the total exposure of RC elements to the atmosphere and its thermal and moisture changes. Figure 1.3 illustrates one railway bridge structure of age 26 years seriously affected by the carbonation and the resulted corrosion of the first layer rebars.

Traffic tunnels and underground structures are also affected by concrete carbonation partially due to the fact that CO_2 from traffic exhaust can accumulate to a high level; for example, a value of three times higher than the normal atmospheric CO_2 content (350–380 ppm) has been reported for the Môquet Tunnel in Paris (Ammoura *et al.*, 2014). A full-scale model was built for the subway stations of Shenzhen City, China, in 1999 during the construction phase to demonstrate the long-term performance of different structural concretes of C30 grade. This model was kept in an outdoor environment after the construction and in-situ tests on the drilled cores have been conducted four times since its construction.

(a)   (b)

(c)

Figure 1.3   Railway bridge after retrofit of the simply supported RC beams (a) and the local corrosion of reinforcement bars by concrete carbonation before the retrofit (b, c). The bridge was constructed in 1976 and the retrofit was finished in 2002. The concrete grade was C50 with OPC as binder, and some local concrete cover was less than 10 mm due to deficient positioning of the reinforcement molds. The bridge is exposed to a typical carbonation environment with an average temperature of 14.6 °C and relative humidity of 56%. *Source:* courtesy of Tingyu Hao.

Figure 1.4 illustrates the inner side of the full-scale model and the carbonation depth measured on the drilled cores from the model walls of age 15 years.

Based on state-of-the-art knowledge of carbonation-induced deterioration and on the experiences of the long-term performance of concrete structures, the favorable conditions for the carbonation deterioration have been well identified: a less-compacted concretes (for carbonation and corrosion), high CO_2 concentrations (for carbonation), and an adequate humidity level (for carbonation and corrosion). It is accepted that the high CH content in concrete is necessary to resist the carbonation of pore solution by infiltrating CO_2, so that a minimum content of 10%, with respect to the binder mass, has been suggested to ensure the carbonation resistance of concrete elements (AFGC, 2007). It is also accepted that the high compactness of concrete helps to limit carbonation by decreasing the CO_2 infiltrating rate and decrease the corrosion risk by increasing the electrical resistivity of concrete. This reasoning favors concretes with both high CH content and high compactness. However, modern concretes increasingly adopt secondary cementitious materials (SCM) in binders, resulting in high compactness of concrete but lower CH content compared with ordinary Portland cement (OPC) binder. For the full-scale model in Figure 1.4, the carbonation depths for OPC concrete ($w/b = 0.56$; binder: OPC 80%, fly ash (FA) 20%) and HPC concrete ($w/b = 0.37$; binder: OPC 36%, FA 36%, granulated blast furnace slag (GGBS) 18%) are respectively 18 mm and 13 mm at an age of 15 years,

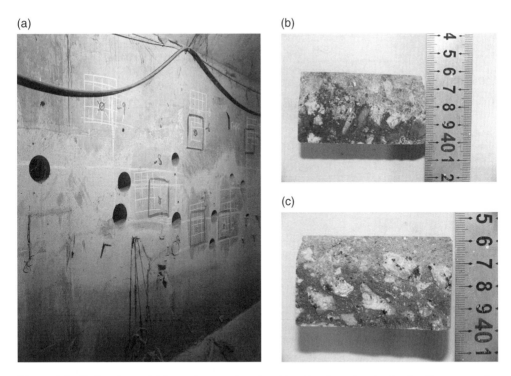

Figure 1.4 Full-scale model for underground station (a) and carbonation depth of wall concrete after 15 years of exposure, 18 mm for OPC concrete (b) and 13 mm for HPC concrete (c). The full-scale model was constructed in 1999 for the long-term observation of different structural concretes used in the subway project of Shenzhen city (average humidity 77%, average temperature 24 °C). *Source:* courtesy of Jianguo Han.

showing that the compactness dominates the CH content with respect to carbonation resistance. Thus, the balance of CH content and the concrete compactness is crucial to making a durable concrete in carbonation environments, particularly for modern concretes incorporating more and more SCM.

1.2 Carbonation of Concrete

1.2.1 Mechanisms

Concrete carbonation includes a series of chemical reactions between the infiltrating CO_2 gas through the material pore space and the liquid interstitial solution in pores. After the hydration of binder materials, the pore solution of hardened concrete contains mainly K^+, Na^+, Ca^{2+} and OH^- ion species and shows high alkalinity, with a pH value around 13.0. Carbonation occurs in the pore solution between the dissolved CO_2 and the aqueous ions species; see Figure 1.5. The preponderant reaction is between CO_2, OH^- and Ca^{2+} as follows:

$$CO_2 + Ca(OH)_2 \rightarrow CaCO_3 \downarrow + H_2O \qquad (1.1)$$

(a)

(b)

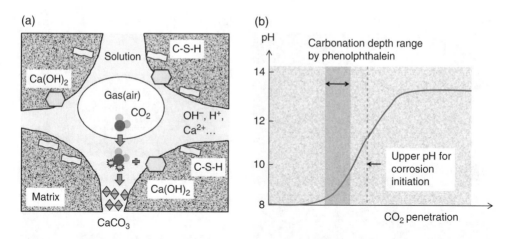

Figure 1.5 Mechanism of carbonation at pore level (a) and pH profile along the CO_2 infiltration direction (b).

Note that this expression masks the dissolution of CO_2 into the pore solution, forming carbonic acid (H_2CO_3) and ionized successively into HCO_3^- and CO_3^{2-}. The results of this reaction are twofold: the consumption of OH^- ions in pore solution and the formation of solid precipitation of calcite ($CaCO_3$) in the pore space. The former decreases the pH value of the pore solution from around 13.0 to below 9.0, while the latter can fill in the original pore space and strengthen the solid matrix of concrete. In the literature, the porosity of carbonated cement pastes with OPC was reported to decrease by 10–15% (Ngala and Page, 1997) and the compressive strength was observed to augment by 30% (Lea, 1970). This knowledge was even used to increase the compactness of hardened concrete by an accelerated carbonation treatment or by introducing CO_2 into the mixing process (Estoup, 1987). However, this compactness effect of CH carbonation is much less in concretes incorporating a large quantity of SCM such as FA or GGBS, and the porosity was even observed to increase. Thus, the beneficial aspect of carbonation on the gain of compactness should be treated with caution.

The CH is not the only phase involved in the reaction with dissolved CO_2 in the pore solution. The C-S-H are the main products from hydration of cement grains and mineral SCM, including actually a group of $(CaO)_x(SiO_2)_y(H_2O)_z$ compounds with Ca/Si ratios varying from 1.2 to 1.7. The solubility of C-S-H is very low in pore solution, and the hydrates can react with the dissolved CO_2, in the form of carbonic acid, forming calcium carbonate and free water (H_2O). According to more detailed research (Dunster, 1989), the carbonation of C-S-H was proposed as follows: the dissolved CO_2 captures calcium ions in C-S-H and leaves anion silicate groups condensed together with lower Ca/Si ratio. In a simplistic way, this reaction was written as (Papadakis *et al.*, 1991b)

$$3CaO \cdot 2SiO_2 \cdot 3H_2O + 3CO_2 \rightarrow 3CaCO_3 \cdot 2SiO_2 \cdot 3H_2O \qquad (1.2)$$

Owing to the very low solubility of C-S-H in the pore solution, the influence of C-S-H carbonation on the pH value of the pore solution is regarded as weak. The C-S-H are the main constituents for the solid matrix of cement pastes; thus, the change in C-S-H properties by carbonation can have consequences in concrete materials. Apart from CH and C-S-H, other minerals in hardened cement paste can also react with dissolved CO_2 in the pore solution, including the unhydrated minerals of tricalcium silicates (C_3S) and dicalcium silicates (C_2S) and aluminates hydrates. And the carbonation of aluminate hydrates is thought to be rapid (Thiery, 2005).

As CO_2 migrates into the pores, the above reactions occur simultaneously with the dissolution of CO_2 from the gas phase into the aqueous phase; see Figure 1.5. In engineering practice, the carbonation depth is usually measured by phenolphthalein ($C_{20}H_{14}O_4$) solution, in 1% mass concentration. The phenolphthalein solution remains colorless for low pH value and turns to a pink–red color for high pH value, and the color change point of pH value is around 9.0. The solution is spayed on the exposed surface, fractured or drilled, and the pink–red zone is assumed to mark the boundary between the carbonated and intact zones. Nowadays, this carbonation depth is widely adopted as the indicator of carbonation extent of concrete. However, the approximate nature of this characterization should be borne in mind: the phenolphthalein solution changes its color over a pH range – that is, between 8.4 and 9.8 (McPolin *et al.*, 2009) – and the pH value of the pore solution along the CO_2 infiltration direction follows a gradual change rather than a steep front; see Figure 1.5. Nevertheless, the concept of carbonation front is still used in the modeling in this chapter owing to its simplicity.

1.2.2 Influential Factors

Concrete carbonation is fundamentally affected by two aspects: the material intrinsic properties involved in the carbonation reactions and the external environmental factors. The material properties include, on the chemical side, the content of reactive hydrates (CH, C-S-H, C_3S, C_2S and aluminates hydrates) in the solid phase, and the resistance of the pore structure to CO_2 infiltration on the physical side. The external factors refer to the atmospheric CO_2 concentration, the temperature and the environmental humidity.

CH Content and CO_2 Diffusivity

The CH content of hardened cement-based materials is the result of the hydration reaction between the binder materials and mixing water, related closely to the chemical composition and hydration extent of the binder materials. Figure 1.6 shows the CH content of cement pastes using OPC as the only binder material for different *w/c* ratios. The CO_2 infiltration is resisted by the tortuous pore structure of concrete and the pore solution as well since CO_2 can only transport in the gas-occupied space. Taking the diffusivity of CO_2 in concrete to characterize the infiltration rate, several analytical expressions were proposed for the diffusivity in terms of the porosity and environmental humidity. Papadakis *et al.* (1991a) proposed the expression

$$D_{CO_2} = 1.64 \times 10^{-6} \phi_p^{1.8} (1-h)^{2.2} \tag{1.3}$$

where D_{CO_2} (m²/s) is CO_2 diffusivity, ϕ_p is the porosity of cement-based materials, and h is the relative humidity in the environment. Using the same data sets, the above relation was

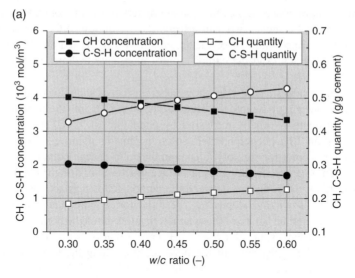

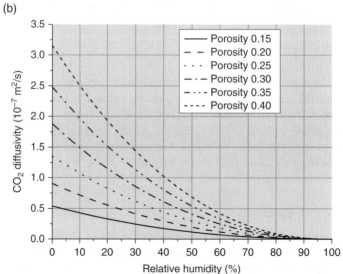

Figure 1.6 CH content for OPC pastes with different *w/c* ratios (a) and CO_2 diffusivity in terms of porosity and environmental humidity using Equation 1.3 (b). The simplified hydration model from Papadakis *et al.* (1991a) is retained and the CH/C-S-H contents are first evaluated for a totally hydrated state. Then, the hydration extent is estimated for different *w/c* ratios using the model from Lam *et al.* (2000), and the stable CH/C-S-H contents are estimated from the total contents and the hydration extents.

calibrated in terms of the CO_2 diffusivity in air $D_{CO_2}^0$, total porosity of concrete ϕ, and the pore saturation s_l as (Thiery, 2005)

$$D_{CO_2} = D_{CO_2}^0 \phi^{2.74} \left(1 - s_l\right)^{4.20} \tag{1.4}$$

with $D_{CO_2}^0$ equal to 1.6×10^{-5} m²/s at 25 °C. The CEB-FIP model code (CEB, 1990) proposed a value range $(0.5–6) \times 10^{-8}$ m²/s for hardened concretes and related the diffusivity to the characteristic value of compressive strength f_{ck} (MPa) through

$$-\log D_{CO_2} = 7.0 + 0.025 f_{ck} \qquad (1.5)$$

This expression was prescribed for concrete elements sheltered from rain, corresponding roughly to a relative humidity $h = 60\%$.

CO_2 Concentration, Temperature and Humidity

The CO_2 concentration in the surrounding environment of concrete structures plays an important role in the carbonation process. If the kinetics of carbonation are assumed to be controlled by the CO_2 diffusion, then the carbonation depth will be proportional to the square root of the CO_2 concentration in the environment (DuraCrete, 2000). In other words, augmenting CO_2 concentration by 25% leads to an increase of carbonation depth by 12% for concrete. The concentration in the atmosphere is normally in the range 350–380 ppm, corresponding to 0.57–0.62 g/m³. Owing to greenhouse gas emissions in recent decades, the CO_2 concentration in the atmosphere has augmented from 280 ppm in 1750 to 380 ppm in 2005 (IPCC, 2007). Compared with the average atmospheric value, the local CO_2 concentration surrounding a particular structural element can be much higher: it can readily double the average value in less-ventilated conditions and adopt even higher values in polluted urban areas.

The temperature is another basic factor for concrete carbonation in the natural environment. Coupled with the greenhouse emission process, the atmospheric temperature is found to have increased from 0.074 °C per decade in the past 100 years to 0.177 °C per decade in the past 25 years (IPCC, 2007). For service lives of 50–100 years, this global warming gives a rather moderate augmentation. Again, the local climate has much more influence on the temperature for concrete structures. A higher temperature is expected to accelerate the carbonation since all the reactions involved are by nature electrolytic and thermally activated. But note that the solubility of CO_2 in the pore solution is also reduced at a higher temperature. Owing to these two opposite effects, the influence of temperature on carbonation rate is considered not important (Chaussadent, 1999). Contrary to the temperature, the environmental humidity has a very distinct impact on the carbonation process: low humidity leaves more pore space for CO_2 infiltration, but the dissolved CH quantity in the pore solution is limited by the low pore saturation; the high humidity increases the pore saturation and thus hinders directly the CO_2 transport; the optimum humidity range was found to be between 50 and 70%. A nominal value of $h_{opt} = 65\%$ was adopted and a factor k_e was proposed for the influence of humidity on the carbonation process (*fib*, 2006):

$$k_e = \left(\frac{1 - h^f}{1 - h_{opt}^f} \right)^g \qquad (1.6)$$

with f and g as model parameters, adopting $f = 5.0$ and $g = 2.5$. Note that this expression gives a conservative prediction for $h < h_{opt}$.

Accelerated Carbonation

Setting the CO_2 concentration and relative humidity level, the carbonation can be accelerated substantially in the laboratory. Actually, accelerated tests have been established to characterize the carbonation resistance of concrete in the laboratory. The principle of these tests is to put concrete specimens under high CO_2 concentration of 2–50% and humidity $h = 60 - 70\%$ and measure the carbonation depth by phenolphthalein solution for a given duration. The measured carbonation depth, or the carbonation resistance deduced from the carbonation depth, is regarded as a useful durability indicator for concrete exposed to a carbonation environment. Theoretically, the difference between the accelerated carbonation (ACC) and natural carbonation (NAC) is due to the environmental conditions; that is, CO_2 concentration and humidity level. However, since the NAC occurs during a time span much longer than the ACC, the intervention of other processes, such as natural drying and the long-term microstructure evolution of concrete, is inevitable and difficult to quantify. Figure 1.7 illustrates the correlation between ACC and NAC at the exposure site in Maurienne, France (Baroghel-Bouny *et al.* 2004). The correlation seems stronger for concrete with strength higher than 25 MPa and weaker for lower strength concretes.

1.2.3   Models

On the basis of theoretical and experimental investigations of concrete carbonation, various models have been proposed in the literature, and the reference (DuraCrete, 1998) provides a comprehensive description for the available models on carbonation depth in its Appendix A.

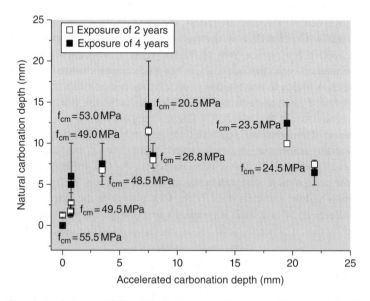

Figure 1.7   Correlation between ACC and NAC of structural concretes. Data were taken from Baroghel-Bouny *et al.* (2004: tables VIII and X). The ACC tests were performed under a CO_2 concentration of 50% for 14 days' duration. The annual average temperature was recorded as between 7.5 and 11.1 °C. f_{cm}: compressive strength.

Here, only two models are introduced: the mechanism-based model (C-1) from Papadakis *et al.* (1991b) and the empirical model (C-2) from *fib* (2006) model code. Both models use the concept of carbonation front and take into account the main influential factors described in Section 1.2.2.

Model C-1

Papadakis *et al.* (1991b) provided the mathematical model for the carbonation depth on the basis of simplified carbonation mechanisms: the cement hydrates that participate in the carbonation reaction include CH, C-S-H and C_3S, C_2S in the unhydrated cement grains. The mass conservation of CO_2 gas considering these carbonation reactions is

$$\frac{\partial}{\partial t}\left(\phi(1-s_1)[CO_2]\right) = \frac{\partial}{\partial x}\left(D_{CO_2}\frac{\partial[CO_2]}{\partial x}\right) - \phi_0 s_w r_{CH} - 3r_{CSH} - 3r_{C_3S} - 2r_{C_2S} \tag{1.7}$$

where ϕ and ϕ_0 are the porosities of concrete after and before the carbonation, $[CO_2]$ (mol/m³) is the CO_2 concentration, s_1 and s_w are respectively the porosity occupied by the liquid phase and the porosity filled with water,[1] r_x (mol/s) (x = CH, CSH, C_3S, C_2S) are the reaction rates of these reactants with CO_2, and the coefficients before these reaction rates are the corresponding stoichiometric constants in the respective carbonation reactions. If the hydrates are stable and no more hydration reactions consume or produce these reactants, these reactions rates are equal to the mass change rate of these compounds:

$$\frac{\partial}{\partial t}[CH] = -\phi_0 s_w r_{CH}, \quad \frac{\partial}{\partial t}[CSH] = -r_{CSH}, \quad \frac{\partial}{\partial t}[C_3S] = -r_{C_3S}, \quad \frac{\partial}{\partial t}[C_2S] = -r_{C_2S} \tag{1.8}$$

where [CH], [CSH], [C_3S], and [C_2S] (mol/m³) are the reactant contents in concrete. Now suppose the carbonation advances into concrete through a carbonation "front"; see Figure 1.8. The position of this front is denoted by x_c. At the right side of the carbonation front, $x > x_c$, all the reactants (CH, C-S-H, C_3S, and C_2S) are assumed to be intact, taking their initial contents [CH, CSH, C_3S, C_2S]⁰; but all these reactants are totally consumed by the carbonation reactions; that is, [CH, CSH, C_3S, C_2S] = 0 as $x < x_c$. In the carbonation range the CO_2 is assumed to be in steady state; that is:

$$\frac{\partial}{\partial t}\left(\phi(1-s_1)[CO_2]\right) = 0 \quad \text{for} \quad 0 < x < x_c \tag{1.9}$$

Under a constant CO_2 concentration imposed on the concrete surface, $x = 0$, this assumption leads to a linear distribution of CO_2 in the carbonation range; that is, $[CO_2] = [CO_2]^0(1 - x/x_c)$.

---

[1] The difference lies in the assumption that, under a certain environmental humidity, the pores with size smaller than the size at equilibrium with a humidity through the Kelvin equation are filled by liquid water, denoted by s_w, and the pores above this size contain a water film on the pore wall. The total liquid saturation s_1 is the sum of the water-filled porosity and the volume of water films.

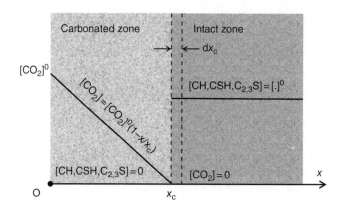

Figure 1.8   Schematic illustration of the C-1 model for carbonation depth.

Now focus on the CO_2 mass conservation as the carbonation front moves from x_c to $x_c + dx_c$. The conservation in Equation 1.7, considering Equations 1.8 and 1.9, becomes

$$-\left(D_{CO_2} \frac{\partial [CO_2]}{\partial x}\right)_{x_c}^{x_c+dx_c} = \frac{dx_c}{dt}\left([CH]+3[CSH]+3[C_3S]+2[C_2S]\right)_{x_c}^{x_c+dx_c} \tag{1.10}$$

Using the linear distribution of CO_2 concentration, this relation turns to

$$x_c \frac{dx_c}{dt} = \frac{D_{CO_2}[CO_2]^0}{[CH]^0+3[CSH]^0+3[C_3S]^0+2[C_2S]^0} \tag{1.11}$$

After integration, the carbonation depth is written as

$$x_c = x_c^0 + \sqrt{\frac{2D_{CO_2}[CO_2]^0}{[CH]^0+3[CSH]^0+3[C_3S]^0+2[C_2S]^0}t}$$

$$= x_c^0 + k_{carb}\sqrt{t} \tag{1.12}$$

where x_c^0 (m) is the initial carbonation depth corresponding to $t = 0$, D_{CO_2} (m²/s) is the CO_2 diffusivity in concrete expressed in terms of the porosity and the pore saturation or relative humidity (see Equation 1.3 or 1.4), and $[CO_2]^0$ (mol/m³) is the atmospheric CO_2 concentration to which the concrete is exposed. This model gives a carbonation depth proportional to the square root of exposure time t. This proportionality has been confirmed by experiments as well as in-field investigations. This model has clear carbonation mechanisms behind it and has been used extensively in research and applications.

However, several aspects are to be considered in applying this model to structural concrete carbonation. First, this model should be applied to concrete in its totally hardened state so that Equation 1.8 can be observed. Second, the concentration of reactants near the carbonation front in Equation 1.10 is greatly simplified, and this simplification, in addition to the carbonation front concept, gives an approximate nature for the carbonation depth predicted by Equation 1.12. Third, the concentrations of CH, C-S-H, C_3S, and C_2S depend much on the cement composition and the hydration extent in concrete; unless accurately quantified, the estimation of these concentrations always introduces errors into the prediction results. Further, the carbonation reaction is also simplified with respect to C-S-H using a constant Ca/Si ratio in Equation 1.2; this assumption can be acceptable for concretes using OPC but is less valid for concretes incorporating SCMs. Finally, the carbonation depth in this model corresponds to the position where all reactant are consumed; thus, the local pH value at the carbonation front will be lower than the color change point of the phenolphthalein solution (i.e., pH 8.4).

The model parameters are summarized in Table 1.1, and some usual values are provided for the initial mineral contents in cement hydrates using the results in Figure 1.6 and assuming a volumetric volume ratio of 0.35 for cement paste in concrete. On the basis of these parameters and relations, simulation results are given in Figure 1.9 for carbonation depths in terms of the concrete porosity and environmental humidity. For comparison, two cases are presented: carbonation with CH and C-S-H, and carbonation with only CH. The results show the dominating impact of humidity on the carbonation depth and the contribution of C-S-H to the carbonation resistance; that is, the carbonation depth for the CH + C-S-H case is 30–50% smaller than for the CH case. With the simplification in Equation 1.2 for C-S-H carbonation, the predictions from these two cases can actually be used as boundary values for engineering applications.

Note that the above results are computed for concretes using 100% OPC as binder materials. Incorporating mineral admixtures, such as pozzolanic materials, can decrease substantially the CH content in hardened concrete. In that case, the CH and C-S-H cannot be taken directly from Table 1.1 or Figure 1.6, but from more specified hydration models or direct experimental characterizations.

Model C-2

Following a very engineering approach, the *fib* (2006) model code proposed an empirical model for carbonation depth on the basis of the ACC results, taking into account the environmental and the curing conditions of concrete. The carbonation depth is expressed through

$$x_c(t) = \sqrt{2k_e k_c R_{NAC}^{-1} C_{CO_2}} \sqrt{t} W_t \qquad (1.13)$$

Table 1.1   Model parameters for the C-1 model and their usual value ranges

Parameter (unit)	Value/relation
CH content in concrete $[CH]^0$ (10^3 mol/m^3)	1.17–1.41 from Figure 1.6
C-S-H content in concrete $[CSH]^0$ (10^3 mol/m^3)	0.590–0.709 from Figure 1.6
C_3S content in concrete $[C_3S]^0$ (mol/m^3)	0 for total hydration state
C_2S content in concrete $[C_2S]^0$ (mol/m^3)	0 for total hydration state
CO_2 concentration in atmosphere $[CO_2]^0$ (mol/m^3)	0.0147–0.0295 (400–800 ppm)
CO_2 diffusion coefficient in concrete D_{CO_2} (m^2/s)	Equation 1.3 or 1.4
Porosity of concrete ϕ	0.09–0.18 ($w/c = 0.30 - 0.60$)
Relative humidity (%)	0–100
Initial carbonation depth x_c^0 (m)	0–0.01

(a)

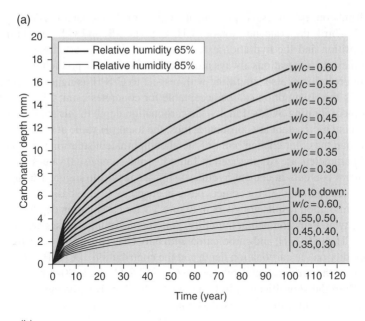

(b)

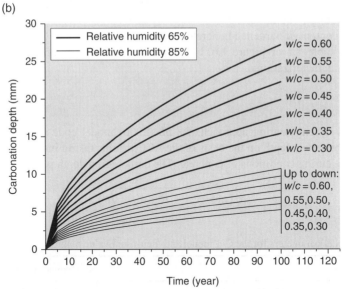

Figure 1.9 Carbonation depth from the C-1 model for concretes under different relative humidity: model results considering CH and C-S-H carbonation (a) and model results considering only CH carbonation (b). The CO_2 concentration is 0.0184 mol/m³, corresponding to 500 ppm.

where k_e is the environmental factor considering the relative humidity and further the pore saturation of concrete (and its value can be evaluated from Equation 1.6); k_c is the curing factor reflecting the hardening extent of concrete as exposed to the atmosphere, taking the form

$$k_c = \left(\frac{t_c}{7}\right)^{b_c} \tag{1.14}$$

with t_c (days) as the curing duration of concrete and b_c the regression exponent. R_{NAC}^{-1} (m²/s per kg CO_2/m³, or mm²/a per kg CO_2/m³) is the resistance of concrete to carbonation in natural exposure under a reference relative humidity of 65%, and can be related to the ratio between the binding capacity of CO_2 in concrete (kg CO_2/m³) and the effective diffusion coefficient of CO_2 in concrete (m²/s or mm²/a) (*fib*, 2006). This quantity is related to the ACC test[2] through the relation

$$R_{NAC}^{-1} = k_t R_{ACC}^{-1} + \varepsilon_t \tag{1.15}$$

where k_t and ε_t (m²/s per kg CO_2/m³, mm²/a per kg CO_2/m³) are regressed parameters. This relation issues from the correlation analysis between the NAC resistance of concrete and the resistance under the ACC test. The resistance R_{ACC}^{-1} was tested for various binders (composition and *w/b* ratio) and the recommended values are given in Table 1.2. C_{CO_2} is the CO_2

Table 1.2 Model parameters for the C-2 model and their usual value ranges

Parameter (unit)	Value/relation	Uncertainty range (%)
Environmental factor f_e	1.0 (65%), 0.31 (85%)	n.d.
Curing condition factor f_c	3.0 (1 day), 1.61 (3 days), 1.0 (7 days), 0.77 (11 days), 0.67 (14 days)	n.d.
Regression parameter k_t	1.25	28
Regression parameter ε_t (mm²/a per kg CO_2/m³)	315.5	15
R_{ACC}^{-1}(*w/b*) (10^3 mm²/a per kg CO_2/m³) Binder: CEM I R42.5	0.98 (0.40), 1.64 (0.45), 2.14 (0.50), 3.09 (0.55), 4.22 (0.60)	4
R_{ACC}^{-1}(*w/b*) (10^3 mm²/a per kg CO_2/m³) Binder: CEM I + FA (20%)	0.95 (0.40), 0.60 (0.45), 0.76 (0.50), 2.05 (0.55), 2.62 (0.60)	40
R_{ACC}^{-1}(*w/b*) (mm²/a per kg CO_2/m³) Binder: CEM III/B	2.61(0.40), 5.32(0.45), 8.38(0.50), 13.95(0.55), 25.20(0.60)	25
C_{CO_2} (10^{-3} kg/m³)	0.82 (500 ppm)	12
Weather exponent w	0.0 (indoors), 0.13, 0.21 (outdoors)	n.d.
Weather factor W_t	1.0 (indoors), 1.16 (outdoors)	n.d.

---

[2]This ACC test is after the experimental procedure from the DARTS project (DARTS, 2004). The concrete specimens undergo 6 days' curing in water and 21 days' curing under 20 °C and 65% humidity. Then the specimens are subject to a carbonation test with 2% CO_2 concentration for 28 days, and the carbonation chamber is controlled to 20 °C and 65% humidity.

concentration in the atmosphere (kg CO_2/m^3). The conventional value can be adopted as 0.82×10^{-3} kg/m³, corresponding to a concentration of 500 ppm in air. Note that the average CO_2 concentration in the atmosphere is between 350 and 380 ppm. W_t is the weather coefficient considering the influence of the wetting by precipitation on the concrete carbonation, expressed as

$$W_t = \left(\frac{t_0}{t}\right)^w \tag{1.16}$$

The reference time t_0 refers to the standard age for concrete under the ACC test (28 days), and the exponent w is determined by the time of wetness and the probability of driving rain (*fib*, 2006). This factor makes the difference between the indoor and outdoor exposure conditions, and some typical values are calculated and presented in Table 1.2.

This model is more adapted to engineering use since no chemical parameters are involved. The compactness of concrete is considered through the inverse carbonation resistance R_{NAC}^{-1} and related to the laboratory ACC test. The influence of humidity on carbonation is addressed through the environmental factor k_e and the weather factor W_t. Moreover, the crucial influence of curing age on concrete carbonation is included through the factor f_c. Some further studies linked the NAC resistance R_{NAC}^{-1} with the concrete compressive strength (Guiglia and Taliano, 2013). Moreover, the uncertainty associated with the parameters in this model was also quantified, which is important in evaluating the reliability of model prediction.

The drawback of this model comes from the relationship in Equation 1.15. Actually the credibility of the model hinges on this equation, which links the NAC resistance with ACC resistance. The ACC test follows a particular procedure, including curing and drying on concrete specimens and the CO_2 concentration/relative humidity level in the carbonation chamber. This particularity limits the possibility of more extensive model calibration for concretes with different binder materials, though a relatively complete study was done on different binders.

The main parameters in the model are summarized in Table 1.2. On the basis of these values, simulations are performed for the carbonation depth for the usual ranges of concrete carbonation resistance and weather conditions in Figure 1.10.

1.3 Steel Corrosion by Carbonation

1.3.1 Mechanism

Carbonation consumes OH^- ions in the pore solution, and the electrochemical stability of embedded steel bars is affected. The electrochemical stability of the reinforcement steel can be described by the Pourbaix diagram of iron in water, which illustrates the thermodynamic equilibrium between the iron and its ion and oxide species; see Figure 1.11. On the figure, several zones are identified for iron in the $Fe-H_2O$ system: immune zone, passivation zone, and corrosion zone. The pore solution of concrete without carbonation usually has a pH value in the range 12.4–13.2, situating the iron in the immune–passivation zones. In other words, a passivation film is formed on the steel surface.

As the pH value of the pore solution is decreased by carbonation, this passivation film is destroyed and the anode reaction of corrosion can occur, turning Fe to Fe^{2+}. This phase is

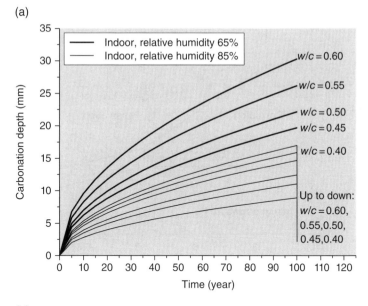

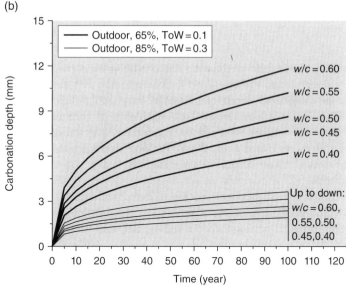

Figure 1.10 Carbonation depth from the C-2 model for OPC concrete with different *w/b* ratios for indoor exposure (a) and outdoor exposure (b). For the outdoor exposure, two cases are studied: one for the time of wetness ToW = 0.1 and driving rain probability p_{SR} = 0.5, and the other case for ToW = 0.3 and p_{SR} = 0.5. The former is representative of a dry climate and the latter of a wet climate.

termed the "corrosion initiation" for carbonation-induced corrosion. As mentioned before, the concept of carbonation depth is used to simplify the engineering judgement of carbonation extent, and the profile of pH value actually follows a gradual change along the carbonation depth; see Figure 1.5. Thus, the position of carbonation depth, determined by phenolphthalein

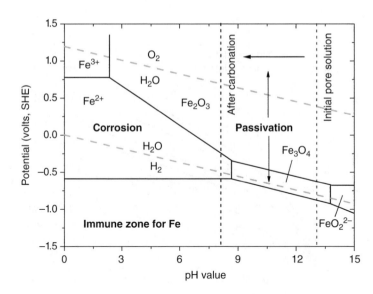

Figure 1.11 Pourbaix diagram for Fe–H$_2$O system at 25 °C (note that the accurate position of the lines drawn depends on the concentration of the ions, and the ion concentration is taken as 10^{-6} M). *Source:* adapted from Kośmider 2011, Wikimedia Commons, Public domain.

solution, cannot accurately capture the pH value condition for corrosion initiation. Actually, both laboratory research and in-field investigations revealed that the steel corrosion can be initiated before the carbonation depth (quantified by the phenolphthalein solution) reaches the steel surface. This distance was termed "residual depth of carbonation" or "carbonation remains," and this value was observed as 5.8–27 mm for a cover thickness of 16.9–50 mm for outdoor concrete elements (Dong *et al.*, 2006). But note that this residual depth adopted negative values for indoor exposure from the same data set; that is, −16 to +5.3 mm for concrete cover of 20–40 mm.

Once initiated, the electrochemical process of corrosion consists of anode and cathode reactions: the Fe atoms lose electrons at the anode and turn into Fe^{2+} ions; the electrons, oxygen and water form OH$^-$ ions at cathode; and the OH$^-$ ions formed flow back to the anode to produce Fe(OH)$_2$ or Fe(OH)$_3$ compounds and deposit as solid phases (rust). The intensity of the electrochemical process can be quantified by the electrical current density, or corrosion rate I_{corr}, between the anode and cathode. The corrosion rate is controlled by both the electrical resistance of the concrete and the cathode reaction: when the pore saturation is low (e.g. the corresponding relative humidity lower than 90%), the corrosion rate is controlled by the concrete resistivity since the OH$^-$ ions must flow back to the anode to close the electrical circuit through the concrete around the steel; when the pore saturation is high (e.g. the relative humidity higher than 95%), the diffusion of oxygen to the cathode controls the whole process; for a pore saturation in between there exists a transition from resistance-controlled mechanism to a cathode-reaction-controlled one (Raupach, 2006). Under carbonation, the corrosion is a general corrosion type; that is, the anodes and cathodes are evenly distributed on the steel surface and the anode and cathode are regarded as having comparable sizes.

The corresponding anode/cathode ratio is usually considered as 1.0 in models for corrosion current evaluation (Yu *et al.*, 2014).

1.3.2   Influential Factors

The target parameter for the steel corrosion is the corrosion current density I_{corr} hereafter. Admittedly, the first influential factor for the steel corrosion is the electrochemical property of steel itself. Incorporating some metal elements like chromium (Cr) and nickel (Ni) will greatly increase the corrosion resistance of steel, but this section treats only reinforcement steel bars made from conventional carbon steel (black carbon steel). Thus, the electrochemical properties of steel are not included as influential factors.

Temperature and Humidity

The influential factors of environmental conditions refer to the relative humidity and the ambient temperature. The relative humidity determines the pore saturation of concrete, and thus the corrosion mode. This influence is determinant for corrosion current: as the pore saturation increases with the relative humidity, the corrosion mode changes from a resistance-controlled one to a cathode-reaction-controlled one. The range of corrosion mode change is estimated as 90–95% in terms of environmental relative humidity (Yu *et al.*, 2014). Accordingly, given the concrete quality and cover thickness, the corrosion current is to increase firstly with the pore saturation due to the reduced electrical resistance, reaching a maximum around $h = 90\%$, and then to decrease with the pore saturation due to the slower oxygen diffusion to the cathode. The ambient temperature is believed to play a role in the corrosion rate of reinforced steel, since both the electrochemical reactions and the oxygen diffusion are influenced by ambient temperature. Tuutti (1982) reported a logarithmic rise of corrosion rate for carbonated concrete specimens for the temperature range −20 to +30 °C. Thus, the ambient temperature should be taken into account for engineering application.

Concrete Cover

The concrete cover has a determinant impact on the corrosion current through its thickness, compactness, and pore saturation. The influence of pore saturation has been considered through the environmental relative humidity. The thickness of concrete cover has a different impact on the two corrosion modes: for the resistance-controlled mode the thickness can contribute to the electrical current path and thus the current increases with concrete cover, but this increase is estimated to be rather moderate; for the cathode-reaction-controlled mode the thickness delays the oxygen diffusion and thus reduces the corrosion current. The compactness of concrete influences the corrosion rate in the resistance-controlled mode through the concrete resistivity, and the corrosion rate in the cathode-reaction-controlled mode through the diffusivity for oxygen. For both modes, high compactness of concrete helps to reduce the corrosion current. In some studies the compactness of concrete was denoted by the *w/c* (*w/b*) ratio of the concrete mixture. Note that the compactness of structural concrete can also evolve due to the further hydration of cementitious materials, particularly for those concretes incorporating a large quantity of SCM.

Concrete–Steel Interface

Another factor involved in the corrosion current is the quality of the concrete–steel interface and its impact on the corrosion process. So far, the steel corrosion induced by the carbonation of concrete cover assumes a perfect adhesion between the concrete and steel. Actually, the interface can contain heterogeneous defects, and these defects can change the corrosion initiation time and corrosion rates (Nasser *et al.*, 2010). The impact of concrete–steel interface quality on the corrosion process is still to be quantified.

1.3.3   Models

The aforementioned knowledge on the initiation of steel corrosion by carbonation and the corrosion current allows the establishing of practical models on corrosion initiation and electrical current for engineering use. Two models are presented for corrosion initiation and corrosion current in the following.

Model C-3: Corrosion Initiation by Carbonation

Corrosion initiation has been conventionally taken as the carbonation depth reaching the surface of reinforcement bars. However, owing to the simplistic nature of the carbonation front concept and the approximate nature of carbonation depth measured by phenolphthalein solution, the initiation condition can be more appropriately expressed as

$$x_c - \Delta x_c \geq x_d \qquad (1.17)$$

with Δx_c (m) the residual carbonation depth and x_d (m) the concrete cover thickness to the reinforcement bars. This residual depth accounts for the fact that the corrosion can be already initiated before the carbonation depth (detected by phenolphthalein solution) reaches the steel surface. The residual carbonation depth can take the following expression, adapted from the CECS technical standard (CECS 2007),

$$\Delta x_c = \left(1.2 - 0.35\sqrt{k_{carb}}\right)x_d - 1.94k_{carb} - 3.46 \quad \text{with} \quad k_{carb} = \frac{x_c}{\sqrt{t}} \qquad (1.18)$$

with x_c, Δx_c, and x_d in millimeters and t in years. The carbonation rate coefficient k_{carb} (mm/a$^{0.5}$) in this expression is defined as the ratio between the carbonation depth x_c and the square root of exposure age at measurement. Note that this model was regressed from 120 RC elements from 18 cities in China with the concrete cover thickness ranging from 20 to 50 mm. No dispersion analysis is available for this model. Taking the measured value range for k_{carb}, the carbonation residual depth is illustrated in Figure 1.12 for different values of cover thickness.

Model C-4: Corrosion Current

Once initiated, the corrosion develops with a certain density of corrosion current. From the mechanisms for steel corrosion, the model should include the resistance-controlled mode and cathode-reaction-controlled mode of steel corrosion in concrete. However, few models in the

(a)

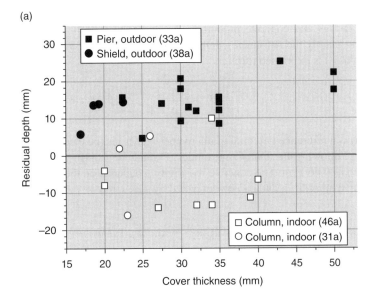

(b)

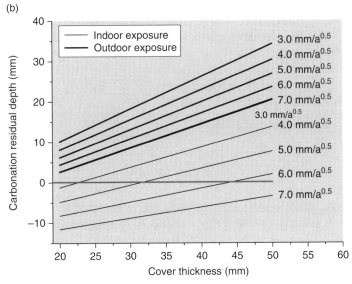

Figure 1.12 Residual carbonation depth measured from in-situ RC elements (a) and predicted from model C-3 in terms of the ratio k_{carb} (3.0–7.0 mm/a$^{0.5}$) and cover thickness (b). Image (a) is generated from data taken from Dong *et al.* (2006: table 2). The residual depth on the figure is for $W_t = 1.0$ and 0.426 for indoor and outdoor conditions, and the outdoor value corresponds to ToW $= 0.1$ and $p_{SR} = 0.5$ in Figure 1.10.

literature can account for both mechanisms and properly describe the influence of both environmental factors and concrete properties. Accordingly, a simplified model is retained here considering mainly the resistance-controlled corrosion. The electrical current density of corrosion I_{corr} (A/m²) is expressed through

$$I_{corr} = \frac{V_0}{\rho} \tag{1.19}$$

where V_0 (V/m) is a reference potential gradient and ρ (Ω m) is the electrical resistivity of the concrete cover. Following the DuraCrete model (DuraCrete, 2000), the reference potential gradient is taken as 0.76 V/m. The concrete resistivity ρ depends on the pore saturation (environmental relative humidity), and can be described by

$$\rho = \rho_0 \left(\frac{t}{t_0}\right)^m k_{RH} k_T \quad \text{with} \quad t \le 1\,\text{year} \tag{1.20}$$

where ρ_0 is the concrete resistivity measured at age t_0, the exponent m is the aging exponent related to the electrical resistivity, k_{RH} is the influence factor of relative humidity on the resistivity, and k_T is the influence factor of temperature. This expression is adapted from the DuraCrete model on corrosion rate. The original model also included the influence factors of curing condition and the chloride concentration. According to Yu et al. (2014), the factor k_{RH} can take the following form:

$$k_{RH} = \left(-1.344h^2 + 3.709h - 1.365\right)^{-1} \quad \text{for} \quad 50\% \le h \le 100\% \tag{1.21}$$

The factor k_T takes the following form:

$$k_T = \frac{1}{1 + \alpha_T(\theta - \theta_0)} \tag{1.22}$$

with α_T (°C⁻¹) a coefficient and θ and θ_0 (°C) the ambient and reference temperatures. Further, as the corrosion is assumed to reduce the cross-section of steel bars uniformly, the corrosion rate (units of mm/a or µm/a) can be estimated through Faraday's law as

$$\dot{r} = 0.0116 I_{corr} \tag{1.23}$$

with $\dot{r}$ (mm/a) the corrosion rate and I_{corr} (µA/cm²) the corrosion current density.

The combined model from Equations 1.19 to 1.23 describes the resistance-controlled corrosion of reinforcement steels and considers the concrete compactness through the electrical resistance. The influence of the humidity and temperature is included in the concrete resistivity, and the time dependence of resistivity is also included to take into account the microstructure evolution of concrete during the service life. This model can be extended without difficulty to include other factors, such as initial chloride content in concrete. Since all parameters are macroscopic the model can be easily used in applications. However, several drawbacks are to be noted for this model. First, the measurement and quantification of corrosion current are subject to significant dispersion and dependent on the test methods, and no dispersion analysis for this model is available. Thus, for a practical case the credibility of model should go to the predicted magnitude than the predicted value. Second, the model considers only the resistance-controlled mechanism; thus, the valid humidity range of this model is below 90%, and the model gives a conservative estimation for the higher humidity range. Finally, the model, from Equations 1.19 to 1.22, was calibrated mostly from short-term tests in the laboratory, and so the validation of long-term in-situ tests is needed.

Nevertheless, this model of corrosion current includes the main mechanism of carbonation-induced corrosion and the main influential factors, and the prediction on corrosion rate can be helpful for engineering application. Figure 1.13 shows the simulation results for the corrosion current in terms of concrete cover, relative humidity, and concrete resistivity. The relevant parameters and their usual value ranges are given in Table 1.3. Note that a corrosion current of less than 0.1 $\mu A/cm^2$ is commonly regarded as the passivated state for steel.

1.4   Basis for Design

1.4.1   Structural Consequence

The corrosion products, identified as a compound of $Fe(OH)_2$ and $Fe(OH)_3$ (Liu and Weyers, 1998), have an expansive nature. With the formation of the products, a local stress can accumulate on the interface between the concrete and steel that is high enough to facture the concrete cover. As the corrosion develops further, the corrosion products can augment the width of the surface cracks until spalling of the concrete cover occurs. Meanwhile, the adhesion between the concrete and steel can be destroyed by the corrosion products, and the bonding strength of the concrete–steel interface is reduced. The direct consequence of this bond loss is the reduction of the RC element stiffness, particularly for RC beams in bending. The aforementioned issues constitute the main concerns for corroded RC elements for the service limit state (SLS); that is, the fracture of the concrete cover, the spalling of the concrete cover due to crack width augmentation, and the reduced element stiffness due to bond loss of concrete–steel interface.[3] The corrosion rate in Equation 1.23 was used as a basic indicator to quantify the impact of corrosion on RC elements. The following model was given for the fracture limit for a corroding steel bar (DuraCrete, 2000):

$$\Delta r_{\mathrm{f}} = a_1 + a_2 \frac{x_{\mathrm{d}}}{\phi_{\mathrm{d}}} + a_3 f_{\mathrm{t,sp}} \qquad (1.24)$$

where Δr_{f} (mm) is the corrosion depth of steel bar needed to fracture the concrete cover, a_x ($x = 1,2,3$) are model parameters (74.4 μm, 7.3 μm, and -17.4 μm/MPa respectively), x_{d} and ϕ_{d} are respectively the concrete cover thickness and the diameter of the steel bars, and $f_{\mathrm{t,sp}}$ (MPa) is the splitting tensile strength of concrete. From this expression, normally a corrosion

---

[3]It is observed from experiments that a slight extent of steel corrosion, up to 4% in mass loss, can help to augment the bond strength (Li C.Q. *et al.*, 2014c). However, this observation should be treated with caution since a general and uniform corrosion extent is always assumed in these studies. In real cases, corrosion, even induced by carbonation, can present nonuniform characteristics, and this threshold can lose its sense.

(a)

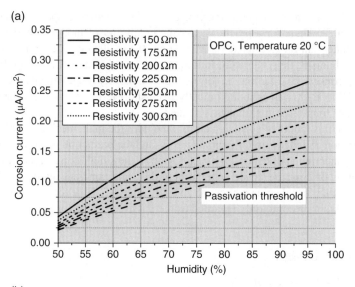

(b)

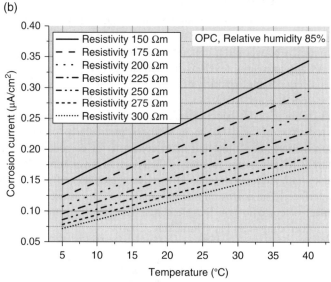

Figure 1.13 Simulation results from model C-4 for corrosion current in terms of relative humidity (a) and ambient temperature (b). The local resistivity data are retained for OPC concrete: resistivity measured at 28 days ranges from 150 to 300 Ω m for a *w/c* ratio from 0.6 to 0.3.

depth of 50—100 μm corresponds to the fracture limit state of the concrete cover. The following expression was proposed to account for the augmentation of surface crack width with the corrosion depth into steel bars (DuraCrete, 2000):

$$w_c = 0.05 + b_{crack}\left(\Delta r - \Delta r_f\right) \tag{1.25}$$

Table 1.3 Model parameters for the C-4 model and their usual value ranges

Parameter (unit)	Value/relation
Reference potential gradient V_0 (V/m)	0.76
Ageing exponent of electrical resistivity m	0.23 (OPC), 0.54 (GGBS), 0.62 (PFA)
Reference time t_0 (year)	0.0767 (28 days)
Resistivity of concrete measured at t_0, ρ_0 (Ω m)	150–300 (OPC), 250–550 (GGBS), 350–500 (PFA) $(w/c = 0.3 - 0.6)$
Humidity influence factor k_{RH}	1.0 ($h = 100\%$) to 6.52 ($h = 50\%$)
Environmental humidity h (%)	50–95
Temperature factor coefficient α_T ($^\circ C^{-1}$)	0.073 ($\theta > 20\,^\circ C$), 0.025 ($\theta < 20\,^\circ C$)
Reference temperature θ_0 ($^\circ C$)	20.0

PFA: pulverized-fuel ash.

where w_c (mm) is the crack width and the parameter b_{crack} is equal to 0.1 and 0.125 for top cast and bottom cast bars respectively. Taking the conventional crack control width for an RC element in design, this expression gives a corrosion depth of 25 μm from cover fracture to crack width attaining 0.3 mm, which is a rather short delay. Accordingly, from a design point of view, these two stages can be conservatively combined into one state, taking Δr equal to 50–100 μm as the criterion.

As the corrosion develops further, the loading capacity of RC elements can be affected through different aspects of deterioration: the spalling of the concrete cover between corrosion cracks, the reduction in cross-section of the steel bars, the reduced ductility and tensile strength of the steel bars, and the loss of bond strength of the steel bars in the concrete section. The spalling of the concrete cover reduces the concrete section and exposes the steels more directly to aggressive environments; the reduced section of steel bars deceases directly the loading capacity of the RC section, while the reduced ductility of steel bars will decrease the resistance capacity of RC elements to dynamic loadings. The loss of bond strength reduces the loading capacity of the RC section mainly through the decrease in the effective area in compression. It was shown that the reduction of bonding strength can be as much as 90% as the bar section loses 20% of its mass (Li C.Q. *et al.*, 2014), and the flexural deflection of corroded RC beams is mainly due to the bond loss (Castel *et al.*, 2011). On the basis of the available knowledge, some standard assessment methods were proposed for loading capacity for the column and beam sections; see DuraCrete (1998). Since the corrosion extent is never allowed to develop further than affecting the SLS, the quantitative aspect of structural performance of RC elements in the post-corrosion phase is not elaborated in this text.

1.4.2 Design Considerations

The durability design is to be presented in a systematic way in Chapter 7, but some aspects specific to carbonation-induced corrosion are discussed here. The durability design of RC elements against carbonation-induced corrosion should be performed at both the materials level and the structural level for a given intensity of environmental action and expected design working life. At the materials level, the compactness of concrete materials can always help because with high compactness concrete has higher resistance to CO_2 diffusion and a higher electrical resistance to control effectively the corrosion current in the normal humidity range. To resist carbonation and provide a highly alkaline environment for reinforcement steel, a high CH content in concrete in its hardened state is appreciated in durability design; that is, OPC is favored as a concrete binder. However, some kind of compromise can be made between the compactness of concrete achieved and its CH content: with the extensive practice of incorporating SCM in concrete binders, the CH can be consumed by the secondary hydration of the mineral SCM but a high compactness can also be achieved as a result. According to state-of-the-art knowledge, high compactness seems to be favored over the CH content; that is, as long as the compactness is achieved the CH content is no longer a concern. On the last point, extensive cracks at an early age should be avoided to assure the compactness of the concrete.

At the structural level, the concrete cover thickness should be adequate to provide sufficient resistance to CO_2 diffusion. Since RC elements can bear working cracks with a certain opening width at the concrete surface, these openings are believed to accelerate CO_2 diffusion as well as the subsequent corrosion process. Normally, the cracks with opening width below 0.3 mm are regarded as harmless for carbonation-induced corrosion (Li K.F. *et al.*, 2014b). According to state-of-the-art concrete technology and experiences, high compactness of concrete, adequate cover thickness, and efficient control of working cracks can provide RC elements and structures with service lives of 50 years or longer without additional protection measures. As RC elements and structures are subject to other environments, carbonation is always the fundamental process for other deterioration processes. In this case, the possible intervention of carbonation in other deterioration processes should be treated in the design.

2

Chloride Ingress and Induced Steel Corrosion

This chapter treats the deterioration of reinforced concrete by chloride ingress. The chlorides are present in seawater, groundwater, and deicing salts. The RC elements exposed to these sources can have the chlorides migrating from external environments into structural concrete. While the chloride ingress into concrete will not necessarily compromise the concrete properties, the chlorides can seriously destroy the electrochemical stability of the embedded reinforcement steel bars, inducing electrochemical corrosion. Actually, the deterioration of RC elements by chloride ingress has become one of the major concerns for durability of concrete structures. This chapter begins with some typical cases of RC elements and the structures affected by chloride ingress. The mechanisms are then introduced successively for chloride ingress and induced steel corrosion, together with an analysis on the main influential factors. Using the state-of-the-art knowledge, the modeling of these processes is treated through mechanism-based or empirical models. The basis for durability design against chloride-induced corrosion is given at the end.

2.1 Phenomena and Observations

RC elements in concrete structures can be seriously affected by chlorides present in the environment. As concrete comes into contact with chlorides, the chloride ions can transport into the material via the pore space by several mechanisms. The electrochemical stability of the embedded reinforced steel can be destroyed as the chloride ions reach the steel surface and accumulate to a sufficient level. The corrosion induced by chlorides assumes a "pitting" pattern, and the local corrosion rate can be more significant than the general corrosion pattern. Thus far, steel corrosion by chlorides has resulted in extensive damage to RC structures worldwide and is the primary concern in durability design.

Durability Design of Concrete Structures: Phenomena, Modeling, and Practice, First Edition. Kefei Li.
© 2016 John Wiley & Sons Singapore Pte. Ltd. Published 2016 by John Wiley & Sons Singapore Pte. Ltd.

(a)   (b)

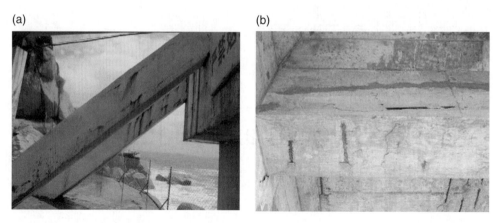

Figure 2.1   Corroded RC elements in a high-pile wharf on the southeastern coast of China. (a) Concrete cracking of 0.5 mm by chloride ingress on an inclined RC beam and (b) general spalling of concrete cover under an RC beam. The structure was put into service in 2001, the concrete grade was C30, and the design value of concrete cover was 50 mm. The in-situ strength of concrete was 35–48 MPa, and the concrete cover thickness inferior to design value was about 50% for these beams. *Source:* courtesy of Shengnian Wang.

The first chloride source is the marine environment, where chlorides exist either as aqueous ions in seawater or as air-borne chlorides contained in tiny droplets of seawater. Thus, marine chlorides can affect RC structures exposed directly to seawater as well as those situated in coastal areas. The usual chloride concentration in seawater is near 20 g/L (for a total salinity 3.5%). The air-borne chlorides can be transported by air currents and deposited on the surface of concrete. The deposited quantity of chloride on a concrete surface depends very much on the local climate and meteorological conditions; for example, RC structures were observed to be affected by air-borne chlorides at a distance of 3 km or further (Neville, 1995a). The intensity of chloride action on RC elements is closely related to the exposure conditions to marine chlorides, among which the splashing by sea waves is considered to be the most severe exposure condition. The RC structures most affected in marine environments include sea link projects, ports and harbors, sea tunnels and building structures in coastal areas. The premature deterioration of RC structures in a marine environment is rather general, and Figure 2.1 illustrates the RC elements affected by chlorides in one coastal high-pile wharf of age 10 years.

The second chloride source is deicing agents (salts) applied to melt ice or snow on road surfaces to ensure traffic safety. These deicing agents can be industrial salts containing mainly the chloride salts ($NaCl$, $CaCl_2$, or $MgCl_2$), or organic agents as ethylene glycol or propylene glycol.[1] As chloride salts are applied, the chlorides melt the ice and form a high-concentration chloride solution on the concrete surface, as high as 14–23% ($NaCl$), 15–25% ($MgCl_2$) and 17–28% ($CaCl_2$) according to Jain *et al.* (2012). The solution formed, rich in chlorides, can be sprayed onto the surface of RC elements by traffic and affect the embedded bars. If the deicing salts are applied on the pavement surface on a bridge, the chloride solution can easily infiltrate into the RC elements underneath. The chlorides can also be transported much farther by

---

[1] Organic deicing agents are more conventionally used in the aviation industry for aircraft.

Figure 2.2   Urban viaduct affected by deicing salts. The solution of deicing salts infiltrated to the pier cap and induced advanced corrosion of the embedded reinforcement steel. The structure was 12 years old at inspection. The concrete grade was C30 for the pier cap, and the deterioration is typical for water-proof failure followed by deicing salts attack. *Source:* courtesy of Peixing Fu.

automobile tires into parking lots and garages, affecting indoor RC elements. This problem was reported to heavily affect parking structures in North America (Trethewey and Roberge, 1995). Note that the deicing salts are also toxic to the aquatic environment in soils. Hence, agents that are more ecological are encouraged for deicing purposes. Figure 2.2 illustrates an urban viaduct affected by the deicing salts.

Apart from these two sources, chlorides can also be present in the soils and groundwater. The soils and groundwater can contain a high content of salts for different reasons: either the water table is connected to seawater and thus the ionic species in seawater are also present in the soils and groundwater, or the soils are subject to a salinization process for geological reasons. For the latter case, soils can have very high content of salts, of which the main ion species are Na^+, K^+, Ca^{2+}, Mg^{2+}, Cl^-, and SO_4^{2-}. Table 2.1 gives some typical values for the ion concentration in seawater, groundwater connected to seawater, and saline soils in the northwestern region of China. For the RC elements exposed to such soils or groundwater, the chloride ions are not the only aggressive species; other deterioration processes, such as sulfate attack or pore crystallization of salts, can dominate over the action of chlorides.

Since chlorides have caused a heavy loss of RC structures, multiple efforts have been dedicated to the prevention of premature deterioration by chloride ingress. The process of chloride transport into concrete involves multiple physical and chemical properties of concrete that are by nature random and difficult to quantify. Thus, engineers give more credit to the long-term observation of chloride ingress in natural environments. Exposure stations were established to this purpose. Nowadays, a number of exposure stations or sites for RC elements subject to chloride action have been set up worldwide; for example, La Rochelle station in

Table 2.1 Typical values for ion concentration and pH value in water and soil in China

Source	Na$^+$	K$^+$	Ca^{2+}	Mg^{2+}	Cl$^-$	SO$_4^{2-}$	pH
Seawater (mg/L), Hong Kong, Macau coast (Li K.F. *et al.*, 2015)	—	—	416	1013	14650	1700	7.64
Saline soil (mg/kg), Turpan Basin (Wei *et al.*, 2014)	52876	377	3861	1620	67199	35715	7.36
Coastal groundwater (mg/L), Tianjin City (Li F.X. *et al.*, 2015)	—	—	—	235	4339	1465	7.53

(a)   (b)

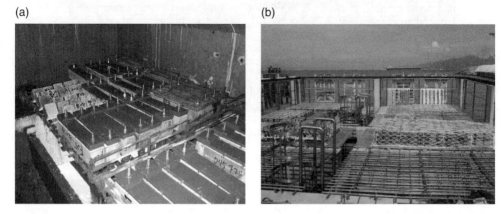

Figure 2.3 Exposure stations on the southeastern coast of China: (a) Zhanjiang exposure station and (b) the exposure station for Hong Kong–Zhuhai–Macau (HZM) sea link project. The Zhanjiang exposure station was established in 1987, located in the wharf of Zhanjiang port (average temperature 22.5 °C and relative humidity 85%), and has accumulated data on chloride ingress for nearly 30 years. The HZM exposure station was newly built in 2013 on one of the project's artificial islands (average temperature 22.3–23.1 °C and relative humidity 77–80%), and served for the long-term monitoring of chloride ingress for the RC elements in the HZM project. *Source:* courtesy of Shengnian Wang.

France (Baroghel-Bouny *et al.*, 2004), Qeshm island site in the Persian Gulf (Valipour *et al.*, 2013) and Träslövsläge harbor on the Swedish west coast (Sandberg, 1998). Figure 2.3 shows two marine exposure stations on the southeastern coast of China for concrete and RC elements.

2.2   Chloride Ingress

2.2.1   Mechanism

The chloride ions can transport into concrete through different mechanisms: diffusion through the connected liquid phase in pores, convection flow of pore liquid, and the binding of chloride ions by the solid matrix of concrete; see Figure 2.4.

Diffusion refers to the transport of chloride in pore solution due to the chloride concentration gradient; that is, from a high concentration region to a low concentration one. The transport rate is scaled by the diffusivity of chloride ions. The diffusivity of chloride ions in water or

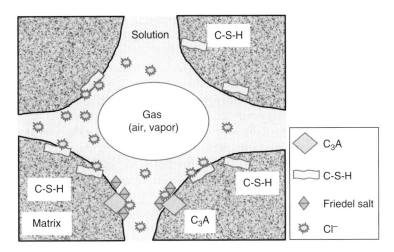

Figure 2.4   Transport of chloride ions through concrete pores by different mechanisms.

dilute solution is 2.03×10^{-9} m²/s (25 °C) while the Cl⁻ diffusivity in pore solution will be lower due to the existence of other ions, positively or negatively charged. For example, for the same solution concentration (0.5 M), the diffusivity of Cl⁻ was measured as 6.25×10^{-12} m²/s in NaCl solution but 10.38×10^{-12} m²/s in CaCl₂ solution on cement paste specimens (Ushiyama and Goto, 1974). From the literature (Allard *et al.*, 1984), the pore solution contained 1500 mg/L Na⁺, 6300 mg/L K⁺, 90 mg/L Ca²⁺, <300 mg/L SO₄²⁻, and 4267 mg/L OH⁻ for freshly hardened cement paste made of OPC (age 10 months, pH 13.5), and these values changed to 1500 mg/L Na⁺, 92 mg/L K⁺, 92 mg/L Ca²⁺, 640 mg/L SO₄²⁻, and 1350 mg/L OH⁻ for a concrete of over 65 years of age (pH 12.9). Apart from this chemistry aspect, the apparent diffusivity of chloride in concrete is also affected by the porosity and the tortuosity of the pore space. Moreover, pore saturation plays an important role in the chloride diffusivity since the chloride ions can only transport via the liquid phase: the lower the pore saturation, the less the transport rate.

The convection transport of chloride is the movement of Cl⁻ in the pore space carried by the flow of the liquid phase. The flow rate of the pore liquid is determined by the pressure gradient and the permeability of concrete, which can be described through Darcy's law as

$$J_{\text{solution}} = -\frac{k}{\eta_l}\,\text{grad}\,(p) \tag{2.1}$$

where J_{solution} (m³/(m² s)) is the volumetric flow rate, p (Pa) is the pressure, η_l (kg/(m s) or Pa s) is the viscosity of the liquid, and k (m²) is the intrinsic permeability, related only to the pore structure of concrete. (Note, as the pressure is noted by water head h_w (m) and the infiltrating liquid is water, Darcy's law is

$$J_{\text{solution}} = -k_S^w\,\text{grad}\,(h_w) \tag{2.2}$$

and the corresponding permeability k_S^w (m/s) is defined as the saturated permeability.) Although termed *intrinsic permeability*, the k values measured for gas flow can be several orders of magnitude higher than the liquid flow, and the underlying mechanism is discussed later, in Chapter 8. The intrinsic permeability of concrete is normally in the range of 10^{-23}–10^{-20} m^2 for liquids (Baroghel-Bouny, 2007).

A particular pattern of liquid flow, termed *surface absorption* or *capillary suction*, occurs as the dry concrete surface comes into contact with liquids. The liquid is driven by the capillary pressure on the concrete surface and flows through the connected coarse pores into the material. If the liquid solution contains chlorides, this transport pattern is much more efficient for intake of chlorides, partially accounting for the high chloride content accumulated on the surface layer of concrete in the marine splash zone. For this process, the physical parameter called water sorptivity, or *water absorption rate*, can pertinently describe the transport capacity of the concrete surface, and this parameter will be used later as one of the durability indicators.

The binding of chloride by the solid matrix of concrete involves physical and chemical mechanisms. The physical binding of Cl^- is due to the ion exchange with the OH^- ions on the adsorption sites, provided mainly by the C-S-H (Nguyen, 2007). The chemical binding of Cl^- is mainly due to the reaction between the cement mineral C_3A, C_4AF and Cl^- ions to form the chloroaluminate calcium hydrates; that is, $3CaO \cdot Al_2O_3 \cdot CaCl_2 \cdot 10H_2O$ (Friedel salt) or $3CaO \cdot Al_2O_3 \cdot 3CaCl_2 \cdot 10H_2O$ (Suryavanshi and Scantlebury, 1996). Since the ion exchange involves OH^- ions the physical binding is sensitive to the pH value of the pore solution. The physical binding can be reversible with respect to the chloride concentration in the pore solution, wheras the chemical binding is irreversible. For concretes, the relation addressing the equilibrium between the chloride ions in the aqueous phase and adsorbed chlorides on solid phases is termed the *chloride sorption isotherm*. This isotherm actually predicts the quantity of chlorides bound to solid phases from the chloride concentration in the pore solution. From the available knowledge, the content of C-S-H, the mineral C_3A, C_4AF in cement grains, and the OH^- concentration (pH value) of the pore solution are considered to be determinant on the adsorption quantity of chlorides.[2]

2.2.2   Influential Factors

Several properties are fundamental to the different transport mechanisms of chloride ingress: the chloride diffusivity for diffusion, the permeability for pore liquid flow, the water sorptivity for the surface absorption, and the binding capacity for the chloride adsorption by solid matrix. These properties can be divided into the transport properties, including chloride diffusivity, permeability and water sorptivity, and the binding capacity. The transport properties are all closely related to the microstructure of concrete and the pore saturation, but the binding capacity is more related to the cement composition and cement hydrates.

---

[2]Owing to chloride binding, the quantity of chlorides in concrete is noted by two different but related terms: the "chloride concentration" refers to the concentration of chloride ions in the aqueous phase of the pore solution (mol/L, or mol/m^3), and the "chloride content" is the mass ratio of total chlorides, including the chlorides in the pore solution and bound by the solid matrix, with respect to binder materials or concrete (wt%), or the mass (quantity) of total chlorides in a unit volume of concrete (kg/m^3, or mol/m^3). These two terms are related through Equation 2.16.

Transport Properties

The chloride diffusion coefficient is related to the microstructure of concrete, pore saturation and the multispecies environment of the pore solution. The following expression accounts for the influence of the pore saturation (Nguyen, 2007):

$$D_{Cl}\left(s_1\right) = D_{Cl}^0 s_1^{\lambda} \quad \text{with} \quad s_1 \geq 0.40 \tag{2.3}$$

with D_{Cl}^0 the chloride diffusivity in totally saturated concrete ($s_1 = 1.0$) and λ the exponent. This expression was adapted from the saturation dependence of ion diffusivity through the spectroscopic impedance technique and calibrated for concrete materials. The suggested value for concrete materials is $\lambda = 6.0$. Note that the valid range for this expression is a pore saturation above 0.40, below which the liquid solution is considered to be no longer connected or percolated in pores.

The permeability of concrete is also related to the microstructure of concrete and the pore saturation. The influence of pore saturation on permeability is expressed by the relative permeability k_{rl}, expressed through the Mualem model as (Van Genuchten, 1980)

$$k_{rl} = \frac{k\left(s_1\right)}{k_0} = \sqrt{s_1}\left[1-\left(1-s_1^{\beta}\right)^{1/\beta}\right]^2 \tag{2.4}$$

with β as the regression parameter in the relative liquid permeability. Note that this expression also gives a very low value, $k_{rl} \approx 0$, as $s_1 < 0.40$, meaning that below this saturation Darcy flow of the pore solution is almost impossible. As for the surface absorption, the water sorptivity S_w (m³/(m² s$^{1/2}$)) is related to the pore saturation through (Hall, 1989)

$$S_w\left(s_1\right) = S_w^0\left(1-1.08s_1\right)^{1/2} \tag{2.5}$$

with S_w^0 the water sorptivity in the totally dried state ($s_1 = 0$). The saturation dependence of the transport properties is illustrated in Figure 2.5.

Binding Capacity of Chlorides

The binding capacity of chloride by concrete is mainly determined by the cement composition and the hydrates and by the chloride concentration in the aqueous phase. The capacity s_{Cl} is the mass or quantity of bound chlorides with respect to the concrete volume or mass. Some empirical laws are available for chloride binding by cement-based materials in the form of linear Langmuir or Freundlich laws (Tang and Nilsson, 1993), but experiments showed that these laws have different valid ranges in terms of the chloride concentration (Li and Li, 2013). A recent mechanism-based model (Nguyen, 2007) considered ion exchange and Fiedel's salt formation and expressed the chloride adsorption quantity s_{Cl} (mol/m³) in terms of the contents of C-S-H and the C_3A and C_4AF minerals in cement, as

$$s_{Cl} = n_{CSH}\frac{\alpha_{Cl}c_{Cl}}{c_{OH}+\beta_{Cl}c_{Cl}} + 2\left(n_{C_3A}+0.5n_{C_4AF}\right) \tag{2.6}$$

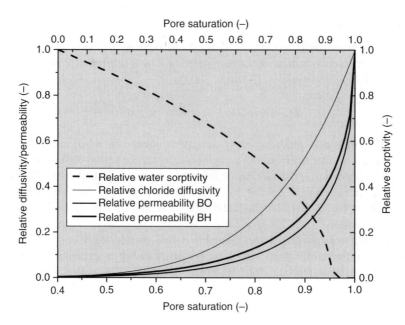

Figure 2.5 Influence of the pore saturation on the relative diffusivity, relative permeability (left-bottom axis) and relative sorptivity of concrete (right-top axis). The main characteristics of BO and BH concretes are the compressive strength $f_c^{BO} = 49.5$ MPa, $f_c^{BH} = 115.5$ MPa; $\beta_{BO} = 2.275$, $\beta_{BH} = 2.06$ from Baroghel-Bouny (2007).

with n_{CSH}, n_{C_3A}, n_{C_4AF} (mol/m³) the molar concentration of C-S-H, C_3A and C_4AF minerals in concrete, c_{Cl}, c_{OH} (mol/m³) the molar concentration of Cl⁻ and OH⁻ ions in pore solution, and α_{Cl}, β_{Cl} the model parameters depending on the composition of the cement hydrates. For concrete incorporating OPC, these parameters were calibrated as $\alpha_{Cl} = 0.0343$ and $\beta_{Cl} = 0.132$. The first term of this expression accounts for the ion exchange on C-S-H and the influence of OH⁻ ions (pH value of pore solution), and the second term represents the formation of salts. In a mathematical sense, this expression is an adapted Langmuir law. Using the same data set of OPC pastes in Figure 1.6, the chloride binding capacity of OPC concrete is illustrated in Figure 2.6 for different w/c ratios. One can see that a low pH value of the pore solution favors chloride binding. However, this statement cannot be applied to the carbonated concrete since the carbonation decreases the pH value and consumes the C-S-H simultaneously.

Temperature, Humidity, and Surface Chlorides

The environmental conditions that influence chloride ingress include the temperature, relative humidity, and the surface chloride concentration. The relative humidity determines the pore saturation of concrete through the moisture equilibrium in the pores. This fundamental relation is explained in Chapter 6. This influence has been reflected in Equations 2.3, 2.4, and 2.5 through the pore saturation levels. The temperature is considered to influence the diffusion and permeation processes. For diffusion, a high temperature tends to increase the molecular

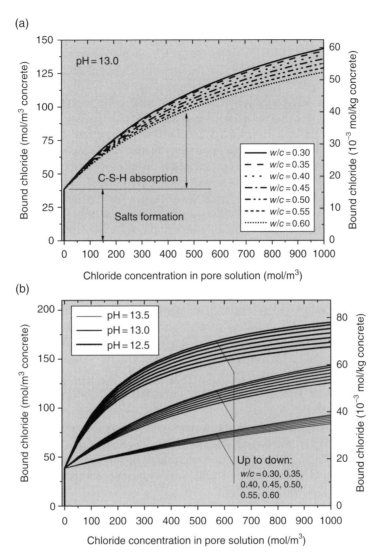

Figure 2.6 Chloride binding capacity for OPC concrete in terms of *w/c* ratio for (a) pH 13.0 and (b) different pH values. The figure uses the same hydration data for C-S-H contents as in Figure 1.6. The C_3A and C_4AF contents are taken respectively as 40 mol/m³ (mmol/L) and 30 mol/m³ (mmol/L) for the cement pastes following Nguyen (2007), and the volumetric ratio of paste in concrete is 0.35.

vibration rate and thus accelerate the diffusion process. This accelerated effect has been quantified through a thermal-activation coefficient b_e (K):

$$D_{Cl}(T) = D_{Cl}(T_0)\exp\left[-b_e\left(\frac{1}{T} - \frac{1}{T_0}\right)\right]$$ (2.7)

where the reference temperature $T_0 = 293\,\text{K}$ and $b_e = 4800 \pm 700$ K are suggested by the *fib* model code (*fib*, 2006). This value range seems to be confirmed by the Nguyen *et al.* (2009)

study on the chloride diffusivity on mortar specimens made of CEM I and CEM V binders. The temperature impacts on the permeation mainly through the viscosity of fluids. Take water for example; the viscosity of water changes from 1.793 Pa s at 0 °C to 0.653 Pa s at 40 °C, resulting in an increase of 175% for the flow rates. The mathematical form is given in Equation 6.43.

Chloride deposition on a concrete surface is one crucial environmental action. The surface chloride content increases with the deposition of air-borne chloride and needs several years to stabilize in a marine environment. Figure 2.7 illustrates the surface chloride accumulation processes collected from the concrete specimens in the Zhanjiang exposure station. The attainable chloride content in concrete cover depends on the exposure condition and the binder composition. This value has been investigated extensively for engineering use, and Table 2.2 summarizes the recommended values for different exposure zones and different binder compositions. The intensity of the chloride deposition will decrease with the distance away from the coast. However, quantifying the chloride deposition on RC elements away from the coast involves the local climate and meteorological conditions such as the wind speed, orientation and the atmospheric precipitation.

For chlorides from deicing salts, the chlorides accumulate on the surface of RC elements more by spray than by air-borne deposition. The chloride content in the concrete cover also has a transition period before achieving its maximum value. This maximum value can be estimated from the chloride sorption isotherm of the concrete and the chloride concentration in the salt solution on the concrete surface (*fib*, 2006). The following expression is given for the chloride content on the concrete surface away from the source of deicing salts:

$$C_s(x_h, x_v) = 0.465 - 0.051 \ln(x_h + 1) - \left[0.00065 (x_h + 1)^{-1.87} \right] x_v \qquad (2.8)$$

where C_s (% concrete) is the chloride mass ratio with respect to concrete and x_h and x_v are respectively the horizontal and vertical distances from the source of deicing salts. This relation is illustrated also in Figure 2.7. Equation 2.8 was established for concrete structures subject to deicing salts spray in German urban and rural areas with concretes containing CEM I cement and having $w/c = 0.45 - 0.60$ (*fib*, 2006).

Drying–Wetting Actions and Ageing of Diffusivity

A particular environmental action is the drying–wetting cycle. This action is the very reason for the high chloride content on the RC surface in the splash zone; see Figure 2.8. During wetting, as the splashing water droplets touch the concrete surface both water and chloride ions transport into the concrete pores; during drying, the water evaporates and leaves the chloride ions in pores. With repeated cycles the chloride ions can accumulate to a very high level; see Table 2.2. More important, this action creates a convection depth on the concrete surface in which the surface absorption and water evaporation dominate over the chloride diffusion. As the profiling method[3] is used to determine the surface chloride

---

[3]The chloride profiling method consists of determining the chloride content at different depths from the concrete surface and regressing this chloride content–depth relation through the empirical Fick's second law and obtaining the apparent surface chloride content and the apparent chloride diffusivity.

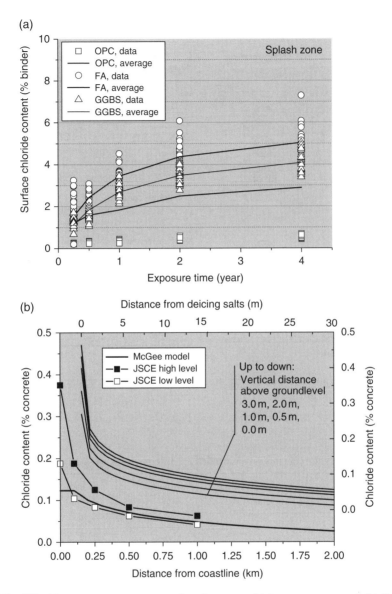

Figure 2.7 Chloride content on concrete surface in terms of (a) exposure age and (b) the distance from the coastline. (a) Data from the exposure tests on OPC, FA, GGBS specimens at Zhanjiang exposure station in the splash zone. (b) Illustration of the decrease of surface chloride content from the McGee (1999) model and the recommended value from Japanese standard JSCE (2010: table C8.2.2, p. 139) (left-bottom axes) and the surface chloride content in terms of distance from deicing salts source (*fib*, 2006) (top-right axes).

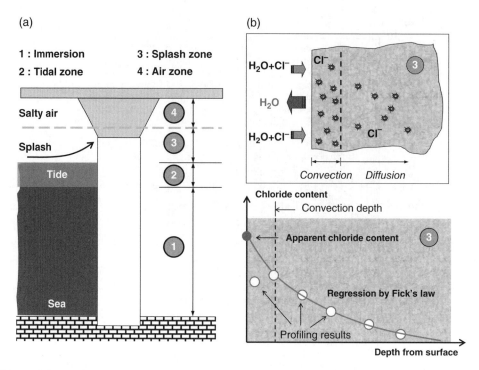

Figure 2.8 Marine exposure zones for RC structures (a) and chloride accumulation on the concrete surface in the splash zone (b).

Table 2.2 Surface chloride contents recommended for different marine exposure zones (%binder)

Source	Atmospheric zone (marine air)	Splash zone	Tidal zone	Immersed zone
DuraCrete (2000)				
OPC	$2.57 \times w/c$	$7.76 \times w/c$	$7.76 \times w/c$	$10.3 \times w/c$
PFA	$4.42 \times w/b$	$7.46 \times w/b$	$7.46 \times w/b$	$10.8 \times w/b$
GGBS	$3.05 \times w/b$	$6.77 \times w/b$	$6.77 \times w/b$	$5.06 \times w/b$
SF	$3.23 \times w/b$	$8.96 \times w/b$	$8.96 \times w/b$	$12.5 \times w/b$
Bamforth (1999)				
OPC	1.5–3.0	4.5	—	—
SCM	1.8–3.6	5.4	—	—
Life-365 (Thomas and Bentz, 2000)	3.6	6.0	4.8	—

SF: silica fume concrete, PFA: pulverized-fuel ash.

content, the regressed value, termed the *apparent surface chloride concentration* (content), will be higher than the actual chloride content; see Figure 2.8. This statement applies to the values in Table 2.2.

Under natural marine environments, the chloride diffusivity, characterized from the chloride profiling method, has been observed to decrease substantially with time. According to the established model (*fib*, 2006), this decrease obeys a power function:

$$D_{Cl}^{app}(t_2) = D_{Cl}^{app}(t_1)\left(\frac{t_1}{t_2}\right)^{n_{Cl}} \quad \text{with} \quad t_2 > t_1, \ n_{Cl} > 0 \quad (2.9)$$

This expression states that the apparent chloride diffusivity at a later exposure age t_2 can be related to an earlier age value at t_1 through a power function with the exponent n_{Cl}. The exponent adopts different values and depends mainly on the binder material composition of concrete, and the conventional value range is between 0.30 and 0.60, with the lower values for OPC concretes and higher values for concretes incorporating mineral SCMs. The durability design benefits considerably from this observation since long-term chloride diffusivity decreases with exposure age: the apparent diffusivity can decrease by 50% from 1 year age to 10 years for $n_{Cl} = 0.3$ and by 75% for $n_{Cl} = 0.6$. The underlying mechanisms for this ageing of apparent diffusivity include both the material ageing and the long-term interaction between the complex environmental actions and the concrete. Actually, continual hydration over long exposure ages can densify the material microstructure, thus decreasing the chloride diffusivity (Andrade *et al.*, 2011). Moreover, as concrete is exposed to real marine environments, the chloride transport is always coupled to the transport of moisture and other ion species in the pore solution, and this decrease can also be the result of the moisture transport from the same side of chloride ions. From the first mechanism, this law should be bounded by a certain age since one cannot expect the densification process to continue forever; but for the second mechanism, a specific study should be performed to determine the valid duration of this decreasing law.

Characterization of Chloride Diffusivity

The last issue concerns the laboratory characterization of the chloride diffusivity. Nowadays, multiple techniques have been developed in the laboratory to measure the chloride diffusivity of concrete. Depending on the experimental principles, these methods can measure the chloride diffusivity in the steady or nonsteady state, or the chloride migration coefficient in the steady or nonsteady state. The diffusivity in Equation 2.9 is actually a nonsteady-state diffusion coefficient D_{Cl}^{nssd}. The rapid chloride migration (RCM) method (Nordtest, 1999) is one of the most used methods, measuring the nonsteady-state chloride migration coefficient D_{Cl}^{nssm} under an electrical potential of 30–60 V. A correlation was found among different chloride coefficients: the migration coefficient D_{Cl}^{nssm} from the RCM method was nearly equal to the D_{Cl}^{nssd} from the immersion test (Chlortest, 2005). This was confirmed recently by laboratory tests on concrete specimens for the HZM project, and a further analysis on the correlation between the D_{Cl}^{nssm} from the RCM method and D_{Cl}^{nssd} from the long-term exposure tests showed that D_{Cl}^{nssm} is nearly two times larger than the D_{Cl}^{nssd} value; see Chapter 9 (Figure 9.5). Here, no analysis is given on the theoretical relations among these coefficients, but further details can be found in Tang (1999).

2.2.3 Models

Nowadays, multiple transport models are available for chloride transport in concrete. In real RC elements, the chloride ions transport via a pore solution containing multiple species and subject to multi-fields, but this multispecies aspect is neglected in this book and the multi-fields aspect will be presented later, in Chapter 6. Two models are introduced for chloride transport with chlorides as the sole transport ion. One model is mechanism-based, considering the different transport processes of chlorides, and the other is an empirical model recommended for engineering use. Both models are related to Fick's second diffusion law. Accordingly, this law is first presented as the modeling basis. Let us consider a pure diffusion process in one dimension. The ion flux J (mol/(m^2 s)) is related to the gradient of ion concentration C (mol/m^3) in the material through Fick's law:

$$J_F = -D\operatorname{grad}(C) \tag{2.10}$$

with D (m^2/s) the diffusion coefficient. The mass conservation of ions becomes

$$\frac{\partial C}{\partial t} = -\operatorname{div}(J_F) = -D\frac{\partial^2 C}{\partial x^2} \tag{2.11}$$

This equation is also called Fick's second law to make the difference with Equation 2.10. Analytical solutions are available for this equation for different boundary and initial conditions. If the boundary and initial conditions are

$$C(x,t=0) = C_0, \quad C(x=0,t>0) = C_S \tag{2.12}$$

then the analytical solution of Equation 2.11 becomes

$$C(x,t) = C_0 + (C_S - C_0)\left[1 - \operatorname{erf}\left(\frac{x}{2\sqrt{Dt}}\right)\right] \tag{2.13}$$

where erf is the error function and takes the following form:

$$\operatorname{erf}(y) = \frac{2}{\sqrt{\pi}}\int_0^y \exp(-z^2)\,dz \tag{2.14}$$

with $\operatorname{erf}(0) = 0$ and $\operatorname{erf}(\infty) = 1.0$. Actually, the analytical solution in Equation 2.13 forms the basis for most models of chloride ingress in concrete.

Model Cl-1: Mechanism-based Model

On the basis of the Fick's second law, this model takes into account the chloride binding and the influence of the pore saturation. The chloride concentration in the pore solution c_{Cl} (mol/m^3) is used as the basic variable. In concrete with a pore saturation s_l, the flux of the chloride ions in the pore solution includes the diffusion flux J_F (mol/m^2/s) and the convection flux J_C (mol/m^2/s):

$$J = J_F + J_C = -D_{Cl}^{eff}(s_l)\operatorname{grad}(c_{Cl}) - c_{Cl}k(s_l)\operatorname{grad}(p) \tag{2.15}$$

with D_{Cl}^{eff} (m^2/s) as the effective chloride diffusion coefficient in the pore solution in concrete. The total chloride content in concrete is

$$C = \phi s_1 c_{Cl} + s_{Cl}(c_{Cl}) \tag{2.16}$$

with the chloride binding capacity s_{Cl} (mol/m^3 concrete) expressed in Equation 2.6. The mass conservation of chlorides in concrete is

$$\left[\phi s_1 + s_{Cl}'(c_{Cl})\right]\frac{\partial c_{Cl}}{\partial t} = -D_{Cl}^{eff}(s_1)\frac{\partial^2 c_{Cl}}{\partial x^2} - k(s_1)\text{grad}(p)\frac{\partial c_{Cl}}{\partial x} \tag{2.17}$$

with p (Pa) the pressure exerted on the pore liquid phase. Neglecting the Darcy flow and using the boundary and initial conditions in Equation 2.12, the chloride profile is

$$c_{Cl}(x,t) = c_{Cl,0} + (c_{Cl,S} - c_{Cl,0})\left[1 - \text{erf}\left(\frac{x}{2\sqrt{\dfrac{D_{Cl}^{eff}(s_1)}{\phi s_1 + s_{Cl}'(\hat{c}_{Cl})}t}}\right)\right] \tag{2.18}$$

The $\hat{c}_{Cl}$ term stands for some intermediate concentration value with which the first term on left side of Equation 2.17 can be treated as constant. Note that this model describes the diffusion process under constant pore saturation, so no moisture transport is involved. If the pore saturation is subject to change during the chloride transport, one has a multi-field problem involving the concentration field and the pore saturation field, and one can refer to Chapter 6 for the relevant description. In Equation 2.18 the chloride diffusivity refers to the effective diffusivity of chloride in the pore solution. The saturation dependence and the thermal dependence of the chloride diffusivity can take the expressions in Equations 2.3 and 2.7, and the ageing law of diffusivity can follow Equation 2.9, but the ageing exponent n_{Cl} should be calibrated to the effective chloride diffusivity in the pore solution.

The Cl-1 model in Equation 2.18 involves some crucial material parameters, including the porosity ϕ, the pore saturation s_1, and the linearized binding capacity s_{Cl}'. The credibility of this model depends very much on the appropriate values adopted for these parameters for a practical case. The clear physical significance of these mechanism-based parameters can be an advantage for this model, while the determination of these parameters does raise a problem for practical engineers. Table 2.3 gives the usual value ranges for this model, and some simulation results are given in Figure 2.9 for constant pore saturation and temperature, without the ageing effect being taken into account. From the results, the impact of the saturation is determinant for the chloride profile, and the chloride binding capacity also plays an important role.

Model Cl-2: Empirical Model

As the counterpart of the mechanism-based Cl-1 model, the Cl-2 model is an empirical one for chloride ingress. The Cl-2 model is also based on the analytical solution in Equation 2.13 but

Table 2.3 Parameters for the Cl-1 model and usual value ranges

Parameter (unit)	Value/relation
Initial chloride concentration $c_{Cl,0}$ (mol/m³)	0.0
Surface chloride concentration $c_{Cl,S}$ (mol/m³)	300–480 (marine)
Effective chloride diffusivity D_{Cl}^{eff} (10^{-12} m²/s)	Equation 2.3
Effective chloride diffusivity at 100% saturation $D_{Cl,0}^{eff}$ (10^{-12} m²/s)	1–10
Pore saturation s_l	0–1.0
Porosity of concrete ϕ	0.09–0.18
Chloride binding capacity of concrete, s_{Cl} (mol/m³)	0–200; see Equation 2.6 or Figure 2.6
Linearized partial binding capacity s'_{Cl}	0.03–0.60; see Equation 2.6 or Figure 2.6
Concrete depth x (m)	Positive value
Exposure age t (s)	Positive value

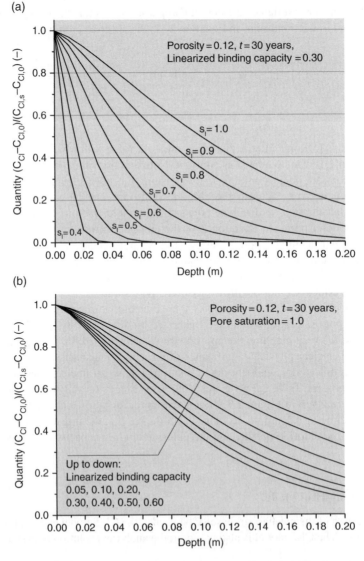

Figure 2.9 Chloride profiles at $t = 30$ years from the Cl-1 model in terms of (a) the pore saturation and (b) the binding capacity of chloride. The effective chloride diffusivity is 5×10^{-12} m²/s and the porosity is 0.12. The saturation dependence of chloride diffusivity follows Equation 2.3 with $\lambda = 6.0$.

uses the chloride contents and the apparent chloride diffusivity as the model parameters. The chloride profile takes the form

$$C(x,t) = C_0 + (C_S - C_0)\left[1 - \text{erf}\left(\frac{x}{2\sqrt{D_{Cl}^{app}t}}\right)\right]$$ (2.19)

where C_S (wt% binder) is the chloride content at the concrete surface, C_0 (wt% binder) is the initial chloride content in concrete, and D_{Cl}^{app} (m²/s) is the apparent chloride diffusivity of concrete.

The surface chloride content C_S is obtained from the aforementioned chloride profiling method for concrete under natural exposure. It can adopt the recommended values in Table 2.2 for marine exposure conditions. For the splash zone, a convection zone exists; thus, the surface chloride content loses its physical sense and becomes a fitting value extrapolated to the concrete surface; see Figure 2.8. To overcome this ambiguity, the *fib* model code (*fib*, 2006) proposed applying the model in Equation 2.19 from $x = \Delta x$ instead of $x = 0$, with Δx representing the depth of the convection zone. The corresponding C_S becomes $C_{S,\Delta x}$, adopting the value from the chloride sorption isotherm curve. This treatment is more reasonable for the chloride transport in concrete cover, but note the convection depth is also a result of concrete–environment interaction. For the action of deicing salts, the chloride content depends very much on the precipitation intensity and the deicing frequency, and a value of 0.564% was proposed (*fib*, 2006) for OPC concrete with $w/c = 0.45 - 0.6$ from German experience. Local statistical data are needed to determine this surface content for other applications. With the help of Equation 2.8 and Figure 2.7b, the dependence of C_S on the distance from the marine coastline and the source of deicing salts can also be taken into account.

The initial chloride content C_0 should include the chloride content in the binder materials, the aggregates, and the chemical admixtures. This parameter is of particular importance as the aggregates are contaminated by chloride-borne water (seawater). The diffusion coefficient D_{Cl}^{app} relates to the long-term chloride ingress under natural exposure. As mentioned previously, this value reflects both the chloride ingress and the concrete–environment interaction. Owing to such a complex nature this parameter is normally *a priori* unknown for a practical case. However, multiple efforts were made to relate this apparent diffusivity to the short-term laboratory characterization and the environmental conditions. The following expression was recommended (*fib*, 2006) for this purpose:

$$D_{Cl}^{app}(t) = D_{Cl}^{app}(t_0)\left(\frac{t_0}{t}\right)^{n_{Cl}}, \qquad D_{Cl}^{app}(t_0) = k_{Cl}^T k_{RCM}^{app} D_{Cl,0}^{RCM}$$ (2.20)

The first part of this equation is the same ageing law as Equation 2.9, and the second part links the chloride migration coefficient in laboratory $D_{Cl,0}^{RCM}$ to the apparent chloride diffusivity at the same age t_0 through two parameters: the temperature parameter k_{Cl}^T, expressed in Equation 2.7, and the transfer parameter k_{RCM}^{app} describing the relation between the chloride migration in the RCM method and the complex transport process in the natural environment. This expression contains two crucial parameters: the ageing exponent n_{Cl} and the transfer parameter k_{RCM}^{app}. Empirical values were recommended for this exponent considering the binder composition

Table 2.4   Parameters for the Cl-2 model and usual value ranges

Parameter (unit)	Value/relation
Chloride migration coefficient $D_{Cl,0}^{RCM}$ at standard age t_0 (10^{-12} m^2/s)	1–25 (*fib*, 2006)
Concrete standard age t_0 (days)	28
Surface chloride content C_S (%binder)	Table 2.2 and Equation 2.8
Initial chloride content C_0 (%binder)	0–0.3
Ageing exponent[a] n_{Cl}	
OPC	0.30 (0.37) [0.65]
PFA	0.69 (0.93) [0.66]
GGBS	0.71(0.60) [0.85]
SF	0.62 (0.39) [0.79]
Transfer parameter k_{RCM}^{app}	0.5–1.0
Temperature parameter k_{Cl}^{T}	Equation 2.7
Reference temperature T_0 (K)	293.15
Thermal-activation coefficient b_e (K)	4200

[a]Ageing exponent in different exposure conditions: immersed (splash/tidal) [atmospheric].

and exposure conditions (DuraCrete, 2000; Thomas and Bentz, 2000; *fib*, 2006). The transfer parameter k_{RCM}^{app} is to be calibrated for specific engineering applications by local data, and this process is to be explained in the durability design for the HZM project in Chapter 9. This transfer parameter was set to 1.0 by the *fib* model code (*fib*, 2006), evidently due to the dearth of data.

The recommended value ranges are given in Table 2.4 for the parameters and some useful simulation results are presented in Figure 2.10. The impact of the chloride migration coefficient and the ageing exponent is determinant on the chloride profile in concrete. Compared with the Cl-1 model, neither pore saturation nor chloride binding is explicitly included in the results; thus, these results are by nature empirical. Nevertheless, the input parameters in this model are all macroscopic parameters related to laboratory characterization and the environmental conditions, so it is more accessible to practical engineers to perform the calculation. However, these parameters were calibrated from specific data sources, which do not necessarily cover the concrete materials or the exposure conditions for other applications. If the model is chosen to perform the durability design for an important structure, local data must be gathered to calibrate the key parameters in this model; that is, the ageing exponent n_{Cl} and the transfer parameter k_{RCM}^{app}.

2.3   Steel Corrosion by Chloride Ingress

2.3.1   *Mechanisms*

Chlorides can destroy the electrochemical stability of the embedded reinforced steel. The electrochemical stability of the reinforcement steel can be described by the Pourbaix diagram for the Fe–Cl–H$_2$O system; see Figure 2.11. This figure reveals the fundamental aspects of steel stability in the presence of chloride in terms of the solution pH value. Different from the diagram in Figure 1.11, no zone on this diagram is totally exempt from corrosion risk for an undisturbed

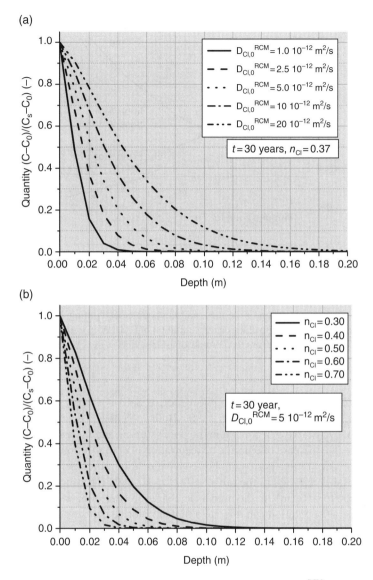

Figure 2.10 Chloride profiles from the Cl-2 model in terms of (a) the $D_{Cl,0}^{RCM}$ values at 28 days and (b) the ageing exponent. The temperature is taken as 293.15 K, and the transfer parameter $k_{RCM}^{app} = 1.0$.

pore solution in concrete with a pH value around 13.0, or for the carbonated concrete with pH below 9.0. For a fixed pH value of pore solution, an increase in chloride concentration tends to move the steel from the protection zone to the imperfect passivation and pitting zones. As the solution pH is decreased by carbonation, the steel has more chance to enter the pitting zone for a given chloride concentration in solution; that is, corrosion is more likely to occur.

Actually, the corrosion risk of steel with chloride is measured by the difference between the electrode potential of steel E_{corr} and the potential for pitting corrosion E_{pit}: where E_{corr} is

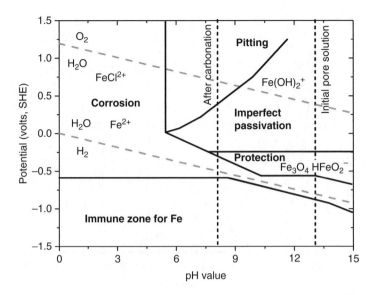

Figure 2.11 Pourbaix diagram for Fe–Cl⁻–H₂O system at 25 °C. (The accurate position of the lines drawn depends on the concentration of the ions, and the boundaries "pitting–corrosion" and "pitting–imperfect passivation" correspond to chloride concentration of 0.001 M.) *Source:* adapted from Moreno *et al.* (2004, Fig. 13). Reproduced from the permission of Elsevier.

less noble than E_{pit} ($E_{corr} - E_{pit} < 0$), then pitting corrosion is not likely to occur, whereas pitting corrosion occurs where $E_{corr} - E_{pit} > 0$. Both potentials depend on the chloride concentration, pH value, oxygen content in the pore solution, and the ambient temperature. At a given pH value, the increase of chloride concentration in pore solution tends to decrease E_{pit}, augmenting the risk of pitting corrosion. For conventional reinforcement bars in concrete, the electrode potential is reported in the range +100 to −200 mV for atmospheric exposure and in the range −400 to −500 mV in the immersed condition (Bertolini *et al.*, 2004). Thus, steel bars in the immersed condition can tolerate a higher chloride concentration in the pore solution before corrosion initiation. Among the various factors contributing to the initiation of pitting corrosion, the ratio between the Cl⁻ and the OH⁻ concentrations, [Cl⁻]/[OH⁻], has been widely used to quantify the initiation of pitting corrosion. For steel bars embedded in concrete, the external chlorides need to penetrate into concrete through the pore space and accumulate to a necessary concentration on the steel surface to initiate pitting corrosion.

Once the corrosion is initiated, the intensity of anodic and cathode reactions can be represented by the electrical current I_{corr}. The electrochemical processes of the anode and cathode reactions are essentially the same as the carbonation-induced corrosion, but the corrosion mode is rather different. The chloride-induced corrosion presents a pitting pattern: the anode reaction is to localize on a pitting point on the steel surface and all the steel around serves as the cathode. Since the area ratio between the anode and cathode plays an important role in the current density, the current density of pitting corrosion is normally higher than the general corrosion induced by the carbonation. Chloride-induced corrosion is also affected substantially by the environmental humidity and temperature. In a moderate humidity range, lower than 80–85%, the corrosion is dominated by the resistance-controlled mode and the current intensity is mainly determined by the concrete resistance; in the high humidity range, higher than 95%, the corrosion is limited by the oxygen availability at the cathode and thus oxygen diffusion controls the current intensity. For the humidity range in between, there exists a transition from the resistance-controlled mode to the oxygen diffusion mode.

2.3.2   Influential Factors

Corrosion Initiation: Pore Solution, Concrete Cover, and Interface

The influential factors for corrosion initiation by chloride ingress involve both the corrosion resistance of steel and environmental factors. The corrosion resistance of steel is related to its chemical composition and metallurgical properties; for example, stainless steel acquires high corrosion resistance by alloying at least 10.5% chromium (Cr) and the chromium oxide can form a passivation film on the steel surface to isolate the steel from the external corrosive agents. This text treats only conventional reinforcement steel bars, made from carbon steel, thus only the environmental factors are addressed here.

The main influential factors from the environmental side include the chloride concentration and pH value of the pore solution at the surface of the steel bar, the compactness of the concrete around the steel bar, the quality of the concrete–steel interface, the moisture and oxygen availability at the steel surface, and the ambient temperature. The electrochemical mechanism of pitting corrosion confirms the inhibitive nature of OH^- in the pore solution. The pH value of the pore solution in concrete is mainly maintained by the portlandite (CH) from the cement hydration and the alkaline ions (Na^+, K^+) in the pore solution. The inhibitive effect of OH^- on pitting corrosion is enhanced by the concentrated distribution of CH on the concrete–steel interface (Page, 1975). Also for this reason, the critical concentration ratio $[Cl^-]/[OH^-]$, quantified from electrochemical studies in a simulated pore solution, is always conservative compared with the real concrete environment (Page *et al.*, 1991a,b). The quality of the concrete–steel interface conditions the electrochemical environment of the pitting corrosion: concentrated CH can effectively inhibit the initiation, while the air voids and cracks at the interface can favor the formation of macrocells and thus accelerate the initiation. The interface is regarded as a major factor in corrosion initiation (Angst *et al.*, 2009).

Since the pores and voids at the concrete–steel interface are not always saturated by pore solutions, the availability of oxygen (gas phase) and the pore solution (liquid phase) impact on corrosion initiation. Both are necessary to start up the cathodic reaction for corrosion. Accordingly, highly saturated and very dry concretes tend to have high resistance to chloride initiation. In practice or laboratory tests, steel bars are regarded as passivated as long as the corrosion current intensity remains below $0.1\ \mu A/cm^2$. From this viewpoint, the electrical resistance can also play an inhibitive role in corrosion initiation; that is, the compactness of the concrete. As for the ambient temperature, chloride ingress is faster and chloride accumulation at the steel surface is more significant at high temperature; thus, increased temperature is believed to accelerate corrosion initiation (Hussain *et al.*, 1995). Other influential factors also include concrete binder type, chloride binding capacity, and the surface state of steel. A comprehensive review of the influential factors on corrosion initiation by chloride ingress can be found in Angst *et al.* (2009).

Two terms are widely used to quantify the risk of pitting corrosion of steel bars in concrete: the concentration ratio $\varphi_{CR} = [Cl^-]/[OH^-]$ and the critical chloride content C_{crit}. The ratio φ_{CR} captures the two most important factors for corrosion initiation, the chloride and the pH value, and pertains to the free chloride in the pore solution. This quantity is favored by laboratory research, especially as a simulated pore solution is used. According to Gouda (1970), the ratio $[Cl^-]^{0.83}/[OH^-]$ is more pertinent to corrosion initiation than the ratio $[Cl^-]/[OH^-]$. The quantity C_{crit} applies to the total chloride content in concrete, noted as the mass ratio of chlorides with respect to the binder or concrete. Though this quantity addresses only the factor of chloride concentration it is more convenient for practical use and in-situ evaluation of corrosion risk. Moreover, this quantity includes the bound chloride, which can serve as a reservoir for free chloride in pore solution. Actually, the two quantities can be related if the chloride binding capacity s_{Cl} is known:

$$C_{crit} = \frac{1}{\rho_c}\left(\phi\varphi_{CR}10^{pH-14.0} + M_{Cl}s_{Cl}\right) \tag{2.21}$$

where C_{crit} (wt%) is the critical chloride content in concrete, ρ_c (kg/m^3) is the concrete density, ϕ is the concrete porosity, M_{Cl} (kg/mol) is the molar mass of chloride, and s_{Cl} (mol/m^3) is the chloride binding capacity of concrete. Considerable effort has been dedicated to the determination of these two quantities, and both quantities can be termed the "critical chloride content (concentration)" or "threshold chloride content (concentration)." Neither the ratio φ_{CR} nor the content C_{crit} reflects all the influential factors, so it is not surprising that rather large scatters were reported for these two quantities in literature. The usual range for φ_{CR} ratio is 0.6–1.0, while the experimental results scattered in the range of 0.1–6.0 (Angst et al., 2009). C_{crit} depends on the exposure condition of the RC elements, and this value is recommended as 0.05 wt% concrete by Life-365 (Thomas and Bentz, 2000) and 1.2 kg/m^3 for absolute chloride content in concrete by JSCE (2010). A more complete synthesis of literature values can be found in Angst et al. (2009).

Corrosion Current: Chloride Concentration

Although different in corrosion patterns, pitting corrosion by chloride ingress shares the same set of influential factors as general corrosion by concrete carbonation: the ambient temperature and humidity, the resistance of the concrete cover, and the concrete-interface quality, which have been reviewed in Section 1.3.2. The only factor specific to the pitting corrosion current is the chloride content (concentration) itself. An increase in chloride content in the pore solution can augment the corrosion current from at least from two aspects: first, the pitting potential is decreased and the driving force for pitting corrosion becomes more important; second, more chlorides in the pore solution help to reduce the electrical resistance of concrete, and thus also help to augment the corrosion current density. Actually, these two aspects cannot be separately readily in the experimental data interpretation. The relevant experiments showed that the corrosion current density (μA/cm^2) scales with the total chloride content (kg/m^3) and the free chloride content (kg/m^3) through a power law with exponents of 0.777 and 0.618 respectively (Liu and Weyers, 1998).

Table 2.5   Parameters for the Cl-3 model and usual value ranges

Parameter (unit)	Value/relation
Critical chloride concentration ratio φ_{CR}	0.6–1.0 (large dispersion/uncertainty)
OH⁻ concentration in pore solution [OH⁻] (10^{-3} mol/m³)	0.0316–0.316 (pH 12.5–13.5)
Critical chloride content C_{crit}, atmospheric (%binder)	0.85
Critical chloride content C_{crit}, tidal/splash (%binder)	0.9 ($w/b = 0.3$), 0.8 ($w/b = 0.4$), 0.5 ($w/b = 0.5$)
Critical chloride content C_{crit}, immersed (%binder)	2.3 ($w/b = 0.3$), 2.1 ($w/b = 0.4$), 1.6 ($w/b = 0.5$)

2.3.3   Models

Model Cl-3: Corrosion Initiation

Corrosion initiation can be described through either the concentration ratio $\varphi = [Cl^-]/[OH^-]$, in terms of the chloride concentration in the pore solution c_{Cl} and the pH value of the pore solution, or the chloride content in concrete C_{crit}. The criterion of corrosion initiation can be expressed as

$$c_{Cl}\left(x = x_d, t\right) = \varphi_{CR}\left[OH^-\right] \text{ or } C\left(x = x_d, t\right) = C_{crit} \tag{2.22}$$

The two criteria can be converted through Equation 2.21 if the binding capacity of concrete is known. These two criteria correspond respectively to the free chloride concentration in pore solution, used usually in a mechanism-based model for chloride ingress, and to the chloride content with respect to binder or concrete, adopted often in empirical models. The first criterion can be jointly used with the Cl-1 model to evaluate the corrosion initiation of steel bars for a given thickness of concrete cover x_d; the second criterion can be combined with the Cl-2 model for corrosion initiation in terms of the total chloride content.

The recommended value ranges of these quantities are given in Table 2.5 for usual exposure conditions. The equivalence between the two quantities for the critical chloride contents is illustrated in Figure 2.12 for OPC concretes.

Model Cl-4: Corrosion Current

The current density of corrosion by chloride adopts essentially the same model (C-4) for carbonation-induced corrosion from Equations 1.19, 1.20, 1.21, and 1.22, except that the presence of chloride in the pore solution should be integrated. For this purpose, a pitting mode parameter $F_{Cl} > 1.0$ is incorporated into Equation 1.19 to account for the larger current density by chloride pitting corrosion than the general corrosion by carbonation:

$$I_{corr} = \frac{V_0}{\rho} F_{Cl} \tag{2.23}$$

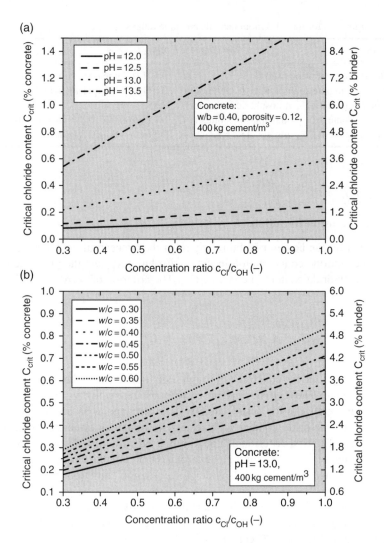

Figure 2.12 Critical chloride content C_{crit} in terms of the critical concentration ratio ϕ_{CR} for (a) different pH levels and (b) different concretes. The concretes contain 400 kg/m³ OPC. The binding capacity is evaluated through Equation 2.6 with the C-S-H contents taken from Figure 1.6. The C_3A and C_4AF contents are 40 mol/m³ and 30 mol/m³ respectively, and the volumetric ratio of paste is 0.35. The porosity values for concretes range from 0.09 to 0.18 for $w/c = 0.30$ to 0.60.

And the decrease of electrical resistance by chloride presence in the pore solution is considered through another factor, $k_{Cl} < 1.0$, into Equation 1.21:

$$\rho = \rho_0 \left(\frac{t}{t_0} \right)^m k_{RH} k_T k_{Cl} \qquad \text{with} \quad t \leq 1 \text{ year} \qquad (2.24)$$

Table 2.6 Parameters for the Cl-4 model and usual value ranges after DuraCrete (2000)

Parameter (unit)	Value/relation
Corrosion mode parameter F_{Cl}	2.63
Electrical resistance parameter k_{Cl}	0.72
Geometric factor α_{Cl},	1.0–5.0, 4.64 (recommended)

Further, the steel consumption by pitting corrosion can concentrate on one local point on the steel section; thus a geometric factor, $\alpha_{Cl} > 1.0$, is introduced into Equation 1.23:

$$\dot{r} = 0.0116\alpha_{Cl}I_{corr} \qquad (2.25)$$

In summary, the corrosion current model Cl-4 is adapted from the general corrosion model C-4 considering the influence of chlorides in the pore solution on the corrosion mode, the electrical resistance of concrete, and the local pitting on the steel surface through the parameters F_{Cl}, k_{Cl}, and α_{Cl}. Table 2.6 gives the recommended values for these additional parameters, and these values are not related to the chloride concentration due to the limited literature data. Using the same set of parameters in Table 1.3 and the supplementary parameters in Table 2.6, Figure 2.13 presents some simulation results for the corrosion current in terms of concrete electrical resistivity and ambient temperature.

2.4 Basis for Design

2.4.1 Structural Consequence

As the corrosion is initiated and develops in pitting mode, the corrosion products at the pitting sites formed expand and can accumulate stress high enough to fracture the concrete cover. The onset of cracking depends on the quantity of corrosion products, the voids at the concrete–steel interface, the diameter of steel bar, the cover thickness, and the fracture properties of concrete. The corrosion in this phase begins to affect the service limit state (SLS) of RC elements and structures via excessive cracking. Some detailed analysis can be found in the literature (Liu and Weyers, 1998), while the corrosion current density can be used as a basic variable to quantify the corrosion extent affecting the SLS. The expressions in Equations 1.24 and 1.25 can be used to evaluate the onset of cracking and the subsequent crack opening at the concrete surface. As the corrosion develops further, both the rigidity and the load-bearing capacity of RC elements can be affected; that is, the corrosion starts to affect the ultimate limit state (ULS). DuraCrete (2000) proposed a standard calculation method for the load-bearing capacity of RC beams and columns affected by corrosion. Although these aspects are essential to evaluate the real structural performance of RC elements and structures, in the design phase the extent of corrosion cannot be allowed to develop further than affecting the SLS. Thus, the performance of RC elements and structures in the post-corrosion phase is not covered by this text, and interested readers can refer to the relevant literature (Andrade and Mancini, 2011).

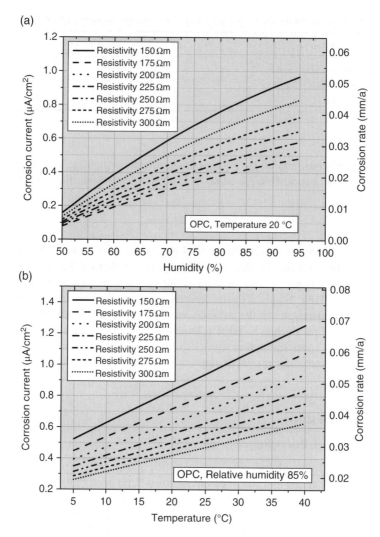

Figure 2.13 Simulation results from the Cl-4 model for corrosion current and corrosion rate in terms of (a) relative humidity and (b) ambient temperature.

2.4.2 Design Considerations

Durability design is presented in a systematic way in Chapter 7, while some aspects specific to chloride-induced corrosion for RC elements and structures are discussed here. Durability design against chloride-induced corrosion should be performed at both material and structural levels. At the material level, a high compactness of concrete and a crack-free concrete are always appreciated and valued. Normally these properties can be acquired through the appropriate use of SCM in concrete binders and sufficient curing at early ages. These concrete materials have low chloride diffusivity and high electrical resistance, which help to delay chloride ingress and reduce the corrosion current. At the structural level, adequate thickness of concrete cover should be achieved together with high compactness of the concrete, which can also contribute to the delay of chloride ingress and the decrease of corrosion current. As RC elements contain, more or less, working cracks in the tension zone of the concrete section, the crack width should be restricted due to their acceleration effect on the external chloride ingress. A detailed study on this effect can be found in Otieno *et al.* (2012) while a surface crack width within 0.3 mm is considered safe with respect to chloride ingress. Note that this knowledge is strictly limited to individual cracks due to mechanical loading, and does not cover the highly dispersed cracks due to early-age shrinkages.

For RC elements or structures locally exposed to severe chloride actions, such as wave splashing in a marine environment or direct spray of deicing salts, additional protection measures should be considered. These measures include applying a protection layer to reinforcement steel bars (epoxy-coated bars), using stainless steel bars, incorporating chemical inhibitors in the concrete to delay the corrosion reactions (corrosion inhibitors), treating the concrete surface with hydrophobic agents (silane impregnation), applying a protective coating on the concrete surface, installing a protection shell outside the elements, and the direct cathodic protection of reinforcement steel bars. These measures can be efficient against local chloride actions without raising the durability requirements for the whole elements. The application of these measures should take into account the life-cycle cost of RC elements and structures, and further aspects can be found in Chapter 7.

3

Freeze–Thaw Damage

This chapter treats concrete damage caused by the freeze–thaw (FT) actions in natural environments. Normally, the hardened structural concretes are rather resistant to low temperature, so concretes can be used in most atmospheric temperature ranges, even in some cryo-storage facilities, such as liquid gas storage tanks. However, this resistance can be compromised by the presence of water, especially as the concrete pores are saturated by water. The external freezing and the subsequent thawing can generate pore pressure associated with the water phase change in pores that high enough to cause damage in the solid matrix of concrete. The damage extent can be further enhanced by the presence of salts on the concrete surface, resulting in significant surface scaling of RC elements. This chapter introduces first the phenomena of FT damage of RC elements and then presents the mechanisms of concrete damage due to pore phase change under freezing, together with the influential factors for the damage process. Then, two models are presented on the basis of the state-of-the-art knowledge for engineering use. Finally, the basis of design is brought forth for the RC elements exposed to FT actions.

3.1 Phenomena and Observations

Structural concrete exposed to natural environments is susceptible to frost action in cold regions.[1] Normally, structural concretes are rather resistant to natural low temperature as long as concrete hardening is properly achieved. That is why concrete infrastructures exist extensively in the countries and regions with heavy frost action, like Canada and northern Europe. Concrete is even

---

[1]Cold region, as a technical term, refers to the geographical zone in which freezing and precipitation in the atmosphere are both available. For durability of concrete, the cold region can be defined by the average atmospheric temperature of the coldest month.

Durability Design of Concrete Structures: Phenomena, Modeling, and Practice, First Edition. Kefei Li.
© 2016 John Wiley & Sons Singapore Pte. Ltd. Published 2016 by John Wiley & Sons Singapore Pte. Ltd.

used for the containers of refrigerated liquid gas in which the service temperature is between +4 °C and −198 °C (ACI, 2013). However, the freezing and the subsequent thawing cycles can induce damage in concretes, especially for concrete elements in contact with liquid water. This damage is related to the water phase change in concrete pores and its mechanical effects on the solid matrix of the material. As early as the 1930s, entrained air was found to increase greatly the concrete resistance against surface scaling by deicing salts. Since then, air-entrainment has become an efficient measure against frost damage of concrete used in FT environments.

The first category of concrete structures subject to frost action concerns hydraulic structures like dams in cold regions. It was reported that 19% of dam deteriorations were attributed to freezing and thawing cycles (ICOLD, 1984). The concrete elements are in constant contact with water, and the concrete is in a high saturation state. Under frost action, liquid water freezes in the concrete surface and thaws back to liquid water with fluctuation of the external temperature. The water freezing increases the pressure in the concrete pores and ice thawing relaxes this pressure. The increase and relaxation of pore pressure induces damage in the concrete. This category of RC elements also includes bridges with concrete piers in flowing water in cold regions. Normally, the change of water level creates a vertical tidal zone on these piers. During frost seasons, these parts of the elements are alternately subject to air freezing and water thawing, resulting in severe damage due to the high saturation degree of concrete. Figure 3.1 shows the concrete piers of a highway bridge across an inland river.

Surface damage of concrete can develop to a very severe extent, as salts are present during the frost action. This occurs due to the use of deicing salts applied to the concrete surface of road pavements in winter, or as a result of the concrete surface being in contact with a salt-bearing water body such as seawater or groundwater with high salinity. Among these cases, the degradation of concrete bridge decks and road slabs was highlighted in the USA (TRB, 2004). For RC elements, the application of deicing salts also raises concern over corrosion of embedded reinforced steel bars; see Chapter 2. Since the role of salts on the surface damage of concrete elements is clearly demonstrated in applications, surface damage by salts is also termed "surface salt scaling" or "salt scaling," and is treated rather differently compared with frost damage without salts. Figure 3.2 illustrates salt scaling on a segment of road pavement.

For the sake of easy terminology, frost damage without salts is termed "internal FT" and damage with salts as "surface scaling." Most observations on concrete damage by frost action have been obtained through laboratory accelerated tests, using either water or salt solution as contact liquids. The laboratory tests can give a clearer effect of FT actions on the surface damage of concrete. Figure 3.3 shows the deterioration of concrete specimens under accelerated FT tests according to the ASTM C666 method. Table 3.1 summarizes the main qualitative observations from these laboratory tests. The most surprising observation for salt scaling is the maximum scaling intensity at an optimum concentration of 3% salts, regardless of salt type (Verbeck and Klieger, 1957). The underlying reason will be explained later.

3.2 Mechanisms and Influential Factors

3.2.1 Mechanisms

Concrete damage from frost action, or FT cycles, originates from the excessive pressure exerted on the solid phases by pore freezing. Research on frost damage has gone on for more than half a century, and multiple mechanisms have been proposed: the hydraulic pressure

Figure 3.1   Concrete piers of a highway bridge across an inland river subject to frost action at the water level. The damage was present as the spalling of concrete cover and the coarse aggregates were exposed by spalling. The minimum temperature in winter reaches −10 to −15 °C in the region, the concrete grade was C30, and the service life of the bridge was 11 years at inspection. *Source:* courtesy of Xiaoxin Feng.

theory (Powers, 1949) attributed the damage to pore pressure that accumulated during the liquid water flow driven by the ice volume expansion of 9% in pores; the osmotic theory (Powers and Helmuth, 1953) explained further that the liquid water is driven to the freezing site in pores due to the local condensation of pore solution, and thus induces the severe scaling with the presence of salts; the micro-lens theory (Setzer, 2001) investigated the vapor–water–ice equilibrium in pores during freezing and confirmed the suction effect of external water during one-sided thawing; the critical pore saturation theory (Fagerlund, 1993) and the low-cycle fatigue theory (Fagerlund, 2002) were proposed for the water absorption and concrete damage process during FT cycles; the pore phase change has been investigated more formally through the pore crystallization theory (Scherer, 1999) and poromechanics theory (Coussy,

Figure 3.2 Salt scaling of concrete surface on road pavement in a cold region in China. *Source:* courtesy of Weizu Qin.

(a)   (b)

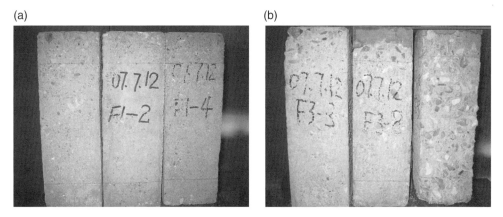

Figure 3.3 Surface spalling of concrete specimens subject to accelerated FT tests: (a) after 100 cycles; (b) after 200 cycles. The specimens were made of OPC concrete with w/c = 0.54 and 28d cubic strength of 37.5 MPa. The test was of ASTM C666 type with pure water as freezing liquid. *Source:* courtesy of Jiaru Qian.

2005). This section attempts to use the soundest elements from these theories to give a realistic picture of pore freezing in concrete. Note that the structural concrete contains 50–80 vol.% aggregates and uses hardened cement paste (HCP) as binder for these aggregates. Both aggregates and HCP are porous media; thus, the following mechanisms apply to both.

Let us begin with some fundamental concepts associated with pore freezing. Under standard atmospheric pressure, liquid water freezes at 0 °C. However, water in a narrow tube with radius r (see Figure 3.4) does not freeze at 0 °C but at a lower temperature since the freezing is hindered by the surface tension at the ice–water interface, or the interface energy.

Table 3.1 Laboratory observations of internal FT and surface scaling damage

Internal FT	Surface scaling
• Damage is in the form of cracking and spalling	• Damage is in the form of surface scaling
• Pore saturation increases with FT cycles	• Pore saturation increases with FT cycles
• Damage occurs with pore saturation above a critical value (0.7–0.9)	• Most severe scaling is observed at 3% concentration of salts
• FT damage decreases concrete strength, elastic modulus, and bond strength	• Surface scaling has no impact on the bulk mechanical properties of specimens
• Damage extent evolves with FT cycles	• Scaling is increased by freezing extent and time
• Damage extent is related to concrete strength and porosity	• Surface scaling is related more to the quality and strength of surface
• Air entrainment can greatly increase the FT resistance	• Air entrainment can greatly increase FT resistance

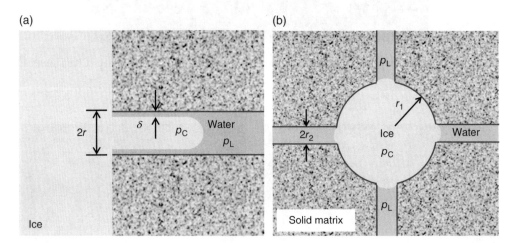

Figure 3.4 Schematic illustration for water freezing: (a) in a tube; (b) in confined pores. The unfrozen water layer is omitted in (b) for ease of illustration.

The extent of this freezing-point depression depends on the radius of the tube, and the relation can be expressed through the Gibbs–Thomson equation (Coussy and Fen-Chong, 2005):

$$T^* = T_f - \frac{2\gamma_{CL}}{r_E S_f}, \qquad r_E = r - \delta \tag{3.1}$$

with γ_{CL} the ice–water interface energy (~0.04 J/m^2), T^* and T_f (K) are the depressed and standard freezing points of water, and S_f is the fusion entropy per unit volume of ice crystal (~1.2 × 10^6 J/(m^3 K)). Here, a liquid layer of thickness δ exists between the ice crystal and pore wall;[2] thus, the contact angle between the crystal and the pore wall is assumed to be 180°;

---

[2]The unfrozen water layer that exists between the solid crystal and the solid pore wall leads to the system being in a lower (stable) energy state. The thickness of this layer is assumed to depend on the freezing temperature (Fagerlund, 1976), and the average value of this thickness is assumed to be 0.8 nm (Sun and Scherer, 2010).

that is, the contact surface between the ice crystal and water is a hemisphere. Using these values and Equation 3.1, a tube with radius 67 nm depresses the freezing point by 1 °C; that is, the liquid water does not freeze until the temperature decreases to −1 °C in the tube, and a radius of 1 nm will depress the freezing point by 67 °C. This result gives the first notion of water freezing in pores: with temperature decrease the water in pores freezes gradually from larger pores to smaller ones, and even at very low temperature (e.g., below −20 °C), water in nanoscale pores is not yet frozen.

Then we look at the mechanical equilibrium as water freezes in pores; see Figure 3.4. In the tube with radius r, the ice pressure p_C and the water pressure p_L are linked by the Young–Laplace equation:[3]

$$p_C - p_L = \frac{2\gamma_{CL}}{r_E} \tag{3.2}$$

According to Scherer (1999), as the ice crystal intrudes into the cylinder pore with radius r with a perfect hemispherical penetrating surface, the additional radial stress exerted on the pore wall is $\sigma_C = \gamma_{CL}/r_E$. Now let us turn to the freezing case in confined pores: a big sphere pore with radius r_1 connected with a group of smaller pores with radius r_2 $(<r_1)$; see Figure 3.4b. Suppose the water in the big pore is frozen at a certain freezing temperature T_1^*. The temperature needs to decrease further to T_2^*, the freezing point for radius r_2, to make the ice crystal penetrate further into the smaller pores. During this process the stress exerted on the solid matrix can be expressed as

$$\Delta\sigma_C\big|_{T_1^* \to T_2^*} = \gamma_{CL}\left(\frac{2}{r_2 - \delta_2}\right) \quad \text{with} \quad T_2^* = T_f - \frac{2\gamma_{CL}}{(r_2 - \delta_2)S_f} \tag{3.3}$$

In other words, as the ice crystal grows into the smaller pores, a higher stress is exerted on the pore wall. This statement assumes the liquid water pressure is held constant during freezing; for example, equal to the external atmospheric pressure p_{atm}. Using this assumption, Equation 3.3 predicts an expansion during freezing, which is usually observed for the freezing strains of free and saturated specimens (Zeng et al., 2014). Note also that the expansion predicted from Equation 3.3 does not need crystal volume augmentation during freezing. This statement was validated experimentally by frost damage of porous materials saturated by benzene, a liquid that shrinks on freezing (Hodgson and McIntosh, 1960). If, alternatively, the solid pressure in the ice crystal is held constant (e.g., equal to p_{atm}), Equation 3.2 predicts a negative liquid pressure. This occurs as the central pore in Figure 3.4b is much bigger and remains void during freezing (e.g., an air void). In this case the ice crystal cannot easily fill up the central pore, resulting in a negative liquid pressure at the ice–liquid interface (i.e., the inlets of the smaller pores). As the temperature is kept above T_2^*, the liquid water flows back to the interface by this local negative pressure, resulting in material shrinkage. This phenomenon is

---

[3]In this chapter, the liquid phase in pores refers to pore water and is denoted by the subscript "L." The same pore liquid phase is denoted by the subscript "l" in the context of gas–liquid equilibrium in pores in Chapter 2 and later in Chapter 6. The two subscripts represent the same pore phase, used in a different context of pore equilibrium.

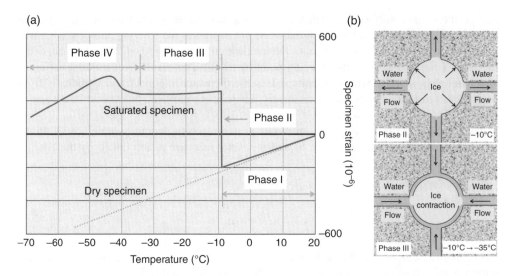

Figure 3.5 Freezing deformation of a saturated mortar specimen (a) and the related mechanisms at pore level (b) for a temperature range from +20 to −70 °C. The figure is drawn from the experimental results of M50 mortar specimens from Penttala (1998); the prism specimens were 10 mm × 10 mm × 55 mm, dried at 105 °C and saturated in water before freezing.

termed "cryo-suction" of freezing, observed in the 1950s (Powers and Helmuth, 1953) and analyzed formally through poromechanics by Coussy and Fen-Chong (2005).

On the basis of these concepts, the mechanical behavior of cement-based porous materials is described for a typical freezing process as follows. Figure 3.5 illustrates the freezing deformation of a saturated mortar specimen and the related mechanisms for a freezing range down to −70 °C. Several typical phases are identified for pore freezing.

Phase I: +20 to −10 °C. During this range the solid matrix and pore water are cooled, and thermal strain (contraction) dominates the global deformation. As the temperature drops below 0 °C, the pore water starts to freeze before overcoming two types of energy barriers: the freezing point depression by the pore confinement expressed in Equation 3.1 and the nucleation energy barrier for ice crystal growth in pores (Fletcher, 1971). Experiments on hardened cement pastes confirm that the heterogeneous nucleation of ice crystals in pores occurs between −6 and −11 °C (Zeng *et al.*, 2015).[4] This supercooling of liquid water in pores persists until the freezing temperature approaches −10 °C. Note that the supercooled pore water has nearly no impact on the thermal deformation of the specimen so that the thermal contractions of saturated and dry specimens are nearly the same.

Phase II: at −10 °C. After sufficient supercooling the pore water starts to freeze massively and the phase change in pores exerts stress on the pore wall or solid matrix, following the mechanisms shown in Figure 3.5. This internal stress causes the global expansion of the saturated specimen. Actually, the pore stress can be generated also from the viscous flow

---

[4]The freezing point depression by heterogeneous nucleation can be reduced, or even disappear, as specimens are exposed to FT cycles because the nucleation sites formed in the preceding freezing cycle can greatly facilitate the crystallization of pore water in the present freezing cycle.

of liquid water driven by the volumetric expansion of ice formation (~9%). This contribution was thought to be totally responsible for the freezing expansion in the hydraulic pressure theory. The ice formation during this phase involves mainly the capillary pores, and the expansion can already damage the material if it is not protected against frost action.

Phase III: −10 to −35 °C. During this phase, several factors interplay with respect to the macroscopic deformation. The ice formed in capillary pores contracts with freezing. Since ice has a larger thermal expansion coefficient than the solid matrix, the ice contraction induces the liquid water refilling into the pores and promotes pore crystallization. This process also releases the hydraulic pressure accumulated in phase II in smaller pores. The pore crystallization contributes to material expansion, while the thermal contraction and the hydraulic pressure release counteract the expansion. The additive effect of pore crystallization and thermal contraction results in a nearly constant deformation of the saturated mortar specimen during this temperature interval.

Phase IV: −35 to −70 °C. Below −35 °C, supercooling overcomes the energy barriers of surface tension and even of homogeneous nucleation (Zeng *et al.*, 2015). Thus, liquid water in the smallest pores, gel pores at the nanometer scale, is expected to freeze. The global expansion around −40 °C marks this process and indicates that the liquid water quantity involved is significant. Afterwards, the specimen, with large part of liquid water frozen in pores, contracts constantly with the decrease in temperature, showing a similar thermal expansion coefficient to the dried specimen.

During this freezing process, the origin of material damage is the pore stress from the crystallization. According to standard poromechanics, the constitutive equation for freezing porous medium is (Coussy, 2010)

$$\sigma = K\varepsilon - \left(b_\mathrm{C} p_\mathrm{C} + b_\mathrm{L} p_\mathrm{L}\right) - 3\alpha_\mathrm{S} K\left(T - T_\mathrm{f}\right) \tag{3.4}$$

where σ (Pa) and ε are respectively the stress and volumetric strain of the material, K (Pa) is the bulk modulus of the material, α_S (1/K or 1/°C) is the thermal expansion coefficient of the solid matrix, p_C and p_L (Pa) are respectively the pressures of the ice crystal and the liquid in the pores, b_C and b_L are respectively the Biot coefficients for the ice crystal and liquids in the pores, and T_f (K or °C) is the reference temperature. Following Coussy's (2010) theory, the mean pore pressure is adopted to describe the overpressure in pores by the crystallization process:

$$\langle\sigma\rangle_\mathrm{p} = s_\mathrm{C} p_\mathrm{C} + s_\mathrm{L} p_\mathrm{L} \quad \text{with} \quad s_\mathrm{C} + s_\mathrm{L} = 1 \tag{3.5}$$

where s_C and s_L are the pore saturations by the ice crystal and liquid water respectively. This overpressure is to be taken by the solid matrix, and the stress level in the solid matrix σ_m is expressed through the volume average of stress over different phases:[5]

$$\sigma_\mathrm{m} = \frac{\phi}{1-\phi}\langle\sigma\rangle_\mathrm{p} = \frac{\phi}{1-\phi}\left(s_\mathrm{C} p_\mathrm{C} + s_\mathrm{L} p_\mathrm{L}\right) \tag{3.6}$$

---

[5]This expression originates from expression (7.33) in Coussy (2010: 160), neglecting the interface stress associated with the formation of solid–liquid interface. This assumption is further validated in Zeng *et al.* (2014), in which the contribution of interface energy in pore stress is two magnitudes lower than other terms for a freezing range to −35 °C.

with ϕ the material porosity. This average matrix stress can be used to characterize the damage of concrete under frost action. If this stress is far below the fracture resistance or tensile strength of concrete, then damage cannot occur; if this stress approaches the facture resistance of concrete, damage can occur in the form of fractures; if the stress is in between, the solid matrix can be damaged probably by low-cycle fatigue action (Fagerlund, 2002). As an engineering approach, the facture criterion of a solid matrix due to pore pressure can be expressed through either the stress intensity for fracture mode I or the cracking strength:

$$\sigma_m = \frac{K_{IC}}{\sqrt{\pi a}} \quad \text{or} \quad \sigma_m = \sigma_{fc} \tag{3.7}$$

where K_{IC} (Pa m$^{1/2}$) denotes the fracture roughness for a mode I crack and σ_{fc} is the stress to initiate crack propagation from the initial flaws in the material. Note that the interface transition zone (ITZ) between the HCP and the aggregates can favor fracture initiation due to its larger porosity and initial defaults and cracks.

So far, the mechanisms are more oriented to the internal FT damage, and the surface scaling is usually thought to be induced by other damage process. During recent decades, several theories were advanced to address the surface scaling with the presence of salts: the osmotic pressure induced by the local condensation of pore solution in the surface layer (Powers and Helmuth, 1953); the thermal shock in the surface layer created by the melting of surface ice with salts (Rostam, 1989); and the pore supersaturation in the surface layer by melting salts (Litvan, 1973). However, the aforementioned internal FT mechanisms were also shown to be capable of accounting for surface scaling under specific boundary conditions (Fabbri et al., 2008). Nevertheless, a recent "glue–spall" mechanism (Valenza and Scherer, 2006) explained the salt spalling in a rather consistent way: (1) the surface salt solution from the melting salts forms a brine-ice layer with mechanical properties dependent on the salt concentration in solution; (2) this brine-ice layer is "glued" to the concrete surface, and the thermal contraction of this layer is larger than that of the substrate concrete; (3) this thermal contraction difference leads to the brine-ice layer fracturing, and the fracture cuts into the concrete surface; (4) isolated islands of brine-ice layer are formed glued to the "cut-out" surface, and with further freezing the substrate surface can be spalled by the mismatch of thermal contraction; (5) fracture analysis showed that a brine-ice layer formed from a salt solution of 3% concentration is weak enough to be fractured by thermal contraction but strong enough to lead to spalling of the substrate concrete surface. This mechanism actually provides new insight into salt scaling.

3.2.2 Influential Factors

The factors that influence frost damage of concrete can be divided into two groups: the material properties of concrete and the environmental conditions. The material properties refer to the pore structure, pore saturation and air entrainment with respect to HCP, and aggregate properties; the environmental conditions include the freezing extent and rate of natural frost action, and the presence of salts.

Table 3.2   Critical pore saturations from literature

Material	Description	Target property	Critical saturation	Source
Mortar	—	RDM	0.77	Fagerlund (2002)
Mortar	$w/c = 0.42$, entrained air 6–14%	RDM	0.88	Li *et al.* (2013)
Concrete	—	—	0.85–0.90	Neville (1995b)
Concrete	—	—	0.80–0.90	Fagerlund (2002)
Concrete	$w/c = 0.4$, entrained air 5.5%	RDM	0.77–0.82	Yang *et al.* (2006)
Limestone	Porosity 45.2%	Porosity change	0.85	Al-Omari *et al.* (2015)
Limestone	Porosity 29.0%	Porosity change	0.85	Al-Omari *et al.* (2015)

RDM: relative dynamic modulus.

Material Properties

Critical Pore Saturation

The most important influential factor for concrete FT resistance is its pore saturation, which applies to both internal FT damage and salt scaling. So far, freezing has been addressed for totally saturated pores. As the pore saturation is below 1.0, the ice formed needs first to fill up the void pore space before exerting stress on the pore wall. As the pore saturation is below a certain value, pore freezing will no longer have a damaging effect on the solid matrix of the material. This value is termed the *critical pore saturation*. The determination of its value is of vital importance for engineering application since only from this saturation does the concrete become susceptible to damage. This value is determined mainly from experimental investigations: specimens are treated to expected pore saturations and subject to laboratory FT tests, and then the critical saturation is judged from the change of material properties. Table 3.2 recapitulates some experimental results from the literature on critical pore saturation.

Sorptivity

In practice, concrete that is subject to frost damage undergoes a saturation process with the freezing and thawing cycles: during the freezing phase the pore water crystallizes and during the thawing phase the thawing in pores tends to absorb more external liquid water. This water intake has been depicted by the micro-lens theory (Setzer, 2001). Accordingly, with the FT cycles, the pore saturation increases gradually, passing the critical saturation and inducing damage to material. This explains why FT cycles are usually needed to damage a structural concrete and also why the FT damage is basically a moisture (water) transport event. To address this crucial aspect of water movement, the water sorptivity can be used as one pertinent property. The water sorptivity S_w (mm/s$^{1/2}$) describes the water intake rate of the concrete surface in contact with liquid water. Some sorptivity data are given in Figure 3.6 for typical OPC concretes. Normally, the water absorption of a concrete surface contains two phases: an initial phase, 6 h according to the ASTM C1585 method, with larger sorptivity, and a secondary phase with smaller sorptivity. The former corresponds to the instant suction of coarse capillary pores and the secondary phase refers to the much slower water movement in gel pores (Martys and Ferraris, 1997) or in closed-form air voids (Fagerlund, 1993).

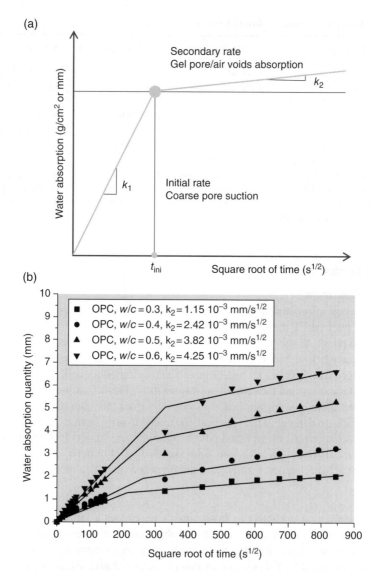

Figure 3.6 Two-phase sorptivity of concrete (a) and usual sorption rates for OPC concretes (b). The concrete specimens were dried to constant weight at 60 °C before absorption, and the absorption procedure conforms to ASTM C1585 (ASTM, 2013c).

Characteristic Curve of Pore Freezing

The pore structure impacts on the freezing through its porosity and the pore size distribution (PSD). The PSD actually determines the pore ice content in terms of freezing temperature, $s_C = s_C(T)$, or the characteristic curve for pore freezing. Only through this relation can the average stress in the solid matrix σ_m be evaluated from Equation 3.6. The pore ice content can be measured experimentally through differential scanning calorimetry or through the electrical capacitance method. Alternatively, the ice content can be estimated from the PSD curve obtained from other pore structure measurement methods, such as mercury intrusion porosimetry (MIP). Mercury intrusion and pore water freezing can both be regarded as a non-wetting phase, mercury or ice crystal, penetrating progressively into pores under the driving force – intrusion pressure (MIP) or freezing temperature (freezing). The basic equation to interpret the PSD through MIP is

$$p_{m} - p_{g} = -\frac{2\gamma_{mg}\cos\theta_{MIP}}{r} \tag{3.8}$$

with p_m and p_g respectively the mercury and air pressures during intrusion; γ_{mg} and θ_{MIP} are respectively the mercury–air surface tension and the contact angle between mercury and the pore wall. Since the MIP test can provide the accumulated intrusion volume v_{MIP} in terms of intrusion pressure p_m, the accumulated volume v_{MIP} can be readily expressed in terms of the radius of pore neck r through Equation 3.8:

$$v_{MIP} = v_{MIP}(r) \quad \text{or} \quad r = r^{-1}(v_{MIP}) \tag{3.9}$$

Now, under a specific freezing temperature, the attainable pore radius can be decided from Equation 3.1. Thus, the frozen pore volume at a freezing temperature T can be deduced as

$$s_{C}(T) = \frac{v_{ice}(T)}{v_{max}} = \frac{v_{MIP}(r_{ice})}{v_{max}} \quad \text{with} \quad r_{ice} = \frac{2\gamma_{CL}}{(T^{*} - T_{f})S_{f}} + \delta(T) \tag{3.10}$$

with v_{max} representing the maximum intrusion volume. The merit of this approach lies in its simplicity and straightforward nature, and the disadvantage is also obvious: Equation 3.10 considers only the physical similarity between mercury and ice intrusion, but neglects the peculiarity related to pore ice formation, such as the nucleation process. Nevertheless, the ice content obtained from Equation 3.10 is considered sufficient here to make an estimation for engineering use. The ice content curves for some OPC concretes are given in Figure 3.7 from MIP results.

Air Entrainment

An important aspect of pore structure related to FT resistance is air entrainment. The technique introduces air entrainment agents (AEAs) into the concrete mixture and the organic molecules of the AEAs can generate spherical air bubbles of diameter about 300 μm and stabilize these bubbles till the hardening of concrete. Normally, the air voids formed are dry, and thus can provide ice nucleation sites without generating crystallization stress in the solid matrix (Valenza and Scherer, 2006). Experience indicates that a volume ratio of entrained air of 4–7%, with respect to concrete, can efficiently protect concrete against frost damage. For the same frost action, concretes with larger aggregate size need less entrained air than those with smaller aggregate size. Alternatively, air entrainment can also be prescribed through the average spacing factor of air voids. This spacing factor can be measured through a standard procedure on a polished concrete surface following the ASTM C457 procedure (ASTM, 2012a), and an entrained air content above 6% can usually decrease the spacing factor below 200 μm. A spacing factor around 250 μm is regarded as a safe value against the frost action, and this factor is usually set to 200–300 μm by technical standards (Hobbs et al., 1998).

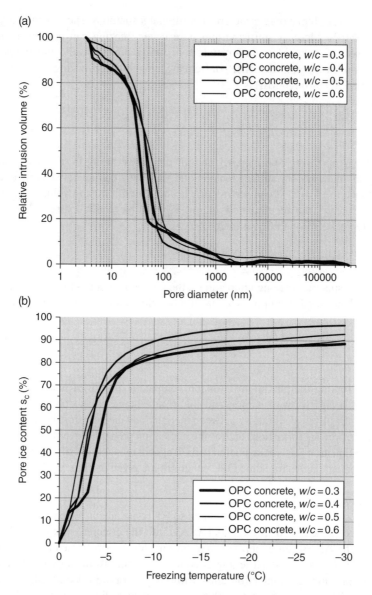

Figure 3.7 Accumulated intrusion volumes from (a) MIP and (b) deduced ice contents for OPC concretes with $w/c = 0.30 - 0.60$. The properties of OPC concretes are given in Table 1.1 and Figure 1.9.

Aggregates

So far, no special attention is given to the aggregates, which occupy a 50–80% volume fraction of concrete. Natural coarse aggregates come from crushed rocks, and normally have lower permeability than the bulk cement paste in concrete. Thus, these aggregates are usually assumed to be impermeable for transport processes in concrete. Actually, the aggregates are also porous media, and subject to the same freezing mechanisms as cement pastes. As indicated in Table 3.2, there also exists a critical pore saturation for rocks. The "D-cracking" damage of concrete pavement joint was attributed to the frost damage of coarse aggregates (Koubaa and Snyder, 2001). The same authors proposed a systematic screening procedure to determine the real frost resistance of natural aggregates, which employed petrography analysis, a laboratory rapid FT test of the ASTM-C666 type and a hydraulic fracture test. For natural fine aggregates, sands, it was shown that, due to the strengthening effect of chemical reaction at ITZ, calcareous and silico-calcareous sands perform better than siliceous sands, and river sands perform better than quarry sands (Girodet *et al.*, 1997). The frost resistance of recycled-aggregate concrete has recently attracted attention: recycled aggregates, coarse or fine, have a higher porosity and larger water sorptivity, and the resulting recycled-aggregate concrete usually has lower mechanical strength and weaker frost resistance (Zaharieva *et al.*, 2004). Blending recycled and natural aggregates and keeping the recycled aggregates saturated or semi-saturated seem to be rational choices for better frost resistance (Yildirim *et al.*, 2015).

Mechanical Properties
Other mechanical properties can have a direct impact on the frost resistance of structural concretes. Taking the frost damage at pore level as a brittle fracture process, it is shown from Equation 3.7 that the fracture toughness and the cracking strength are pertinent to the mechanical damage of the solid matrix by pore pressure accumulation. The fracture toughness of concrete depends on both the aggregate packing and the strength of the HCP (Zhang *et al.*, 2010).

Environmental Conditions

Freezing Range and Rate
The environmental temperature conditions include mainly the freezing temperature (range) and the freezing rate. The lowest temperature determines the attainable pore size of freezing, and thus the quantity of ice formation, which is reflected through the fundamental property $s_C = s_C(T)$. Larger freezing ranges involve more pores in freezing. Note that natural frost action has a freezing temperature rarely below $-20\,°C$. The freezing rate was believed to play a role in the frost damage: faster freezing is supposed to cause more damage than slower freezing. It was even predicted from the hydraulic pressure theory that the magnitude of the hydraulic pressure in pores is proportional to the freezing rate. The results in the literature seem to confirm, if not the proportionality between frost damage and freezing rate, larger spacing factors of air voids could be tolerated at lower freezing rates (Pigeon *et al.*, 1985). The freezing rate involves certainly the kinetics aspect of pore freezing and the viscous flow of liquid water as well. Note that the natural freezing rate is rarely above $2\,°C/h$, but the rate in laboratory accelerated tests can reach $20\,°C/h$. Figure 3.8 illustrates the attainable pore size in terms of the freezing range and some experimental results on the role of freezing rate. In the figure, a higher freezing rate, $12\,°C/h$, induces a larger freezing expansion, but the most severe damage, in terms of residual strain after each FT cycle, occurs at the lowest freezing rate, $2.5\,°C/h$ (Li and Zeng, 2009).

Salts
The last issue of environmental condition is the presence of salts during frost action. The presence of salts is regarded as a severe condition for frost action, inducing severe damage of surface scaling. The related mechanisms were treated in Section 3.2.1, and the most important observation is the optimum concentration of 3% for the most severe scaling regardless of the

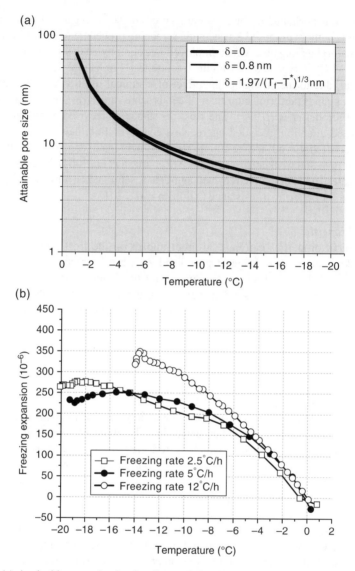

Figure 3.8 (a) Attainable pore size by freezing and (b) role of freezing rate on the expansion of a mortar specimen. In (a), the unfrozen layer thickness $\delta = 1.97/(T_f - T^*)^{1/3}$ comes from Fagerlund (1976). In (b) the mortar has $w/c = 0.60$ and MIP porosity is 15.9%. *Source:* Li and Zeng, 2009, Figure 3(b). Reproduced with the permission of Springer.

salt type (Verbeck and Klieger, 1957). For concrete pavement and RC elements exposed to deicing salts, the salt concentration depends on the application frequency and quantity of salts, and also the position of the concrete surface. For a concrete element in contact with seawater, frost damage was also reported to be more severe compared with the FT action by pure water (Achintya and Prasad, 2003), knowing that the salinity of seawater is in the range 3.1–3.8%, approaching the optimum concentration for frost action.

Accelerated Freeze–Thaw Tests

Laboratory accelerated tests for frost action have been established during the past decades on the basis of the aforementioned mechanisms and influential factors. These tests include mainly the ASTM C666 method (ASTM, 2003) for water freezing and the ASTM C672 method (ASTM, 2012b) for salt scaling, the RILEM TC-176 IDC slab test method (RILEM, 2001) for internal FT, and the RILEM TC-117 CDF method (RILEM, 1996) for FT and deicing resistance of concrete. The main features of these tests will be given later, in Chapter 8. These tests use standard freezing ranges, accelerated freezing rates, and specific water contact conditions to exert an artificial frost action on concrete specimens, and the frost damage is evaluated from the change of material properties or simply by visual appearance of the specimen. These tests are classified into durability performance tests, and the possible use of the experimental results as durability indicators will be discussed further in Chapter 8.

3.3 Modeling for Engineering Use

Though the main mechanisms and influential factors involved in the frost damage of concrete are rather clear, the quantitative modeling to predict the damage extent is far from accomplished. This is partially due to the random nature of the concrete pore structure and the complex nature of the damage process of concretes as heterogeneous composites. On the basis of this state of the art, this section provides two preliminary models for frost action: the FT-1 model accounts for only the pore saturation, and uses the absorption property and the critical pore saturation to predict the frost resistance of structural concrete under specific environmental conditions; the FT-2 model further takes into account the pore pressure by the freezing of pore water, and uses a simple criterion of fracture initiation to judge the onset and rate of material damage. Both models remain at the conceptual level, and more work on validation and refinement is needed.

3.3.1 Model FT-1: Critical Saturation Model

This model is adapted from the internal damage model in DuraCrete (1998) proposed by Fagerlund (1993). The premise of this model is that concrete is susceptible to frost damage once the critical pore saturation s_{CR} is reached. The model is developed particularly for concrete with a certain content of entrained air and can be used to design the needed entrained air content with respect to the frost action.

Suppose the total porosity of concrete is ϕ and the entrained air content is ϕ_a. Further, the air voids are assumed to contain no water at the beginning of wetting. Then, the concrete is exposed to a long-term wetting process during which the pore saturation is gradually increased. This increase is described by the water sorption process in Figure 3.6, and the secondary part of water sorption is attributed to the water invasion into the entrained air voids. The pore saturation increase is described as (DuraCrete, 1998)

$$s_L(t) = \frac{1}{\phi}\left[(\phi - \phi_a) + 0.13V_p(w/c)\log t\right] \tag{3.11}$$

Table 3.3 Parameters for FT-1 model for pore saturation evaluation

Parameter	Value/relation
Critical pore saturation s_{CR}	0.85
Water-to-cement ratio for concrete w/c	0.3, 0.4, 0.5, 0.6
Porosity of concrete $\phi - \phi_a$	0.09 ($w/c = 0.3$), 0.12 ($w/c = 0.4$), 0.15 ($w/c = 0.5$), 0.18 ($w/c = 0.6$)
Entrained air porosity ϕ_a	0.04–0.07
Cement paste volume ratio in concrete V_p	0.35

where w/c is the water-to-cement ratio of concrete, V_p is the paste volume ratio in concrete, and t (h) is the wetting time. The potential service life is attained as

$$s_L(t) = s_{CR} \tag{3.12}$$

The parameters involved in this model are given in Table 3.3, and the simulation results are illustrated for typical OPC concretes in Figure 3.9.

From these results, the impact of the air entrainment on the wetting time is the determinant for a target critical pore saturation. The wetting time in the figure should be interpreted as the accumulated effective wetting time for a concrete surface. In practice, the concrete surface susceptible to frost action is located in the tidal zone or near the water level; see Figure 3.1. For these cases, the concrete surface is alternately subjected to a wetting period and a drying period with atmospheric temperature above zero and a freezing period with temperature below zero. Take the frost action as an annual event and note the periods respectively as t_{wet}, t_{dry}, and t_{freeze}. If the surface has a short wetting period compared with the drying period, the concrete surface globally loses water, then frost damage can be exempted; otherwise, the annual cycle tends to increase the pore saturation, and its liability to frost damage can be measured by the accumulated wetting time through the FT-1 model. To judge the water intake or loss under specific wetting and drying periods, a model is developed in Chapter 6 establishing the rule for water storage and loss of concrete surface in terms of the drying and wetting durations and physical properties of concrete. The interested readers can strengthen the FT-1 model with the drying–wetting model in Section 6.3.

3.3.2   Model FT-2: Crystallization Stress Model

This model aims to incorporate the pore stress accumulation and the facture of the solid matrix of concrete by pore stress into the frost damage model. To this end, the basic equations for pore crystallization in Section 3.2 are recalled. Here, instead of a single pore with a radius, the continuous PSD has to be included into the stress expression in Equation 3.6. Consider a concrete with the PSD of the pore structure characterized by Equation 3.9 and the corresponding ice content $s_C(T)$ determined through Equation 3.10. The average stress in the solid matrix can be obtained through the integral

$$\sigma_m\big|_{T_f \to T^*} = \frac{\phi}{1-\phi} \int_{T_f}^{T^*} \left(p_C \frac{\mathrm{d}s_C}{\mathrm{d}T} + p_L \frac{\mathrm{d}s_L}{\mathrm{d}T} \right) \mathrm{d}T \tag{3.13}$$

Assume the liquid pressure is constantly equal to the atmospheric pressure; that is, pore freezing occurs in the condition of free drainage of pore water. Considering further the

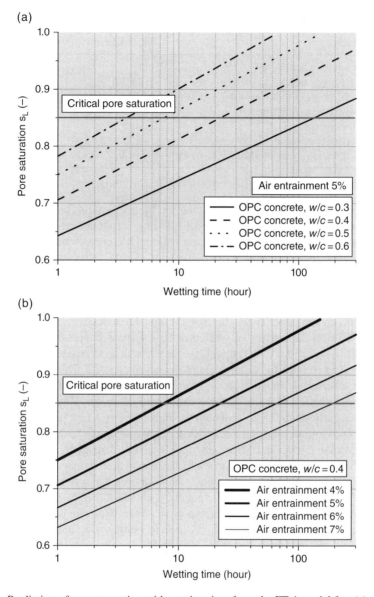

Figure 3.9 Prediction of pore saturation with wetting time from the FT-1 model for: (a) typical OPC concretes with air entrainment of 5%; (b) OPC concrete of $w/c = 0.4$ with different air entrainments. The properties of OPC concretes are given in Table 1.1 and Figure 1.9.

capillary equilibrium condition in Equation 3.2 and ice content expression in Equation 3.10, Equation 3.13 can be arranged as

$$\sigma_{\mathrm{m}}\big|_{T_f \to T^*} = \frac{\phi}{1-\phi} \int_{T_f}^{T^*} \frac{2\gamma_{\mathrm{CL}}}{r_{\mathrm{E}}(T)+\delta(T)} s_{\mathrm{C}}'(T)\,\mathrm{d}T \qquad (3.14)$$

Table 3.4 Parameters for FT-2 model for solid matrix stress evaluation

Parameter	Value/relation
Ice–water interface energy γ_{CL} (J/m^2)	0.04
Fusion entropy of ice S_f (J/(m^3 K))	1.2×10^6
Layer thickness of unfrozen water δ (nm)	$1.97/(T^* - T_f)^{1/3}$ (Fagerlund, 1976)
PSD of concrete pore structure	Figure 3.8
Pore ice content $s_c(T)$	Figure 3.8
Porosity of concrete ϕ	See Table 3.4
Crack initiation stress, σ_{fc} (MPa)	0.8–1.0 (Zhang et al., 2004)

With a known PSD of the pore structure and the freezing extent, the average stress in the solid matrix can be evaluated from Equation 3.14. The model parameters are given in Table 3.4, and the simulation results are illustrated in Figure 3.10 for OPC concretes with $w/c = 0.30 - 0.60$.

From the results, the mean pore pressure remains at almost the same level for the OPC concretes, reaching a magnitude of 5 MPa for freezing temperature down to −30 °C. This pore pressure is a local one and creates tensile stress in the solid matrix; and the solid matrix stress has a notable difference for different concretes, reaching 0.5 MPa, 0.75 MPa, 1.0 MPa, and 1.2 MPa for OPC concretes with $w/c = 0.3, 0.4, 0.5$, and 0.6 respectively. This difference comes mainly from the absolute values of total porosity of these concretes, ranging from 0.09 ($w/c = 0.3$) to 0.18 ($w/c = 0.6$). This stress is the very reason for concrete damage, and its value should be compared with the fracture-related properties of concrete. This fracture-related property is chosen as the crack initiation stress, or cracking strength, of concrete, and the value in literature ranges from 0.9 to 1.8 MPa for conventional concretes (Zhang et al., 2004). This value is schematically set as 0.8–1.0 MPa on the figure to give a first judgement on the stress level susceptible to initiate cracks in the concrete. Using this range as a criterion, one can see that concretes with $w/c = 0.5$ and 0.6 are susceptible to frost damage by pore freezing.

If one wants to use the FT-2 model to evaluate the potential damage extent for a concrete surface of RC elements in frost environments, the saturation depth should then be quantified. As in Section 3.3.1, consider a typical situation: a concrete surface exposed alternately to drying, wetting and freezing actions with respective annual exposure duration. The model on the drying–wetting action in Chapter 6 can be used to estimate the saturation depth or drying depth in terms of the drying and wetting periods and the concrete transport properties. If wetting dominates over drying (i.e., the water intake is more than water loss), a saturation depth can be evaluated. Then, for concrete within this saturation depth, the FT-2 model can be applied to determine the tensile stress in the solid matrix. If the stress level is judged high enough to initiate cracks, the saturation depth is susceptible to spalling damage and this saturation depth can be used to estimate the annual spalling rate of the concrete surface.

The FT-2 model contains the essential information of pore freezing, including the pore ice content, porosity, PSD, the freezing range, and the crack initiation criterion. However, the FT-2 model remains a conceptual model since the pore liquid pressure is neglected. In other terms, the liquid water in pores is assumed to flow freely and keep equilibrium with atmospheric pressure instantaneously. Though the unfrozen water layer can always provide channels for liquid water flow, the release rate of liquid pressure will depend on the freezing rate as well as the connectivity of pores. For a saturated concrete surface exposed to freezing temperature, the pores at the

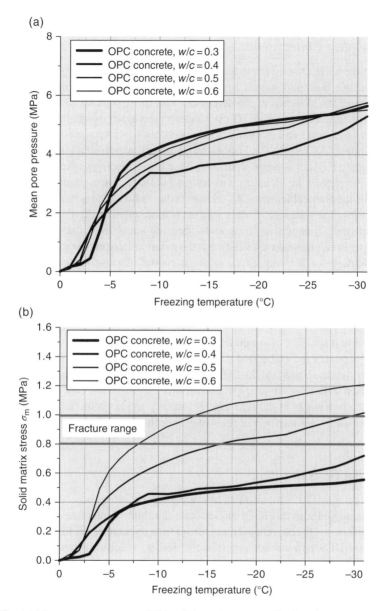

Figure 3.10 (a) Mean pore pressure and (b) solid matrix stress under freezing temperature for OPC concretes with $w/c = 0.3 - 0.6$.

surface freeze first and the subsequent pore freezing is to drive the liquid water into the inner part of the materials. Liquid pressure, similar to that described in the hydraulic pressure theory, can arise as the pore connectivity is poor. This aspect is neglected in the FT-2 model, and a more elaborate poromechanics modeling is needed to account for this aspect. Since the complexity of such modeling is beyond the scope of this textbook, interested readers are suggested to refer to the recent development of poromechanics for pore freezing (Coussy, 2010).

3.4   Basis for Design

3.4.1   Structural Consequence

Concrete damage by frost action is induced by the pore pressure during freezing that is high enough to fracture the solid matrix of the material. As a consequence, the physical and mechanical properties of concrete can deteriorate and the related structural performance can be reduced. Structural concrete after FT cycles can have a lower dynamic elastic modulus, and this change is often used in laboratory accelerated tests to characterize the extent of frost damage in concrete (ASTM, 2003). Further, in laboratory accelerated tests, RC elements with concrete subject to FT actions were reported to have lower loading capacity and less deformation capacity compared with elements without frost damage (Duan *et al.*, 2011).

However, the mechanical failure of RC elements due to frost action has rarely been observed in practice because frost damage of RC elements is more often limited to scaling and spalling of surface concrete. In natural environments, freezing starts from the concrete surface and usually a temperature gradient exists between the external freezing temperature and the inner parts of elements, so a freezing depth exists from the element surface. Moreover, as the concrete surface is saturated by direct contact with external water, the inner part of the material is usually less saturated. Accordingly, freezing damage can only occur at depth from the concrete surface where both the temperature condition and pore saturation condition are met for frost damage. This explains the extensive surface damage by frost action in real cases. Surface scaling and spalling of concrete raise engineering concern for the protection of the inner part of material or the embedded steel bars rather than the altered mechanical resistance of RC elements. It was shown that the diffusivity and permeability of structural concretes were notably increased with frost damage (Wittmann *et al.*, 2010; Wang and Ueda, 2014). Thus, a concrete surface bearing frost damage can no longer provide protection against the external aggressive agents for the inner part of the material or the embedded reinforcement steel bars.

3.4.2   Design Considerations

The durability design of RC elements in a frost environment is to protect the structural concrete against frost damage. From this point of view, two durability limit states (DLSs) can be specified for RC elements: (1) concrete exempted from frost damage or (2) limited spalling/scaling extent. The first DLS is a stricter criterion for design, specifying that no frost damage is allowed for the structural concrete of RC elements during the service life. This DLS applies to RC elements for which frost damage of surface concrete can have a notable effect on the expected performance; for example, the concrete surface is expected to provide protection for reinforcement steel bars, or the element is expected to be leakproof and surface damage can lead to deterioration in the water-tightness. Thin-wall RC elements also belong to this category, since the section is small and any loss of concrete is liable to affect the structural performance. The second DLS refers to a less strict state of which a specified scaling/spalling extent is tolerated. This situation applies to most RC elements for which the loss of surface concrete, if limited to some specified extent, does not compromise the structural performance. The FT-1 model in this chapter can be used to help the design with respect to the first DLS with the focus on the critical pore saturation. The design for the second DLS can use the FT-2 model to estimate the possible spalling of the concrete surface of RC elements.

From the analysis in this chapter, the risk of frost damage of structural concrete is enhanced by the following factors: high saturation, large porosity, low freezing temperature, and the presence of salts. It is on this basis that the measures for durability design are formulated. There are basically two measures to be taken against internal frost damage on a material level: the air-entrained technique or high-strength concrete. Air entrainment can efficiently decrease the pore pressure during freezing, and the necessary content of entrained air is normally between 4 and 7% in terms of the aggregate size, the freezing range, and the water contact conditions. This requirement can be alternatively replaced by specifying the average spacing of air voids in hardened concrete. High-strength concrete achieves high compactness, and thus has low porosity and permeability. These characteristics make the concrete very difficult to be totally saturated. Even saturated, the concrete shows high frost resistance since the absolute quantity of water is very low and the tensile stress created in the solid matrix is not high enough to initiate cracks. From experience, structural concretes with strength grade higher than C50 are rarely observed to have internal frost damage. The presence of salts is a rather special condition. Although the damage mechanisms are not yet unanimously agreed, two measures are proven efficient: air entrainment and treatment of the concrete surface. Both methods help to increase the surface strength of concrete. For RC elements locally exposed to salts, local protection layers can be adopted to isolate physically the concrete surface from contact with the salts.

Apart from these measures, the hardening of structural concrete in frost environments merits the attention of designers. Normally, in-situ concreting avoids severe frost climates, since in low (negative) temperature the hydration of cement is retarded and the concrete hardening can be affected. This is termed "cold weather concreting" in concrete technology, and more details can be found in the relevant specification (ACI, 1988). For durability consideration, the concrete strength is expected to reach a certain level before the first exposure to negative temperature. This minimum strength is specified as 10 MPa, and the delay between the end of concrete curing and contact with liquid water is not allowed to be shorter than 30 days (CNS, 2008).

4

Leaching

This chapter presents the leaching process of structural concretes by soft water in the environment. Leaching refers to the dissolution of solid phases of porous materials through pore solution transport. For cement-based materials, leaching is usually related to the dissolution of calcium in the solid matrix. The stability of cement hydrates is affected by the leaching process and the properties of structural concrete are compromised. Leaching is a long-term process for concrete, and contact with soft water, water with low salinity, is an indispensable condition. This chapter begins with the engineering cases of leaching deterioration by flowing water and then introduces the mechanisms and the influential factors of the process. On the basis of state-of-the-art knowledge, two models are given for calcium leaching of concrete. Some useful indicators, especially the leaching depth of portlandite (CH), are identified for engineering use. Finally, considerations for durability design are developed.

4.1 Phenomena and Observations

Concrete consists of a solid matrix, including hydrates and aggregates, and porosity, which is the void space not occupied by the solid phases. The pores of concrete are more or less occupied by a liquid solution. In the hardened state, the solid phases are in dissolution equilibrium with the liquid phase in pores. The solid phases, especially the hydrates from cements and mineral admixtures, are stable at this chemical equilibrium. This solid–liquid equilibrium can be changed as transport processes occur in the pore liquid phase. Leaching belongs to such processes, and usually refers to the dissolution of calcium contained in solid phases and the subsequent transport of calcium, in the form of aqueous ions, to the external environment via the liquid phase in pores. By this process the calcium in solid phases is gradually dissolved into pore solution and transported out of the material. Thus, leaching is also termed "calcium

Durability Design of Concrete Structures: Phenomena, Modeling, and Practice, First Edition. Kefei Li.
© 2016 John Wiley & Sons Singapore Pte. Ltd. Published 2016 by John Wiley & Sons Singapore Pte. Ltd.

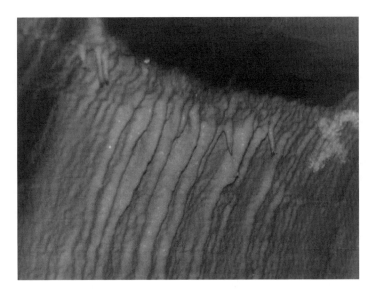

Figure 4.1 Speleothems (CaCO$_3$) formed on the internal wall of a concrete gallery in a dam structure in China. Calcium ions are brought to the wall surface by long-term leaching and precipitate in a reaction with atmospheric CO$_2$ in the gallery to form speleothems. *Source:* courtesy of Gaixin Chen.

leaching" in the literature for concrete. In the long term, calcium leaching can weaken the solid matrix of the material and decrease the performance of concrete elements in structures.

Normally, the leaching process is rather slow, especially for compact concretes, and the leaching is rarely an issue for concrete structures with short service lives (e.g., less than 50 years). The first category of concrete structures liable to calcium leaching is hydraulic structures, such as dams, waterways, and water pipelines. The RC elements in these structures have direct contact with flowing water in which the salinity and calcium concentration are relatively low. Such water is termed "soft water." In such an environment, the concrete surface can be saturated by the water and the concentration gradient of ions between the external water and the pore solution drives the pore ions to diffuse out to water and decrease the concentration of ions in pore solution, which, in turn, promotes the dissolution of solid phases. Under long-term action of leaching, the alkalinity of pore solution is decreased by OH$^-$ transport and the porosity of concrete surface is enlarged by the dissolution of solid calcium. Figure 4.1 illustrates the leaching of a concrete gallery in a dam structure.

Another engineering concern for concrete leaching pertains to concretes used in the long-term disposal of radioactive waste, including the engineered barriers and containment structures in near-surface disposal sites (IAEA, 2001) and the concrete facilities in deep geological disposal sites (Sellier *et al.*, 2011). These concrete barriers and facilities are expected to ensure the containment function of disposal for more than 300 years (near-surface disposal) and more than 500 years (deep geological disposal) (IAEA, 2006). Under such circumstances, the concrete materials can come into contact with groundwater and calcium leaching can develop to an advanced extent considering the very long disposal life. Actually, calcium leaching has been considered as one major control scenario for the safety assessment of these structures (IAEA, 1999).

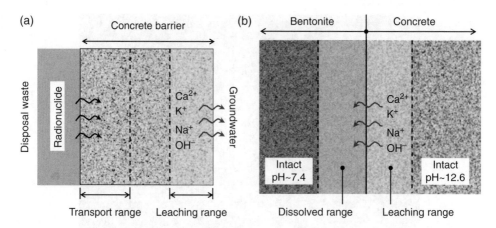

Figure 4.2 Critical scenario of concrete leaching in (a) near-surface disposal and (b) deep geological disposal. The design criterion for concrete barriers (a) is the thickness of barrier should cover the transport depth of radionuclide and the leaching depth by groundwater during the disposal life. The dissolution of smectite in bentonite by concrete leaching (b) creates a leached zone in concrete and a dissolved zone in bentonite, in which the performance of both barriers is reduced.

Figure 4.2 illustrates the problems related to concrete leaching in radioactive disposal facilities. In near-surface disposal the concrete barrier is expected to resist both the internal radionuclide transport and the external leaching process. In geological disposal the adverse effect of concrete leaching on the adjacent bentonite should be limited. The bentonite has a rather neutral internal environment (pH 7.4) and contains salts of very low concentration (Fernández *et al.*, 2004), while conventional OPC concretes have much higher alkalinity (e.g., pH ~12.6) (Allard *et al.*, 1984). As the two materials come into contact, the ions in concrete pores, including Ca^{2+}, K^+, Na^+, and OH^-, are transported across the interface into the bentonite, and the OH^- ions will dissolve the main constituent mineral of bentonite, the smectite, and form secondary minerals (Neretnieks, 2014).

4.2   Mechanisms and Influential Factors

4.2.1   Mechanisms

As the concrete is completely hardened, the hydrates from cement and mineral admixtures are in equilibrium with the ion species in the pore solution. The pore solution contains mainly the ions of Na^+, K^+, Ca^{2+}, OH^- and SO_4^{2-}, and the solid phases in the hydrates establish their respective dissolution equilibrium in this ionic environment. The soluble solid calcium phases are mainly the portlandite (CH) and C-S-H. The dissolution reaction of CH writes

$$Ca(OH)_2 \leftrightarrow Ca^{2+} + 2OH^- \tag{4.1}$$

The dissolution reaction in Equation 4.1 is controlled by the dissolution constant K_{CH}:

$$Q_{CH} = \left[Ca^{2+}\right]\left[OH^-\right]^2 \leq K_{CH} = \left[Ca^{2+}\right]_{eq}\left[OH^-\right]_{eq}^2 \tag{4.2}$$

with $[\cdot]_{eq}$ being the ion concentration in equilibrium. The C-S-H have lower solubility than CH, and the dissolution reaction can be schematically written as

$$C_xS_yH_z \leftrightarrow xCa^{2+} + 2xOH^- + yH_4SiO_4 + (z-x-y)H_2O \qquad (4.3)$$

The variables x, y, and z are chemical stoichiometric numbers associated with CaO, SiO$_2$, and H$_2$O in C-S-H. Such an expression reflects the fact that different groups of C-S-Hs co-exist with different stoichiometric numbers (Atkinson *et al.*, 1989). The dissolution equilibrium is described by the constant K_{CSH}:

$$Q_{CSH} = \left[Ca^{2+}\right]^x \left[OH^-\right]^{2x} \left[H_4SiO_4\right]^y \leq K_{CSH} \qquad (4.4)$$

The dissolution constants K_{CH} and K_{CSH} are also temperature dependent. In addition to CH and C-S-H, other hydrates, such as ettringite (3CaO·Al$_2$O$_3$·3CaSO$_4$·32H$_2$O), can also dissolve into pore solution and produce Ca^{2+} ions. Owing to the limited content of these hydrates in concrete their contribution is assumed to be minor, and thus neglected in this text. As concrete surface comes into contact with soft water, the ion species in pore solution, including Ca^{2+} and OH$^-$, start to transport out to the external water. The transport of the pore species can be due to a pure diffusion process or flow of the whole pore solution. By such transport processes of pore species, the dissolution equilibriums in Equations 4.1 and 4.3 are destroyed and the solid phases of CH and C-S-H start to dissolve further into the pore solution to observe Equations 4.2 and 4.4. The mechanisms involved are illustrated in Figure 4.3.

Since the solubility of CH is higher than that of C-S-H – for example, $\log K_{CH} = -5.18$ and $\log K_{CSH} \approx -11.3$ at 25 °C (Thomas *et al.*, 2003) – CH is always the first phase to dissolve in the pore solution. As the Ca^{2+} and OH$^-$ transport out, the CH phase dissolves into pore solution to maintain the dissolution equilibrium, serving as the solid buffer for calcium in pore solution. The dissolution–transport process can continue until the total consumption of CH in the solid matrix. The alkalinity and Ca^{2+} concentration of pore solution can be maintained at a constant level as long as the solid phases still contain CH. Afterwards, the C-S-H start to dissolve to maintain the solid–liquid equilibrium. Owing to the low solubility of C-S-H, the buffer capacity of C-S-H is much lower and the Ca^{2+} concentration decreases substantially in pore solution. Meanwhile, the dissolution of C-S-H weakens the solid matrix of concrete substantially. If the dissolution–transport process develops further (i.e., the calcium concentration continues to decrease), the C-S-H can be totally dissolved, and this concentration is set to 1.5 mmol/L in the literature (Yokozeki *et al.*, 2004). Certainly, it is an extreme situation for the leaching process, deduced from the dissolution stability of C-S-H.

This combined solid–liquid equilibrium is described by a global calcium equilibrium diagram in Figure 4.3b. The diagram expresses the calcium content in solid phases (CH, C-S-H) in terms of the Ca^{2+} concentration in pore solution. The diagram should be read form the right to left and from up to down: as the leaching starts, the dissolved phase is only CH and the Ca^{2+} concentration is maintained at its initial level, about 20 mmol/L for pH 12.6, but the solid calcium decreases; as the CH is totally dissolved, the aqueous Ca^{2+} concentration cannot be maintained at its initial level and begins to decrease with the C-S-H dissolution. This process is bounded by a Ca^{2+} concentration of 1.5 mmol/L, supposed to dissolve totally C-S-H. Thus,

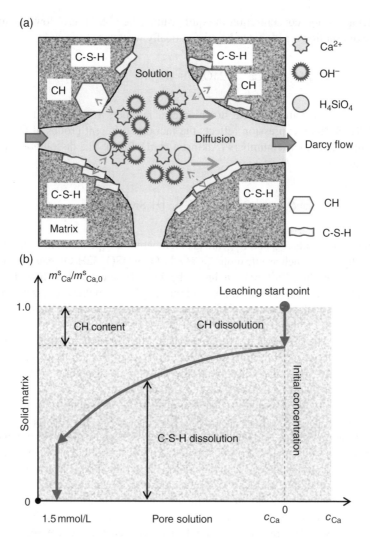

Figure 4.3 (a) Dissolution–transport process of leaching and (b) the calcium equilibrium between solid matrix and pore solution of hardened concrete. (b) *Source:* adapted from Yokozeki *et al.* 2004, Fig. 4. Reproduced with the permission of Elsevier.

this calcium equilibrium diagram is the very identity of the leaching property of concrete, and the precise mathematical form of the diagram depends on the composition of the hydrates.

The effect of leaching on the concrete is twofold. First, the leaching transports Ca^{2+} and OH^- out of concrete to the external environment, decreasing substantially the pH value of the pore solution, especially during the dissolution of C-S-H. Thus, like carbonated concrete, the protection capacity of leached concrete for the embedded reinforcement steel is decreased against corrosion. It is also implied from Equation 2.6 that the binding capacity of chlorides of the solid matrix can also be decreased by the dissolution of C-S-H. Thus, for RC elements exposed directly to seawater the leaching of surface concrete can decrease the chloride binding capacity and change the chloride profiles; see Figure 2.8. Second, the leaching dissolves the solid phases, enlarging the pore space in materials, and thus weakens the solid matrix of the concrete. This results in the deterioration of the relevant properties of the structural concrete (Heukamp *et al.*, 2001; Nguyen *et al.*, 2007). So, for RC elements with surface concrete subject to leaching, long-term deformation and damage can be expected and should be considered in the durability design.

4.2.2   *Influential Factors*

The factors that influence concrete leaching can be divided into three groups: the chemical composition of the hydrates, the transport properties of the concrete, and the external environmental conditions. The chemical composition of the hydrates determines the calcium equilibrium diagram in Figure 4.3, and a high content of CH in concrete tends to provide high resistance against leaching. Also, the composition of C-S-H, reflected usually by the Ca/Si ratio, is also a determinant of the dissolution equilibrium with the Ca^{2+} in aqueous solution (Kulik and Kersten, 2001). The transport properties refer to the ion diffusivity and the liquid permeability of concrete. As the concrete is highly compact, through incorporating a large quantity of mineral admixture in the binder and using a high-efficiency superplasticizer, the porosity and the connectivity of pores can be very low, resulting in very low diffusivity for ions and low permeability for liquids. Thus, the diffusion rate of the ions (Ca^{2+}, OH^-) and the flow rate of pore solution can be greatly limited. This is the reason why concretes incorporating large quantities of fly-ash or slag, compared with OPC concretes, contain low CH but are still considered to have a high leaching resistance (AFGC, 2007).

Environmental Conditions

The environmental conditions for the leaching process include the ambient temperature, the chemistry of the external soft water, and the flow state of the external water. The temperature has an impact on the solubility of the hydrates and the transport rates as well. The CH phase belongs to those rare solids that become less soluble with increased temperature, and the dissolution constant in the pore species environment can be expressed as (Yokozeki *et al.*, 2004)

$$\log K_{CH} = \log\left[Ca^{2+}\right]\left[OH^-\right]^2 = 556\frac{1}{T} - 6.28 \tag{4.5}$$

with the temperature T on the Kelvin scale. And the dissolution constant of Jennite, C-S-H mineral with the stoichiometric numbers $x = 1.5$, $y = 0.9$, and $z = 2.4$, is adapted from Kulik and Kersten (2001) as

$$\log K_{CSH-J} = \log\left[Ca^{2+}\right]^{1.5}\left[OH^-\right]^{3.0}\left[H_4SiO_4\right]^{0.9} = 192.5 - \frac{9906}{T} - 30.7\ln T \tag{4.6}$$

Regardless of the inverse effect of temperature on the CH solubility, the global effect of temperature is to accelerate the leaching rate, and the leaching depths[1] of OPC pastes in pure

---

[1]The leaching depth used here refers to the zone of surface with porosity and properties substantially changed by the leaching process, usually corresponding to the zone where the CH phase has been completely dissolved.

water were observed to be 1.5 mm at 25 °C for 123 days and 3.5 mm at 85 °C for 85 days (Kamali *et al.*, 2003). The chemistry of external water refers to the ion species in the aqueous phase. External water with high pH values shows an inhibitive effect on leaching of concrete because the high concentration of OH^- ions decreases the OH^- gradient between the pore solution and the external water. A pH value above 12.4 was shown capable of completely stopping the leaching of cement pastes (Kamali *et al.*, 2003). Other species also have an effect on the leaching rate through their influence on ion diffusivity.

The flow state of external water is a physical factor for the dissolution–transport leaching process. Consider two flow states: static (stagnant) water and flowing water. As the concrete surface is exposed to static water, the ion species from the pore solution transport to the concrete surface and can accumulate at the near-surface region. With time this local accumulation decreases the concentration gradients of ions and thus decelerates the transport rate, which, in turn, limits the dissolution rate of solid phases. In contrast, flowing water can efficiently evacuate ion species at the concrete surface and maintain the transport rate as well as the dissolution rate. That is why in most laboratory leaching tests the leaching solution (water) is constantly stirred or recycled to obtain an accelerated effect. In practical cases, flowing water on a concrete surface, at high speed, can also possibly create a negative pressure at the water–concrete interface and accelerate the outflow of the pore solution. These effects are quantified later in the modeling section.

Accelerated Leaching Tests

The leaching of concrete, in the natural state, is very slow. For research purposes, some accelerated tests were conceived for cement-based materials in the laboratory to obtain measurable leaching effects within reasonable duration. Two kinds of methods are available: chemical and electrochemical acceleration tests. The chemical acceleration method uses ammonium nitrate (NH_4NO_3) solution instead of pure (deionized) water as experimental fluid, and then the following reaction occurs:

$$NH_4NO_3 + Ca(OH)_2 \leftrightarrow Ca(NO_3)_2 + NH_3(gas) + 2H_2O \tag{4.7}$$

Leaching of calcium is greatly accelerated due to the much larger solubility of solid CH in NH_4NO_3 solution; that is, 2.9 mol/L in NH_4NO_3 solution of 6 mol/L compared with 22 mmol/L in deionized water (Heukamp *et al.*, 2001). The electrochemical method imposes an electrical potential, via two electrodes, on the two surfaces of the concrete specimen, and the ion species in pore solution migrate to different electrodes driven by the electrical potential. In this method, the cations Ca^{2+}, H^+, Na^+, and K^+ migrate to the negative electrode, while the anions OH^- transport to the positive electrode. Both accelerated methods distort, to some extent, the natural process of concrete leaching, so the leached concrete specimens from accelerated leaching are usually used to characterize the alteration of the properties of interest in terms of leaching extent. The kinetics of natural leaching are not expected to be predicted from these accelerated tests, though the accelerated effect of 6 mol/L NH_4NO_3 was estimated as 300 times faster than leaching by deionized water (Heukamp *et al.*, 2001). So far, no method has been accepted universally as the standard test for accelerated leaching of cement-based materials.

4.3 Modeling for Engineering Use

Multiple models are available for the leaching process of concrete materials, either based on the dissolution–transport mechanism or in empirical form regressed from laboratory tests. Two models are presented in this section for engineering use. The first model, L-1, uses the simplified mechanism of leaching and only the dissolution of CH and the subsequent Ca^{2+} diffusion are considered, giving an analytical solution for the dissolution front of CH in concrete. The second model, L-2, considers the dissolution of both CH and C-S-H phases and the subsequent transport of Ca^{2+} through the pore solution, incorporating more realistic aspects of the leaching mechanism. With the help of the L-2 model, the influence of some important environmental conditions on the leaching rate of concrete is investigated, including the evacuation rate of Ca^{2+} at the concrete surface and the surface negative pressure crerated by flowing water.

4.3.1 Model L-1: CH Dissolution Model

This model is adapted from the general dissolution–diffusion reference model from Mainguy and Coussy (2000). The concrete is idealized as a porous medium with porosity ϕ. The leaching process includes only the dissolution of CH in the solid phase and the subsequent diffusion of Ca^{2+} to the concrete surface. The problem is illustrated in Figure 4.4 for the one-dimensional case. The initial content of CH in the solid phase is noted as $m^s_{Ca,0}$ (mol/m^3) and the equilibrium concentration of Ca^{2+} in pore solution is noted as c^0_{Ca} (mol/m^3). As the concrete surface comes into contact with a soft water with $c_{Ca} = 0$ at the boundary $x = 0$, the leaching process starts. With the diffusion of Ca^{2+} and the dissolution of solid CH, a dissolution front of CH forms; see Figure 4.4. The position of the dissolution front x_L depicts the leaching depth in concrete. For modeling purposes, it is further assumed that the CH content is zero in the range of leaching depth ($x < x_L$) and the CH phase remains intact out of the leaching range ($x > x_L$).

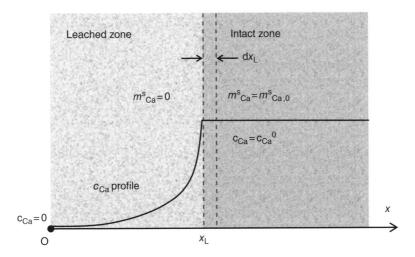

Figure 4.4 Simplified leaching mechanism by the dissolution–diffusion process of calcium in the CH phase.

The mass conservation of calcium in the leaching range is written thus:

$$0 < x < x_L: \quad \frac{\partial(\phi c_{Ca})}{\partial t} = \frac{\partial}{\partial x}\left(D_{Ca}^{eff}\frac{\partial c_{Ca}}{\partial x}\right) \tag{4.8}$$

where D_{Ca}^{eff} is the effective diffusivity of Ca^{2+} in concrete.[2] In the intact range, one has

$$x > x_L: \quad c_{Ca} = c_{Ca}^0, \quad m_{Ca}^s = m_{Ca,0}^s \tag{4.9}$$

At the dissolution front x_L, the mass conservation of calcium across this front is

$$m_{Ca,0}^s\left(\frac{dx_d}{dt}\right) = -D_{Ca}^{eff}\left(\frac{\partial c_{Ca}}{\partial x}\right)\bigg|_{x=x_L} \tag{4.10}$$

Actually, the conservation in Equation 4.10 states the calcium generated from the advancement of the dissolution front x_L is totally transported into the leaching range through diffusion flow. Equations 4.8–4.10 constitute a moving boundary problem. With the boundary condition $c_{Ca} = 0$ at $x = 0$, the dissolution front can be expressed in the form (Gebhart, 1993; Mainguy and Coussy, 2000)

$$x_L = 2\lambda_L\sqrt{\frac{D_{Ca}^{eff}}{\phi}t} \tag{4.11}$$

and the coefficient λ_L is the solution of the following equation:

$$\lambda_L \mathrm{erf}(\lambda_L)\exp(\lambda_L^2) = \frac{\phi c_{Ca}^0}{m_{Ca,0}^s} \tag{4.12}$$

The effective Ca^{2+} diffusivity takes the following form (Yokozeki et al., 2004):

$$D_{Ca}^{eff} = \frac{1 - c_{agg}v_g}{1 - d_{agg}v_s}v_p\phi f\left(\phi_{p,c}\right)D_{Ca}^0 \tag{4.13}$$

where v_g and v_s are respectively the volume fractions of coarse and fine aggregates, c_{agg} and d_{agg} are model parameters, D_{Ca}^0 is the Ca^{2+} diffusivity in water, f is the tortuosity factor of capillary porosity for diffusivity, and $\phi_{p,c}$ is the capillary porosity of the cement paste in concrete. The tortuosity factor f adopts the following expression (Garboczi and Bentz, 1992):

$$f\left(\phi_{p,c}\right) = 0.001 + 0.07\phi_{p,c}^2 + 1.8\left(\phi_{p,c} - 0.18\right)^2 H\left(\phi_{p,c} - 0.18\right) \tag{4.14}$$

---

[2]In the original model of Mainguy and Coussy (2000), the diffusivity was simply noted as ϕD with D as the Ca^{2+} diffusivity in the pore solution. The difference between the diffusivity values in the pore solution and in the concrete was only denoted by the porosity ϕ. This expression greatly overestimates the diffusivity in concrete, and thus the diffusivity here takes directly the value of the calcium diffusivity in concrete.

with $H(\cdot)$ being the Heaviside function. The capillary porosity of hardened OPC pastes can be evaluated after Powers (1960):

$$\phi_{p,c}^0 = \frac{w/c - 0.36\alpha_c}{w/c + 0.32} \tag{4.15}$$

with α_c the hydration degree of cement and w/c the water-to-cement ratio. Note that the capillary porosity can be enlarged by the dissolution of the CH phase. Thus, in the leaching range the capillary porosity of cement pastes can be expressed as

$$\phi_{p,c} = \phi_{p,c}^0 + \Delta\phi_{p,c}, \qquad \Delta\phi_{p,c} = \frac{M_{CH}}{\rho_{CH} v_p} m_{Ca,0}^s \tag{4.16}$$

Here, M_{CH} (kg/mol) is the molar mass of $Ca(OH)_2$, ρ_{CH} (kg/m³) is the mass density of $Ca(OH)_2$, v_p is the volume fraction of cement paste in concrete. Accordingly, the concrete porosity in leached zone is written as

$$\phi = \phi_0 + \Delta\phi, \qquad \Delta\phi = \frac{M_{CH}}{\rho_{CH}} m_{Ca,0}^s \tag{4.17}$$

The L-1 model is described through Equations 4.11–4.14, 4.16, and 4.17; the model parameters and the usual value ranges are given in Table 4.1, and some simulation results are presented in Figure 4.5. The OPC concrete with $w/c = 0.3$ has the smallest leaching depth, attaining 22 mm for 100 years, due to its higher CH content and lower initial porosity.

The virtue of the L-1 model lies in its simplicity, and the model describes the essential mechanism (if not all the mechanisms) of calcium leaching for concrete materials. The CH leaching depth, described in Equation 4.11, contains three main parameters: the (leached) porosity, the effective diffusivity of Ca^{2+}, and the coefficient λ_L. The coefficient λ_L, solution of Equation 4.12, depends on the porosity, the equilibrium calcium concentration, and the initial calcium content. The results in Figure 4.5 fit quite well with the engineering judgement on the magnitude of leaching depth of OPC concretes with time (Haga *et al.*, 2005). It will be showed later that these values give a conservative estimation for the CH leaching depth.

4.3.2   Model L-2: CH + C-S-H Leaching Model

This model takes into account the calcium leaching from both CH and C-S-H. The solid–liquid calcium equilibrium diagram in Figure 4.3 is used as the fundamental relation for calcium leaching. For the one-dimensional problem, the mass conservation of calcium in concrete writes

$$\frac{\partial\left[m_{Ca}^s(c_{Ca}) + \phi c_{Ca}\right]}{\partial t} = \frac{\partial}{\partial x}\left[D_{Ca}^{eff}\frac{\partial c_{Ca}}{\partial x} + c_{Ca}k^{eff}\,\mathrm{grad}(p)\right] \tag{4.18}$$

where c_{Ca} (mol/m³) is the Ca^{2+} concentration in the concrete pore solution, m_{Ca}^s (mol/m³) is the calcium content in the solid phase (related to the Ca^{2+} concentration in the aqueous phase),

Table 4.1   Parameters for the L-1 model and their usual value ranges

Parameter	Value/relation
Initial content of solid calcium (CH phase) in concrete $m_{Ca,0}^s$ (mol/m^3)	2149 ($w/c = 0.3$), 1850 ($w/c = 0.4$), 1623 ($w/c = 0.5$), 1447 ($w/c = 0.6$)
Initial porosity of concrete ϕ_0	0.09 ($w/c = 0.3$), 0.12 ($w/c = 0.4$), 0.15 ($w/c = 0.5$), 0.18 ($w/c = 0.6$)
Dissolution front coefficient λ_L	Equation 4.12
Initial pH value of pore solution	12.65
Initial Ca^{2+} concentration in pore solution c_{Ca}^0 (mol/m^3)	20.9, see Equation 4.5
Effective Ca^{2+} diffusivity in concrete D_{Ca}^{eff} (m^2/s)	Equation 4.13
Ca^{2+} diffusivity in pore solution at 293.15 K $D_{Ca,0}$ (m^2/s)	1.37×10^{-9}
Capillary porosity of cement paste $\phi_{p,c}$	Equation 4.12
Hydration degree of cement α_c	0.65 ($w/c = 0.3$), 0.73 ($w/c = 0.4$), 0.78 ($w/c = 0.5$), 0.81 ($w/c = 0.6$)
Water-to-cement ratio of concrete w/c	0.3, 0.4, 0.5, 0.6
Molar mass of CH M_{CH} (kg/mol)	0.074
Mass density of CH ρ_{CH} (kg/m^3)	2230
Volume fractions of coarse and fine aggregates v_g, v_s	0.390, 0.260
Volume fraction of cement paste v_p	0.350
Concrete tortuosity parameters c_{agg}, d_{agg}	1.5, 0.86 (Yokozeki *et al.*, 2004)

ϕ is the material porosity, D_{Ca}^{eff} (m^2/s) is the effective diffusivity of Ca^{2+} in concrete, k^{eff} (m^2/(Pa s)) is the effective permeability of liquid in concrete, and p (Pa) is the liquid pressure. The central relation in this model is $m_{Ca}^s = m_{Ca}^s(c_{Ca})$, and the relation suggested by Yokozeki *et al.* (2004) takes the form

$$\frac{m_{Ca}^s}{m_{Ca,0}^s} = A_{CSH} \left(\frac{c_{Ca}}{c_{Ca}^0} \right)^{1/n} \qquad \left(c_{Ca}^1 < c_{Ca} < c_{Ca}^0 \right) \tag{4.19}$$

where $m_{Ca,0}^s$ (mol/m^3) is the initial calcium content in the concrete, c_{Ca}^0 (mol/m^3) is the initial calcium concentration in the aqueous phase, c_{Ca}^1 (mol/m^3) is the limit concentration for calcium corresponding to the total dissolution of cement hydrates (C-S-H), n is the power exponent depicting the equilibrium relation between m_{Ca}^s and c_{Ca} during the calcium dissolution of C-S-H, and A_{CSH} denotes the ratio of calcium in C-S-H with respect to the total solid calcium content; that is:

$$A_{CSH} = \left. \frac{m_{Ca}^s}{m_{Ca,0}^s} \right|_{c_{Ca} = c_{Ca}^{0-}} \tag{4.20}$$

The initial calcium concentration c_{Ca}^0 can be evaluated from the dissolution constant of Ca(OH)$_2$, expressed in terms of ambient temperature in Equation 4.5, and the pH value of the

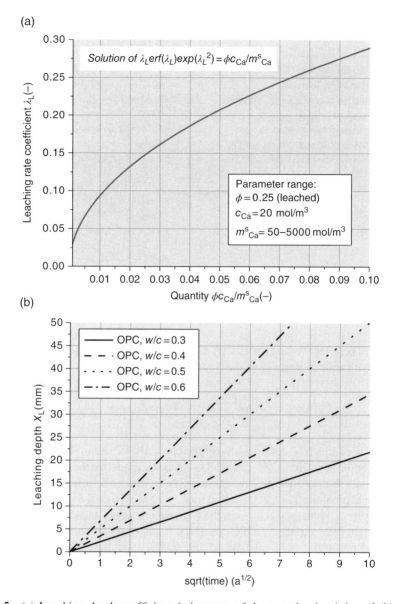

Figure 4.5 (a) Leaching depth coefficient λ_L in terms of the quantity $\phi c_{Ca}/m_{Ca}^s$ and (b) leaching depth for different OPC concretes using the L-1 model. In (a), the Ca^{2+} concentration in the pore solution, $c_{Ca}^0 = 20$ mol/m^3, corresponds to pH 12.6, and the solid calcium content m_{Ca}^s varies from 50 to 2000 mol/m^3. The higher bound corresponds to OPC concretes, while the lower bound corresponds to concretes with a high content of mineral admixtures, such as fly ash and silica fume. The relevant parameters in Table 4.1 are used for the calculation of leaching depth for the OPC concretes ($w/c = 0.3, 0.4, 0.5, 0.6$) in (b).

Table 4.2 Parameters for the L-2 model and their usual value ranges

Parameter	Value/relation
Initial content of solid calcium (CH + C-S-H phases) in concrete $m^s_{Ca,0}$ (mol/m³)	5397 ($w/c = 0.3$), 4645 ($w/c = 0.4$), 4077 ($w/c = 0.5$), 3633 ($w/c = 0.6$)
Solid calcium content in concrete m^s_{Ca} (mol/m³)	Equation 4.9
C-S-H calcium content ratio A_{CSH}	0.60
Dissolution equilibrium law exponent n	4.52 (Yokozeki et al., 2004)
Ca²⁺ concentration in pore solution c_{Ca} (mol/m³)	Variable
Ca²⁺ concentration for total C-S-H dissolution c^l_{Ca} (mol/m³)	1.5 (Yokozeki et al., 2004)
Effective water permeability of concrete k^{eff} (10^{-12} m/s)	1 ($w/c = 0.3$), 5 ($w/c = 0.4$), 10 ($w/c = 0.5$), 20 ($w/c = 0.6$), see Figure 8.2
Pore pressure on liquid phase p (Pa)	Variable
Exchange coefficient λ_{Ca} for boundary layer (m/s)	3.8×10^{-10} (clay, 1 m thickness) 1.37×10^{-9} (water, 1 m thickness) 2.74×10^{-9} (water, 0.5 m thickness) 1.37×10^{-8} (water, 0.1 m thickness) 2.74×10^{-8} (water, 0.05 m thickness)

pore solution. The effective diffusivity D^{eff}_{Ca} can be evaluated from Equations 4.13 and 4.14, and the enlarged capillary porosity of the cement paste is written

$$\phi_{p,c} = \frac{w/c - 0.36\alpha_c}{w/c + 0.32} + \frac{M_{CH}}{\rho_{CH} v_p}\left(m^s_{Ca,0} - m^s_{Ca}\right) \tag{4.21}$$

The model is completely described by Equations 4.18–4.21, and Table 4.2 provides the value ranges for the new parameters in model L-2 and the values for the duplicated parameters can be found in Table 4.1. The L-2 model is solved by the finite-difference method for the one-dimensional problem. The Ca²⁺ concentration is set to zero on the concrete surface as the boundary condition. Figure 4.6 illustrates the CH leaching depth and the quantity of calcium outflow at the concrete surface for OPC concretes. It can be seen that the CH leaching depth varies from 18 mm (OPC, $w/c = 0.3$) to 45 mm (OPC, $w/c = 0.6$) at 100 years. Again, the smaller CH leaching depth of OPC concrete with $w/c = 0.3$ is attributed to its higher CH content and smaller initial porosity.

One can compare the CH leaching depths predicted from the L-1 model in Figure 4.5 and the depths from the L-2 model in Figure 4.6: the CH leaching depth from the L-2 model is systematically smaller than the depths from the L-1 model. In terms of the magnitude, the L-1 leaching depths are 23% ($w/c = 0.30$) to 52% ($w/c = 0.60$) larger than the L-2 depths. The underlying reason is that the L-2 model considers calcium leaching from both the CH and C-S-H phases: as CH is leached out the C-S-H continue to buffer Ca²⁺ into the pore solution. Compared with the L-1 model, the concentration gradient in the CH-leached range is smaller, and thus the Ca²⁺ diffusion rate is reduced. Evidently the CH leaching depth from the L-2 model is more rational than the value from the L-1 model. Nevertheless, both models predict the same order of magnitude for CH leaching depth. On the basis of this observation,

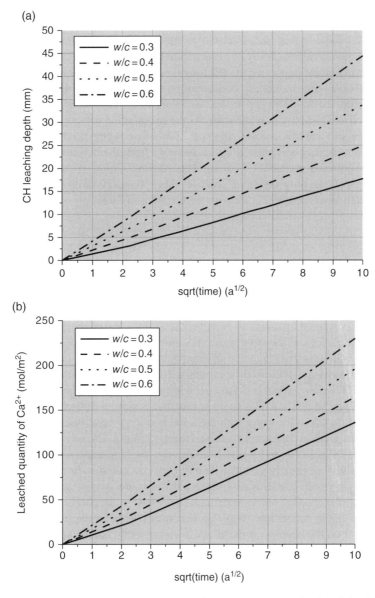

Figure 4.6 Simulation results from the L-2 model for (a) the leaching depths of the CH phase and (b) the leached quantity of Ca²⁺ from both the CH and C-S-H phases. The initial calcium content in Table 4.2 refers to the Ca²⁺ contained in both the CH and C-S-H phases.

the L-1 and L-2 models can be used for different purposes in practice: the L-1 model, owing to its simple and analytical nature, can be used to estimate the order of magnitude of CH leaching depth for concrete elements in contact with flowing soft water, while the L-2 model, together with the numerical solution scheme, can be employed to solve the more accurate leaching depth.

4.3.3   Further Analysis of Surface Conditions

Using the L-2 model, some issues related to more realistic boundary conditions can be investigated quantitatively. Actually, the evacuation rate of leached calcium from the concrete surface has an important impact on the leaching process and CH leaching depth. The simulations in Figure 4.6 assume a zero Ca^{2+} concentration at the concrete surface; that is, the leached Ca^{2+} ions are carried away from the surface instantaneously. This occurs only for the concrete surface exposed to constantly flowing water. As the surrounding water is less free to flow, even stagnant, or the soft water is contained in some porous geotechnical materials such as soils, the leached Ca^{2+} cannot be evacuated instantaneously from the concrete surface but diffuses into the water or water-contained medium. This picture is more realistic for the boundary condition of most leaching cases in practice. The boundary-layer theory addresses this situation through an exchange condition on the concrete surface:

$$J_{Ca} = \lambda_{Ca} \left(c_{Ca} - c_{Ca,env} \right) \tag{4.22}$$

with J_{Ca} (mol/(m² s)) the flow rate of Ca^{2+} ions through concrete surface, c_{Ca} and $c_{Ca,env}$ (mol/m³) the concentration of Ca^{2+} transported to the concrete surface and the Ca^{2+} concentration in the environmental medium, and λ_{Ca} (m/s) the exchange coefficient. Actually, the physical significance of this exchange coefficient is the ratio between the Ca^{2+} diffusivity in this medium (water or water-contained medium) and the thickness of the boundary layer of this medium or the diffusion distance of Ca^{2+} into this medium. Some typical values of λ_{Ca} are given in Table 4.2 for water and water-saturated clay with different values for boundary-layer thickness, and the corresponding results are presented in Figure 4.7 for OPC concrete with $w/c = 0.4$. From the figure, the CH leaching depth is substantially influenced by the exchange coefficient λ_{Ca}: for the less permeable clay the leaching depth is less than 5 mm for 100 years (OPC, $w/c = 0.4$), nearly one-third the depth for water with the same boundary-layer thickness (1 m); for water the CH leaching depth at 100 years increases from 12 mm to 25 mm as the boundary-layer thickness decreases from 1 m to 0.05 m. Note that the leaching depth with a 0.05 m boundary layer of water is very near to the results with null Ca^{2+} concentration at the concrete surface.

Another boundary issue concerns the surface water pressure. According to Equation 4.18, this pressure can contribute to Ca^{2+} transport through Darcy flow of the pore solution, which has been neglected so far in the simulations. A practical concern arises as the flowing water creates a negative pressure on the concrete surface, which can possibly accelerate the calcium leaching process. For hydraulic structures like discharging spillways in dams, this negative pressure due to high-speed water flow was estimated at the order of -10 kPa (Zhang $et\ al.$, 2005). Instead of numerical simulation, the relative importance of the diffusion flow J_{Ca}^{F} (mol/(m² s)) and convection flow J_{Ca}^{D} in Equation 4.18 is compared here. During the leaching range, $0 < x < x_L$, one has

$$\frac{J_{Ca}^{F}}{J_{Ca}^{D}} = \frac{D_{Ca}^{eff}}{k^{eff}} \frac{\nabla c_{Ca}}{c_{Ca} \nabla p} \ \text{and}\ c_{Ca} \leq c_{Ca}^{0}, \quad \nabla c_{Ca} \doteq \frac{c_{Ca}^{0}}{x_L} \tag{4.23}$$

Thus, a lower bound estimate for the flow ratio is given by

$$\xi_L = \frac{D_{Ca}^{eff}}{x_L k^{eff}} \frac{1}{\nabla p} < \frac{J_{Ca}^{F}}{J_{Ca}^{D}} \tag{4.24}$$

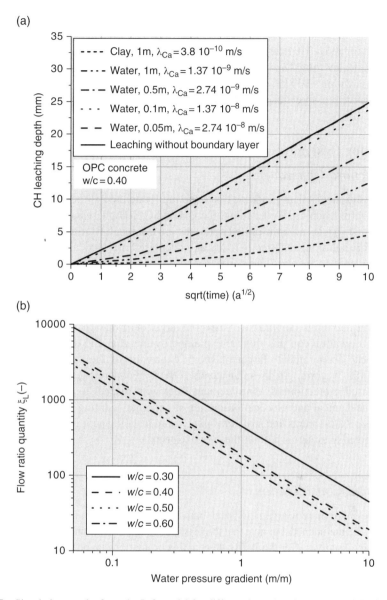

Figure 4.7 Simulation results from the L-2 model for different boundary layers (a) and the flow rate ratio between calcium diffusion and pore solution convection for surface negative pressure (b). The OPC concrete of $w/c = 0.4$ in Table 4.1 is retained for calculation. For boundary layers, the exchange coefficients are calculated for pure water and soil (Shackelford, 1991) with the boundary-layer thickness from 0.05 to 1.0 m.

Figure 4.7 presents the results for ξ_L for OPC concretes assuming $x_L = 1\,\mathrm{cm}$ and the pressure gradient ∇p from -0.05 m/m (0.5 kPa/m) to -10 m/m (100 kPa/m) in terms of water head. The diffusivity of Ca^{2+} adopts the same values as in the simulations in Figure 4.6 for leached concretes, and the permeability values are given in Table 4.2. The results show that the ratio is quite large at low negative pressure, and decreases with the negative pressure gradient. Note that even at the largest gradient level, -10 m/m (-100 kPa over 1 m thickness of concrete), this low bound ratio is still above 10. In other words, the diffusion flow is always at least one magnitude higher than the Darcy flow. Thus, for conventional hydraulic structures where the surface negative pressure is within -100 kPa and the concrete element is around 1 m thickness, the impact of surface negative pressure created by high-speed flow cannot be important.

4.4 Basis for Design

4.4.1 Structural Consequence

The physical and mechanical properties of structural concrete can be substantially deteriorated by long-term leaching. By leaching, the CH and C-S-H phases are gradually dissolved and the dissolved species are transported out to the environment, resulting in a decrease of pH value in the pore solution, enlarged porosity and decreased mechanical resistance. CH dissolution is considered to increase the porosity and affect the stiffness of concrete, while the dissolution of C-S-H changes comprehensively the intrinsic properties of solid phases and induces chemomechanical softening of concrete (Ulm *et al.*, 2003). Relevant experiments on the stress–strain relationship showed that, compared with sound material, the leached specimens had lower maximum compressive stress but larger deformation capacity (Nguyen *et al.*, 2007). This observation can possibly be explained by fact that CH dissolution enlarges the porosity and can accommodate larger plasticity expansion of the solid matrix of concrete (Ulm *et al.*, 2003). Accordingly, the main adverse effects of concrete leaching raise two issues: (1) the decreased protection of reinforcement steel with respect to corrosion, and (2) the deteriorated mechanical and physical properties of concrete resulting from the solid calcium dissolution. For RC elements and structures exposed to leaching actions, the durability design should be targeted at either or both of these two aspects.

4.4.2 Design Considerations

To protect the internal reinforcement steel bars from corrosion, the alkalinity of the pore solution should be maintained, but note that the pH value of the pore solution is to be reduced once the CH is totally dissolved. For durability design, this protection can be assured through a specification of a minimum residual CH content over a specified depth of concrete; for example, the concrete cover to steel bars. This CH content can adopt a specified absolute value in moles per cubic meter or kilograms per cubic meter, or a relative content with respect to the initial CH content. Such a specification is actually a performance criterion because the specified CH content is determined by the initial CH content in concrete, concrete compactness in terms of diffusivity and permeability, and the environmental conditions related to the surface boundary layer. Once this content is fixed, the design can be performed with the help of leaching models. Both models presented in this chapter can provide the quantity of calcium outflow for a specified service life and thus one can deduce the residual CH content from its initial value.

To limit the deterioration of concrete properties by leaching, the CH dissolution depth can be used as a durability indicator. From state-of-the-art of knowledge, the CH dissolution enlarges the material porosity but the solid skeleton remains continuous; thus, the integrity of the material remains intact. The CH leaching depth can be specified with respect to the target mechanical performance or functional requirements of the structure or elements. Admittedly, more knowledge on the altered properties of concrete and altered performance of elements is needed to support this quantitative specification in terms of CH dissolution depth. Take concrete facilities used in radioactive disposals, for example. The leaching depth should be specified together with consideration of the possible accelerated effect of the transport rate of radionuclides; see Chapter 9. Both models in this chapter can help to determine the CH dissolution depths.

From the analysis in this chapter, the leaching kinetics are closely related to the CH and C-S-H contents of concrete, the transport properties, and the environmental conditions. Accordingly, at the material level, concretes with high CH and C-S-H contents or with very low Ca^{2+} diffusivity ($<10^{-12}$ m^2/s) and permeability ($\sim10^{-12}$ m/s) are favored by durability design against leaching action. However, modern concretes tend to incorporate more SCM, resulting in much higher compactness but lower CH content. The general view on this aspect is that the high compactness can counteract the unfavorable effect of low CH content. No further quantitative discussion is given on this point, while the models in this chapter can help to quantify the respective contribution of compactness, through diffusivity and permeability, and the CH content in concrete. As for the environmental effect, flowing water with a low pH value is the most severe condition, and surface treatment or protection should be adopted if necessary.

5

Salt Crystallization

This chapter treats the damage of structural concrete by salt crystallization. This phenomenon is also termed "salt weathering" for other porous building materials, such as stones and bricks. As the external salts are deposited on or transported to the surface of concrete elements, the pore solution of concrete can be supersaturated with respect to the salts under the temperature fluctuations or drying–wetting actions in environment. The pore pressure can accumulate as salt crystals form in pores, high enough to fracture the solid matrix and induce damage to concrete. Though the resulting damage is usually manifested as surface scaling, the mechanical resistance of concrete can be affected in extreme situations. This chapter begins with the introduction of the phenomena of salt crystallization for concrete elements and structures, then it details the mechanisms of salt crystallization in confined pores. Since both pore freezing and salt crystallization involve crystallization in confined pores, some common concepts and expressions from Chapter 3 are adopted. Salt crystallization in concrete is still an ongoing subject with much knowledge to be explored, so the models in this chapter are conceptual, attempting to capture the main mechanisms of pore crystallization. Finally, the basis for durability design against salt crystallization is provided.

5.1 Phenomena and Observations

Salt crystallization belongs to weathering actions of the natural environment on porous materials, which is known for stones and bricks under the term "salt weathering." As building materials have pores open to external environments, the airborne salts, soil-borne salts, or dissolved salts in groundwater can penetrate into material in the form of a solution. Owing to a change of external humidity or temperature, salts in pores can crystallize out from the solution and form crystals, filling up the pore space and exerting stress on the solid matrix of material. As the stress level can be high enough to induce damage, salt crystallization becomes

Durability Design of Concrete Structures: Phenomena, Modeling, and Practice, First Edition. Kefei Li.
© 2016 John Wiley & Sons Singapore Pte. Ltd. Published 2016 by John Wiley & Sons Singapore Pte. Ltd.

a detrimental process to the long-term durability of building materials and structures. However, external salts are not a necessary condition for this process because dissolved salts in pores can originate from soluble minerals in the solid matrix. Drying and cooling can also trigger the crystallization of these dissolved salts in pores. The weathering of limestone in the Great Sphinx, Egypt, was attributed to the long-term pore crystallization from the dissolution of halite, a mineral of sodium chloride in natural rocks, into the condensation water as the ambient temperature cooled below the dew point (Gauri *et al.*, 1990).

Salt crystallization was noted as a detrimental process for building materials as early as the 1930s (Neville, 2004). The first published case for structural concrete was the extensive damage of residential foundations by the crystallization of sulfate salts in southern California, USA (Novak and Colville, 1989). Then, the detrimental nature of salt crystallization was recognized for structural concretes exposed to different salts, including sodium chloride (NaCl) and sodium and magnesium sulfates (Na_2SO_4, $MgSO_4$). Owing to its large solubility at ambient temperature, Na_2SO_4 is regarded as the most aggressive salt for crystallization damage.[1] Damage of concrete elements due to salt crystallization is rather extensive nowadays (Mehta, 2000), so should be taken into account in the durability design.

The first salt source for crystallization is the airborne salts in marine or similar environments, and all the concrete elements with exposed surfaces can be affected. In these environments, the airborne salts deposit on the concrete surface over time. Simultaneously, the surface is subject to natural drying–wetting cycles: during the wetting period the deposited salts are dissolved (e.g. by natural precipitation) and absorbed into the concrete surface; during the drying period the water in pores evaporates and condenses the pore solution. These cycles cause crystallization to occur in the pores as the condensation reaches the necessary degree. In a marine environment, the salt involved is mainly NaCl, and the accumulation of chlorides on a concrete surface was addressed in Chapter 2; see Figure 2.7. The surface deposition of airborne salts also occurs for concrete elements in areas of saline soils or inland salt lakes. Airborne salts can have other sources: the surface weathering of structural concrete in a 40-year-old tunnel was attributed to the deposition of air pollutants on the concrete surface and the subsequent crystallization of gypsum ($CaSO_4$) (Marinoni *et al.*, 2003).

The second source of salts comes from saline soils or groundwater with high salinity. Concrete elements in direct contact with these media can have severe damage from salt crystallization, and this case is more frequently encountered in practice. The damage particularly affects those elements partially buried in salt-rich soils and partially exposed to air, such as the basement and footings for residual structures (Yoshida *et al.*, 2010). Figure 5.1 illustrates such a concrete electrical pole and the mechanisms involved. The buried part of the electrical pole is in contact with salts in the saline soil, and the concrete serves as capillary channels to draw the salt solution up to the air-exposed part when the salts in the soil are dissolved by groundwater or natural precipitation. Afterwards, the water evaporates on the upper part and leaves behind the condensed salt solution to crystallize. The concrete damage is surface spalling and is characterized by a damage zone about 0.5 m above ground level. This damage pattern is rather typical for such concrete elements. Na_2SO_4 in the soils is usually assumed to be responsible for the damage.

---

[1]Note that sulfate salts can also induce chemical reactions with cement hydrates, and the ettringite and gypsum formed can damage the concrete by the expansion of these products. This chapter treats only the physical process of pore crystallization, not covering the possible chemical reactions between the salts and cement hydrates.

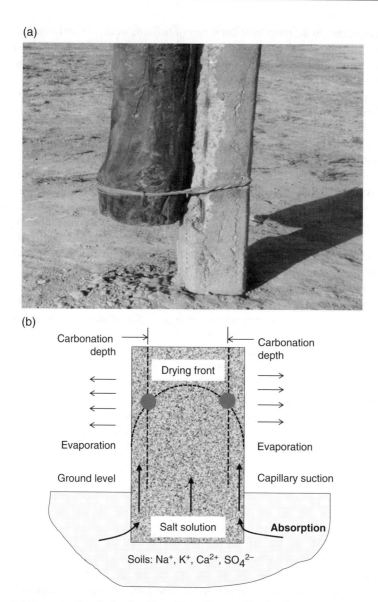

Figure 5.1 (a) Concrete electrical pole affected by salt crystallization in an area of saline soil and (b) the related mechanisms. (a) *Source:* courtesy of Hengjing Ba. (b) *Source:* adapted from Scherer, 2004, Fig. 1. Reproduced with permission of Elsevier.

Compared with other deterioration processes, salt crystallization has been much less investigated so far. A considerable amount of knowledge on salt crystallization was acquired from laboratory tests. A partial immersion test, aimed at reproducing the in-situ damage of Figure 5.1, is of particular interest: specimens were partially immersed into a solution of Na_2SO_4 or $MgSO_4$ and partially exposed to air with the relative humidity controlled. The literature provides several typical observations (Rodriguez-Navarro *et al.*, 2000; Haynes *et al.*, 2008): (1) Na_2SO_4

causes much greater damage than $MgSO_4$; (2) the most severe damage is induced by the alternative formation of thenardite (Na_2SO_4) and mirabilite ($Na_2SO_4 \cdot 10H_2O$); and (3) concrete compactness seems to be the determining factor in damage extent. The correct interpretation of these observations will be helpful for durability design against salt crystallization damage.

5.2 Mechanisms and Influential Factors

5.2.1 Mechanisms

Material damage by salt crystallization arises from the pore stress accumulated during the phase change process. We begin with some fundamental concepts for salt crystallization before turning to the stress issue. Consider a single-salt solution with c_{sat} (mol/L) as the solubility of this salt. In the bulk state, the salt does not crystallize for concentration $c < c_{sat}$. Crystallization begins to be possible as the concentration is increased above c_{sat}, and the state $c > c_{sat}$ is termed *supersaturation* of the solution. For ease of description, the molar fraction z,[2] instead of the molar concentration c, is used to describe the salt concentration in the following. Actually, the supersaturation degree z/z_{sat} is the driving force for crystallization, and the phase equilibrium between the aqueous solute and the solid crystal can be expressed through the Correns equation:

$$p_C - p_L = \frac{RT}{V_C} \ln\left(\frac{z}{z_{sat}}\right) \tag{5.1}$$

where p_C and p_L are the crystal and liquid pressures,[3] R (J/(mol K)) is the ideal gas constant, T (K) is the temperature, and V_C is the molar volume of the crystal. Note that the solubility of salts is temperature dependent; that is, $z_{sat} = z_{sat}(T)$. If the liquid pressure p_L is assumed constant, the equilibrium condition in Equation 5.1 states the following: crystals under higher pressure p_C will be in equilibrium with a higher supersaturation degree z/z_{sat}.

If the crystal is a sphere, with radius r, and the only stress comes from the surface energy of the crystal–liquid interface γ_{CL} (see Figure 5.2a), Equation 5.1 can be expressed as

$$\frac{2\gamma_{CL}}{r_C} = \frac{RT}{V_C} \ln\left(\frac{z}{z_{sat}}\right) \tag{5.2}$$

This equation describes the equilibrium condition between the z/z_{sat} ratio and the crystal size. Now let us look at the crystal formation in a cylinder pore (see Figure 5.2b). Similar to pore freezing, a liquid film of thickness δ exists between the crystal and the pore wall. The tip of the crystal is assumed to be a hemisphere having a curvature of $2/(r - \delta)$, while the curvature of the cylinder side of the crystal is $1/(r - \delta)$. Let the supersaturation in solution z/z_{sat} be in

---

[2]The molar fraction is the ratio between the mole quantity of a component in a mixture and the total mole quantity of this mixture, and has no units.

[3]In this chapter, the liquid phase in pores refers to pore solution and is denoted by the subscript "L." The same pore liquid phase is denoted by the subscript "l" in the context of gas–liquid equilibrium in pores in Chapter 2 and later in Chapter 6. The two subscripts represent the same pore phase, used in a different context of pore equilibrium.

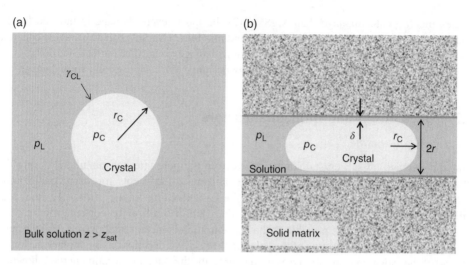

Figure 5.2 (a) Phase equilibrium between the solute and crystal and (b) crystallization pressure in a cylindrical pore.

equilibrium with the crystal tip (Equation 5.2). Since the cylinder side can only provide half of the crystal pressure through its own interface energy, the pore wall will exert a surplus stress onto the crystal:

$$\Delta\sigma = \frac{\gamma_{\mathrm{CL}}}{r_{\mathrm{C}}} = \frac{RT}{V_{\mathrm{C}}}\ln\left(\frac{z}{z_{\mathrm{sat}}}\right) \text{ with } r_{\mathrm{C}} = r - \delta \tag{5.3}$$

This stress must be equally taken by the pore wall, and Equation 5.3 gives the origin of the crystallization pressure. In real pores of complex geometry, the expression for pressure can be different but the principle remains the same. If we take the pore structure in Figure 3.4b to deduce the incremental stress on the pore wall as the crystal (salt here) penetrates into the smaller pores, the magnitude of pressure can reach $2\gamma_{\mathrm{CL}}/r_{\mathrm{C}}$.

Now let us look at the pore crystallization under a drying process. The pore structure is illustrated in Figure 5.3 through an idealized representative elementary volume (REV) where the pores are totally interconnected. Initially, the pores are saturated completely by a solution with salt concentration z_0, noted in molar fraction. This state can be the end of the absorption of the dissolved salt at the concrete surface. As $z_0 < z_{\mathrm{sat}}$, no crystal forms and all pores are filled with salt solution; only as $z_0 > z_{\mathrm{sat}}$ and the supersaturation degree z_0/z_{sat} satisfies Equation 5.2 for the largest pore radius r_{max} can crystals form in the REV.

As the saturated REV undergoes a drying process, subject to a relative humidity $h_{\mathrm{R}} < 100\%$, the liquid–gas interface begins to retreat by water evaporation from the interface until the liquid–gas equilibrium is reached; that is, the curvature of the interface is in equilibrium with h_{R}. (This pore radius in equilibrium can be deduced from the Kelvin equation applied to the liquid–vapor equilibrium across the liquid–gas interface and the Laplace equation for gas and liquid pressures:

$$p_{\mathrm{L}} - p_0 = \frac{RT}{V_{\mathrm{W}}}\ln\left(h_{\mathrm{R}}\right), \, p_{\mathrm{G}} - p_{\mathrm{L}} = \frac{2\gamma_{\mathrm{GL}}}{r} \tag{5.4}$$

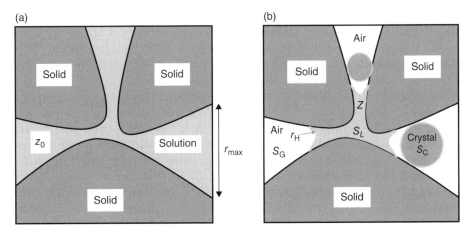

Figure 5.3 Pore crystallization under drying: (a) initial state of saturated pores and (b) pore phases in equilibrium with the external humidity h_R.

with V_w the molar volume of water and γ_{GL} the liquid–gas interface energy. Letting the reference pressure $p_0 = p_G$, one obtains the attainable pore radius r_H corresponding to the external drying h_R:

$$\frac{1}{r_H} = -\frac{RT}{2\gamma_{GL}V_w}\ln\left(h_R\right) \tag{5.5}$$

Note that in Equations 5.4 and 5.5 the pore liquid phase is assumed to be pure water, neglecting the influence of solutes. The complete description considering the contribution of solutes can be referred to Coussy (2006a).) During drying the pore solution is constantly condensed by evaporation, increasing the pore supersaturation z/z_{sat}. Once z/z_{sat} satisfies Equation 5.2 during the receding of the liquid–gas interface, pore crystallization can occur. And this pore crystallization also decreases the supersaturation in the residual solution. At the final equilibrium state, the smallest pores are still occupied by the liquid phase, illustrated in the central part on Figure 5.3. Pores in which the liquid has receded are occupied either by crystals or the gas phase (air). In Figure 5.3, the volumes of these phases are noted by their respective saturations, s_L, s_G, and s_C. If the concrete is subject to drying–wetting cycles and during each wetting period some new salt is introduced into the pore solution, one can expect that the crystal-occupied volume of pores increases in both the saturated state and the dried state.

The evaluation of pore pressure from the crystallization follows the same concept of mean pore pressure in Equation 3.5, except that the gas-phase term contribution should be added:

$$\langle\sigma\rangle_p = s_C p_C + s_L p_L + s_G p_G \text{ with } s_C + s_L + s_G = 1 \tag{5.6}$$

As only two phases are present (e.g., the saturated state in Figure 5.3), the expression in Equation 5.6 becomes Equation 3.5. The pressure in the solid matrix can be calculated from Equation 3.6, and Equation 3.7 can be still used as a first criterion to judge the liability of concrete to fracture damage.

Finally, the damage of partially buried elements deserves more attention. The damage, illustrated in Figure 5.1, is generally attributed to salt crystallization, and of sodium sulfate in particular. However, some recent investigations showed rather complex mechanisms are involved in the damage. First, the damage above the ground is confirmed due to crystallization, while the surface carbonation is also relevant. According to Yoshida *et al.* (2010), the crystals, thenardite or mirabilite, can only form within the carbonation depth of the concrete surface, and the formation of ettringite is favored in concrete not yet carbonated. Second, the typical pattern of damage at about 0.5 m above the ground is actually the balance between the capillary suction of salt solution from the ground and the pore evaporation above the ground. As the evaporation occurs at the surface, the salt crystallization is visible at the surface, termed "efflorescence," while "subflorescence" occurs as the evaporation deepens into the material (Scherer, 2004); see Figure 5.1a. Since the crystals can grow freely at the surface, the damage is more relevant to the subflorescence. Based on this knowledge, it is reasonable to assume that the most likely damage zone by crystallization is bound by the subflorescence line and the carbonation depth.

5.2.2   Influential Factors

Based on the aforementioned mechanisms, two groups of factors are relevant to the crystallization and the resulting damage to concrete: material properties and the environmental conditions. The material properties include the pore structure, the transport properties, and the fracture-related resistance of concrete; the environmental conditions are the salt deposition/intake rate from the environment, the salt type, and the environmental temperature and humidity.

Material Properties

Pore Structure
The pore structure impacts on the crystallization process through the total porosity, the pore size distribution (PSD), and the maximum pore size. Through Equation 5.2, the minimum attainable pore radius can be determined for a given supersaturation z/z_{sat}. With the help of PSD, Equation 5.2 can actually provide the liquid saturation in pores in terms of the supersaturation; that is, the characteristic curve of pore crystallization. Taking the PSD curve from mercury intrusion porosimetry (MIP) as the basic data, the liquid saturation in pore s_L, under a given supersaturation z/z_{sat}, is

$$s_L = \frac{v_L(z/z_{sat})}{v_{max}} = \frac{v_L(r)}{v_{max}} \quad \text{with } r = \delta + \frac{2\gamma_{CL}V_C/RT}{\ln(z/z_{sat})} \tag{5.7}$$

with v_{max} the maximum intruded volume in the MIP test and v_L the liquid volume in pores. Note that in the saturated state in Figure 5.3 the crystal saturation $s_C = 1 - s_L$, but this does not hold for the dried state in Figure 5.3 since one also has $s_G > 0$. Figure 5.4 provides the attainable pore size (radius) of crystallization for two salts, NaCl and mirabilite, and the characteristic curves of crystallization for OPC concretes using Equation 5.7. For a same supersaturation degree, the NaCl crystal can penetrate into much smaller pores than the mirabilite due to the smaller molar volume of NaCl crystals.

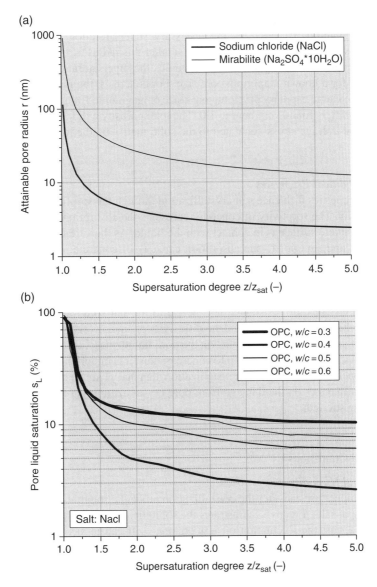

Figure 5.4 (a) Attainable pore size of crystallization for sodium chloride and mirabilite and (b) characteristic curves of pore crystallization of OPC concretes of sodium chloride. The parameters retained for the evaluation are as follows: γ_{CL} (sodium chloride, mirabilite), 0.1 J/m^2; δ, 1.0 nm; V_C, 27 cm^3/mol (sodium chloride), 220 cm^3/mol (mirabilite); z_{sat} (20 °C), 0.0996 (sodium chloride), 0.02235 (mirabilite). The PSD curves of OPC concretes are taken from Figure 3.7. Note that the crystallization of mirabilite (Na$_2$SO$_4$·10H$_2$O) induces a volume contraction compared with the total volume of its constituents of solid thernadite (Na$_2$SO$_4$) and water molecules. This volume change has an impact on the mechanical condition in Equation 5.1. This volumetric change associated with the mirabilite crystallization is recorded as between −7.01 and −3.28% (Flatt, 2002), and an averaged value of −4.97% was adopted by Coussy (2006a). Here, this volumetric contraction is neglected for the evaluation of attainable pore size.

The maximum pore size governs the minimum z/z_{sat} needed to start the crystallization in pores. Again, Equation 5.2 provides the necessary supersaturation corresponding to the maximum pore size r_{max} on the PSD curve. The smaller this radius is, the higher is the supersaturation needed to start the crystallization. Actually, the supersaturation needed can be read from Figure 5.4 for a known maximum pore size. For concretes, the maximum pores pertain to the air voids and large capillary pores, having a size of magnitude 100 μm. The corresponding supersaturation z/z_{sat} is found very near to 1.0. The total porosity will have an influence on the magnitude of the average stress generated in the solid matrix by the local pore pressure; see Equation 3.6.

Transport/Mechanical Properties

The transport properties influence salt crystallization through the sorptivity, the permeability, and the diffusivity. The sorptivity of concrete represents the absorption capacity of salt solution once the concrete surface is in contact with the liquid. As the surface supply of dissolved salt is abundant and superior to the absorption capacity of concrete, only some of the dissolved salts can be taken in. Otherwise, if the surface supply is lower than the absorption capacity, the sorptivity is no longer important since all dissolved salts can be absorbed. The latter case corresponds to a dry environment and is more liable to pore crystallization damage. The permeability scales the resistance of concrete to the salt solution penetration once the solution is absorbed into the surface, and the diffusivity of aqueous salt species determines the transport rate of salts into material inside via the pore liquid phase. Both properties are pertinent to the long-term transport of salts into concrete; thus, it is important to estimate the range or depth of concrete susceptible to crystallization damage. The mechanical properties of concrete have influences on the damage from pore crystallization via the fracture toughness and the cracking strength.

Environmental Conditions

Temperature and Humidity

The ambient temperature can alter the crystallization process through the thermal dependence of the solubility of salts. Figure 5.5 illustrates the solubility of several salts in term of temperature. In other terms, the fluctuation of temperature itself can induce crystallization in the pores. The ambient humidity governs the drying extent for pore crystallization. As the drying humidity is low (see Figure 5.3), the crystals can form in smaller pores and generate higher pore pressure. Also, it should be noted that lower humidity will decrease the liquid pressure in pores and the crystal pressure p_C can be decreased according to Equation 5.1. This offsets to some extent the mean pore pressure in Equation 5.6. However, this statement is theoretical and should be taken with caution: if the crystals formed hinder liquid evaporation and the low humidity can make the liquid film δ crystallize between the crystal and the pore wall, then the local crystallization pressure generated can be very high (Scherer, 2004). Pore crystallization in natural environments can result from the combined thermal–hydraulic effect without precipitation: the temperature drops below the dew point during the night and the vapor in air condenses into liquid droplets on the concrete surface, dissolving the deposited salts and absorbing into the concrete pores. This process, quite slow in itself, does increase the salt concentration in pores in the long term.

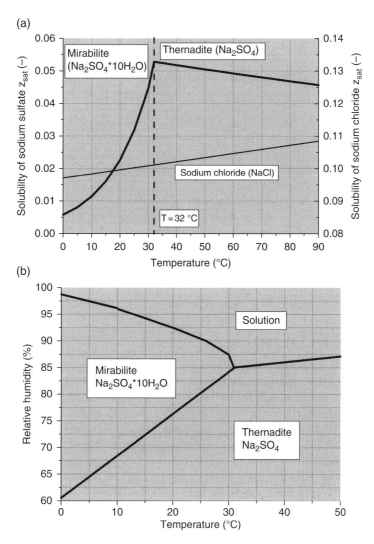

Figure 5.5 (a) Temperature dependence of the solubility of salts and (b) phase diagram of Na_2SO_4–H_2O. (b) *Source:* adapted from Flatt, 2002, Fig. 1 and Fig. 9. Reproduced with permission of Elsevier.

Salt Type

The salts responsible for crystallization damage are mainly NaCl, Na_2SO_4, and $MgSO_4$. NaCl is typical for concrete elements exposed in marine and similar environments, while the sulfates exist typically in saline groundwater and soils. Among these salts, Na_2SO_4 is assumed to be the most aggressive due to its two crystal types: thenardite (Na_2SO_4) and mirabilite ($Na_2SO_4 \cdot 10H_2O$). Both crystals can form from the supersaturated solution of Na_2SO_4, but each has its own stable thermal–hydraulic conditions; see Figure 5.5. Compared with thenardite, mirabilite has a much larger molar volume due to the 10 water molecules combined into the crystal, being 214% larger than thenardite. Owing to these properties, Na_2SO_4 is the salt most used in laboratory crystallization tests (e.g., the partial immersion test). The significant spalling in partial immersion above the solution level was attributed to the precipitation of mirabilite crystals from the solution supersaturated with respect to mirabilite but not to thenardite (Haynes *et al.*, 2008). From the same set of experiments, the most severe damage was observed in the scheme to shift the exposed environment between the stable ranges for thenardite and mirabilite.

Salt Deposition Rate

The last environmental factor is the intake rate of salts into the concrete elements, closely related to the salt intake mechanism. For airborne salts, it is usually the salt deposition rate on the surface that determines the intake rate. For salts in saline soils or groundwater, the intake rate involves the salt content in water or soils, the sorptivity and permeability of concrete, and the evaporation rate as well; see Figure 5.1b. The intake rate of airborne salts can be estimated from the surface deposition rate, while the intake rate from soils and groundwater needs to consider the concrete–environment interaction. On the basis of the surface deposition of chloride in Figure 2.7, the NaCl intake rate can be estimated as 0.254 mol/m² per year in the first year of exposure and 0.141 mol/m² per year for the subsequent 3 years for OPC concrete with $w/c = 0.35$. Note that these values of deposition rate correspond to the concrete surface exposed directly to the splashing of seawater; concrete elements away from the splash zone will have a much lower deposition rate.

5.3 Modeling for Engineering Use

Pore crystallization in concrete is a subject far from being thoroughly investigated. This is partly due to the crystallization process in confined pores itself, and partly due to the complex microstructure of concrete materials. The current knowledge on this subject is not yet capable of supporting reliable modeling for engineering use. Bearing such a premise in mind, this section attempts to propose two models capturing the most important mechanisms of pore crystallization: the CT-1 model addresses the onset of pore crystallization through the critical supersaturation, and the CT-2 model estimates the stress in the solid matrix of the material generated from pore crystallization. Both models are at the conceptual stage and further validation and elaboration are needed.

5.3.1 Model CT-1: Critical Supersaturation Model

This model aims to predict the onset of crystal formation in pores. It is an early stage for crystallization damage, since crystal formation in pores does not necessarily induce damage in the solid matrix of concrete. Imagine a concrete surface with initial liquid (water) saturation s_L^0 exposed to the invasion of external salts. The deposited salts are dissolved by the natural precipitation, forming a solution with equivalent concentration z_{in}, noted in molar fraction. The concrete surface is then exposed to drying and wetting cycles: during the wetting period the salt solution is absorbed into the concrete and saturates the surface to some depth; during the drying period the pores are dried back to the initial saturation. By such cycles, the salt concentration in the pore solution keeps increasing, obeying the salt mass balance for each cycle:

$$s_L^0 \frac{z_n^1}{v_L\left(z_n^1\right)} + \left(1 - s_L^0\right)\frac{z_{in}}{v_L\left(z_{in}\right)} = \frac{z_{n+1}^0}{v_L\left(z_{n+1}^0\right)} \text{ and } s_L^0 \frac{z_{n+1}^1}{v_L\left(z_{n+1}^1\right)} = \frac{z_{n+1}^0}{v_L\left(z_{n+1}^0\right)} \qquad (5.8)$$

where z_n^1 and z_{n+1}^1 are the pore concentrations in the dried state at cycles n and $n+1$, z_{n+1}^0 is the concentration at the beginning of cycle $n+1$ in the saturated state, and v_L is the molar volume of the salt solution. The molar volume of salt solution depends on the salt concentration and the temperature as well, and the regressed expression from Flatt (2002) is retained for the molar volume (in cm³/mol) for NaCl solution at 20 °C:

$$v_L(z) = 18.069 - \frac{20.618z}{1+4.215z} \tag{5.9}$$

If we set $z_0^1 = 0$ and z_{in} as constant as the initial conditions, using Equations 5.8 and 5.9 can give the pore saturation evolution with the number of cycles. Figure 5.6 illustrates the concentration evolution in terms of the cycles for different z_{in}. Here, the invading concentration z_{in} is noted as a ratio of saturated fraction z_{sat}. The saturated fraction z_{sat} is set as the crystallization criterion since the supersaturation degree z/z_{sat} needed for pore crystallization in concrete is very near to 1.0; see Figure 5.4.

The parameters involved in the simulation are given in Table 5.1. From the results, the higher the invading concentration is, the sooner the pore concentration reaches the critical value; the lower the initial liquid saturation is, the quicker the pore concentration will increase. The initial liquid saturation s_L^0 is assumed to be in equilibrium with the environmental humidity, with a lower s_L^0 indicating a drier environment. This result confirms that the crystallization is accelerated by the drying extent of the concrete surface. In this analysis, no explicit time scale is attributed to the cycles, but the following reasoning can be used: if the equivalent concentration z_{in} is deduced from the annual surface deposition quantity of salts and the annual environmental precipitation, the cycles can be interpreted on a yearly basis.

5.3.2   Model CT-2: Crystallization Stress Model

This model attempts to quantify the stress in concrete from the pore crystallization process. Let us follow the logic line of the CT-1 model and suppose the pore supersaturation z/z_{sat} reaches the critical level after n cycles. To simplify the expression of pore pressure, the stress analysis focuses on the saturated state (see Figure 5.3a), since only two phases, s_L and s_C, exist in the pores. Using Equation 5.2, the attainable pore radius r_{n+1} can be determined as

$$\frac{1}{r_{n+1}-\delta} = \frac{RT}{2\gamma_{CL}V_C}\ln\left(\frac{z_{n+1}^0}{z_{sat}}\right) \tag{5.10}$$

It is assumed that all pores smaller than r_{n+1} are still liquid saturated, while pores greater than this are occupied by salt crystals. Following Equation 5.7 with the help of PSD measurement, the saturations are

$$s_L^{n+1} = s_L(r_{n+1}), \quad s_C^{n+1} = 1 - s_L^{n+1} \tag{5.11}$$

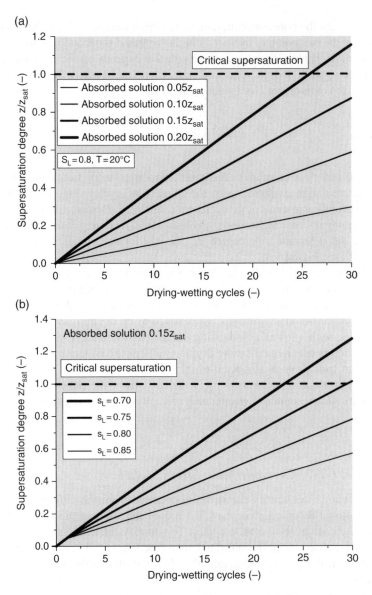

Figure 5.6 Increase of pore concentration of salt (NaCl) in terms of drying–wetting cycles for (a) different invading concentrations and (b) different initial liquid saturations.

Once the crystal saturation is determined, the mean pore pressure can be evaluated from Equation 3.5 or 5.6 with $s_G = 0$. Following the same integral scheme in Equations 3.13 and 3.14, the stress in the solid matrix can be evaluated as

$$\sigma_m\big|_{s_C=0 \to s_C^{n+1}} = \frac{\phi}{1-\phi} \int_{r_{max}}^{r_{min}} \left(p_C \frac{ds_C}{dr} + p_L \frac{ds_L}{dr} \right) dr \qquad (5.12)$$

Table 5.1 Parameters used in the CT-1 and CT-2 models

Parameter (unit)	Value/relation
Initial pore saturation s_L^0	0.7–0.85
Invading salt solution concentration z_{in}	$0.05z_{sat}$ to $0.20z_{sat}$
Saturated salt solution concentration for NaCl at 20 °C z_{sat}	0.0996
Molar volume of NaCl solution at 20 °C v_L (cm³/mol)	16.62–18.07, see Equation 5.9
Interface energy of water–crystal γ_{CL} (J/m²)	0.1
Thickness of liquid film between crystal and pore wall δ (nm)	1.0 (Scherer, 2004)
Ideal gas constant R (J/(mol K))	8.314
Temperature in environment T (K)	293.15 (20 °C)
Porosity of concrete ϕ	0.09 ($w/c = 0.3$), 0.12 ($w/c = 0.4$), 0.15 ($w/c = 0.5$), 0.18 ($w/c = 0.6$)
Crack initiation stress σ_{fc} (MPa)	1.0 (Zhang et al., 2004)

Assuming $p_L = p_{atm}$ and using the sphere-channel model in Figure 3.4b to estimate the crystallization pressure for a specified pore radius r, the above expression can be written as

$$\sigma_m\big|_{s_C=0\rightarrow s_C^{n+1}} = \frac{\phi}{1-\phi}\int_{r_{max}}^{r_{n+1}}\left(\frac{2\gamma_{CL}}{r-\delta}\frac{ds_C}{dr}\right)dr \tag{5.13}$$

The expressions in Equations 5.10, 5.11, and 5.13 constitute the CT-2 model to evaluate the accumulated stress in the solid matrix during pore crystallization subject to drying–wetting cycles. Using this model, the mean pore pressure and the solid stress are evaluated for OPC concretes subject to NaCl solution invasion. The results are illustrated in Figure 5.7 with the relevant parameters given in Table 5.1. In the figure, the fracture criterion is simply the comparison of solid stress with the cracking strength of concrete.

The results give the stress in the solid matrix in terms of the supersaturation z/z_{sat}, but not explicitly in terms of drying–wetting cycles. This can be done using the CT-1 model: given drying extent s_L^0, invading solution concentration z_{in}, and drying–wetting cycles, the supersaturation z/z_{sat} can be evaluated and the stress level can be read on Figure 5.7 in terms of cycles. The relevant simulations are omitted here. From the results, it can be seen that the PSD curves of concrete do not have a significant impact on the local crystallization pressure, though the pore pressure for $w/c = 0.6$ concrete is systematically lower than the others. This is because the OPC concrete with $w/c = 0.6$ contains more coarse pores than others. The total porosity is determinant for the transfer from mean pore pressure to solid matrix stress, and the concrete with $w/c = 0.3$ has the lowest stress due to its smallest total porosity. Note that all concretes are liable to fracture as z/z_{sat} attains 1.25 using 1.0 MPa as the cracking strength. This indicates that the concretes cannot withstand a high degree of supersaturation of NaCl solution. This model provides the quantified stress but not the damage extent (depth) from pore crystallization. This damage extent is related to the wetting depth of the invading salt solution, and can be deduced from the precipitation intensity and the sorptivity of the concrete surface. Interested readers can elaborate this depth with the help of the modeling of drying–wetting actions in Chapter 6.

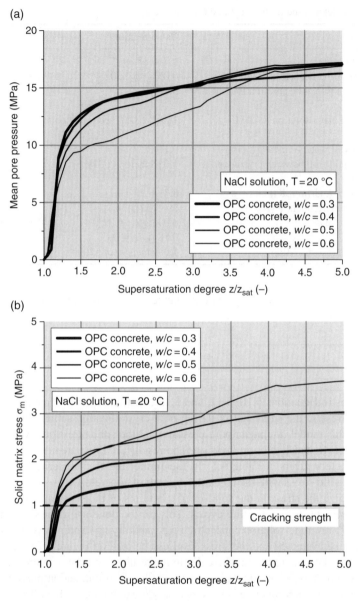

Figure 5.7 (a) Mean pore pressure and (b) solid stress of OPC concretes subject to NaCl solution invasion. The characteristic curves $s_L(z/z_{sat})$ are given in Figure 5.4.

5.4   Basis for Design

As the solid matrix of concrete is damaged by the crystallization pressure, the mechanical properties of concrete deteriorate. Some concrete specimens, subject to salt crystallization damage by partial immersion tests in the laboratory, showed complex damage patterns under compression, and the fracture initiation was found not to be correlated to the crystallization damage zone (Hartell *et al.*, 2011). Certainly, the alteration of concrete mechanical properties needs further investigation, but experience in practical cases confirms the principal damage pattern by salt crystallization is the surface spalling of concrete elements. For the elements partially buried into saline soils, surface spalling is limited to a local zone above ground level; see Figure 5.1. Thus, the target of durability design of concrete elements against salt crystallization is to protect the structural concrete from surface spalling by the invasion of external salts.

Similar to frost damage, two durability limit states (DLSs) can be specified for durability design against salts crystallization: (1) concrete exempted from pore crystallization, and (2) concrete with a limited extent of surface spalling. The first DLS is a strict one, corresponding to the early stage of the increase of salt concentration in pores by external salt invasion. The invasion is allowed but the accumulated concentration in pore solution should be less than the critical supersaturation degree. The CT-1 model can help to quantify the allowed degree of supersaturation. The second DLS is a less strict state and corresponds to a later stage of pore crystallization: the salt concentration surpasses the critical supersaturation, salt crystals form in pores, and spalling occurs by pore crystallization. This DLS just imposes a limit on the spalling extent; for example, the annual spalling rate of surface concrete. The CT-2 model, together with the correct estimate of wetting depth of invading salt solution, can help in estimating the spalling extent and the spalling rate. In this sense, the first DLS applies to concrete elements with external aesthetic requirements, or concrete elements with a very limited section dimension for which any loss of material can affect the mechanical resistance. The second DLS applies to concrete elements with no aesthetic requirements for which surface spalling does not affect the mechanical resistance.

From the analysis in this chapter, damage by salt crystallization can be enhanced by high salt deposition rate, low environmental humidity, large porosity, and weak fracture resistance of concrete. Accordingly, the design measures against crystallization damage can be formulated on this basis. The principal measure is to adopt concrete with high compactness, having very low (connected) porosity and permeability, and high fracture toughness. These properties help to decrease the invasion rate and wetting depth of the external salt, and to reduce the fracture risk as the pore crystallization occurs. Though not discussed so far, air entrainment also helps to decrease the damage risk of crystallization. The role of air voids is similar to that in frost action: the voids can reduce the crystallization pressure by letting crystals grow freely. The volume of entrained air is recommended to be 4–5.5% (CNS, 2008).

Besides these measures at the material level, measures can be taken at the element and structural levels. Surface treatment can be adopted for concrete elements to separate the surface from salt deposition (epoxy-coating) or render the surface hydrophobic (silane impregnation). Thus, the salts and the external water cannot easily penetrate into the pore space of the concrete. Some experimental results in the laboratory have confirmed their protection capacity against salt crystallization (Suleiman *et al.*, 2014). However, the long-term effect of these treatments needs further investigation. For elements partially buried in saline soils, a waterproof isolation of the buried part of elements from the soils will be efficient in cutting the source of salts for capillary absorption into the concrete (Yoshida *et al.*, 2010).

Part Two

From Materials to Structures

6

Deterioration in Structural Contexts

This part contains only one chapter, treating the transition from knowledge at the material level, through ageing modeling, to the performance of structural concretes. "Structural concrete" refers to concrete material used in structures resisting various service loadings and actions during its service life. Different deterioration processes of concrete materials are investigated following their respective ageing mechanisms covered in Part 1 (Chapters 1–5). Actually, the service conditions of concrete structures, including the mechanical loads and environmental actions, can greatly alter the deterioration kinetics of the concrete materials. Further, as multiple ageing processes occur simultaneously, the deterioration mechanism can be different from a single process. It is from this perspective that this chapter finds its reason of being. In other terms, the ageing laws and models introduced in the previous chapters should be calibrated by the mechanical loadings and the actual environmental actions. This chapter treats successively the effect of loading and cracking, multi-fields problems, and drying–wetting actions.

6.1 Loading and Cracking

The first aspect of the structural context is the mechanical loading. Structural materials, including concrete and steel, are subject to mechanical loading of different natures during their service life. The mechanical loadings have different effects on concrete and steel: concrete can exhibit brittleness under loading, and excessive loadings often lead to cracking damage, while steel in RC elements usually remains within its elastic limit. Accordingly, the effects of mechanical loading on the deterioration kinetics of concretes involve both the loading effect, without material damage, and the intervention of cracks, in the damage phase of concrete. Moreover, concrete materials can bear cracks from origins other than mechanical loading, such as autogenous or drying shrinkage. This section focuses on the loading effect on

Durability Design of Concrete Structures: Phenomena, Modeling, and Practice, First Edition. Kefei Li.
© 2016 John Wiley & Sons Singapore Pte. Ltd. Published 2016 by John Wiley & Sons Singapore Pte. Ltd.

concrete materials and attempts to mark the difference between the "pure" loading effect and the effect of cracks. These elements provide a basis for deterioration kinetics calibration with respect to loading and cracking for RC elements.

6.1.1   Mechanical Loading

Mechanical loading alters the durability processes, if any, through its action on the solid matrix of the concrete. The consequence of mechanical loading on the solid matrix of concrete is either elastic deformation, at low stress level, or irreversible deformation and cracking, at a higher stress level. The impact of macroscopic stress on the deterioration processes, physical or chemical, has to be quantified. To this purpose, a simple analysis is performed on the concrete in its elastic phase to quantify the magnitude of deformation at the pore level under service loading. The concrete is considered as a poroelastic medium, and the macroscopic stress of material can be related to the porosity and pore pressure for an isothermal case through (Coussy, 2006b: 75)

$$\sigma - \sigma_0 = \frac{K}{b}(\phi - \phi_0) - \left(\frac{K}{bN} + b\right)(p - p_0) \tag{6.1}$$

where σ and σ_0 (MPa) are stress and reference stress, ϕ and ϕ_0 are porosity and reference porosity, p and p_0 (MPa) are pressure and reference pressure in pores, K and N (MPa) are the bulk modulus and Biot's tangent modulus, and b is the Biot coefficient. For drained cases, $p \equiv p_0$, and the stress change is only related to the porosity change through K/b. Take the usual value ranges of K and b: $K = 3000$ to $15\,000$ MPa, and $b = 0.04$ to 0.35 for mortars (Coussy, 2006b). The macroscopic stress needed for porosity change of 1% is estimated as

$$p \equiv p_0: \quad \Delta\sigma = \frac{K}{b}(\phi - \phi_0) = 85.7 - 3750\,\text{MPa} \tag{6.2}$$

Thus, the stress needed is much higher than the realistic service stress in structural concretes, or the elastic change of porosity can be actually neglected under service stress at the order of a megapascal. This simple analysis makes it clear that the alteration of deterioration processes by mechanical loading must involve cracking or/and irreversible deformation of the solid matrix of concrete.

So far, considerable studies have been dedicated to the impact of stress level on the durability-related properties, and a comprehensive compilation of data can be found in Reinhardt (1997) and Yao *et al.* (2013). Basically, the available data were obtained from specimens either under loading or after loading. Some data are compiled in Figure 6.1 for loading and unloading cases. Although these data were obtained from different experimental procedures for different concretes, a direct observation is that the stress level has only very limited influence on the transport properties within $(0.7–0.8)f_c$ for compression, and this threshold value was observed to be lower for tension cases. Given that the allowable compressive stress of concrete can rarely surpass $0.5f_c$ in structural design, the stress level in structural concretes under service condition is too low to cause tangible alteration of transport properties according to these observations. However, it cannot be concluded that the compressive stress within $0.5f_c$ has no impact on transport properties since the long-term effect of service stress is not yet

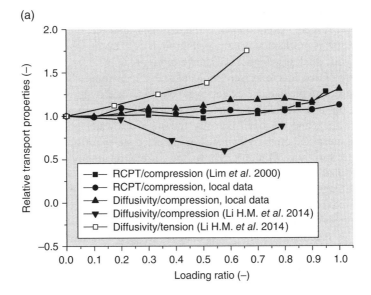

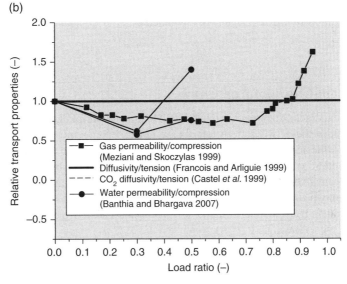

Figure 6.1 Literature data on impact of loading levels on the transport properties of structural concretes: transport properties measured at (a) unloaded conditions and (b) sustained loading conditions.

included in these investigations. The long-term loading can lead to considerable transient deformation of concrete (i.e., the creep), and no results are available yet on its impact on the transport properties. This is a key aspect to be explored in the future, and can be crucial for a rational judgement on the long-term effect of mechanical level on the transport properties of structural concretes.

So, the stress level does not seem to be the pertinent variable to address the impact of mechanical loading on transport properties. The damage or cracking extent can be a better parameter since the loading alters the transport properties through the solid matrix at a rather high stress range. A recent experimental campaign measured the altered transport properties of concrete specimens mechanically damaged through axial cyclic loadings. The experimental setup and the cyclic loading schemes are shown in Figure 6.2.

The specimens of two concretes were subject to a cyclic loading scheme to achieve different extents of damage, and the damage extents were characterized by the residual strains of specimens (in the axial and lateral directions), the damage factor from ultrasonic pulse velocity (UPV) measurement, and the open porosity measured by water absorption. The transport properties were measured on the specimens in unloaded states for open porosity, gas permeability, and water sorptivity. Figure 6.3 shows the correlation among the UPV damage factor D, the axial residual strain ε_{33}^r, and the volumetric residual strain $\varepsilon_v^r = \varepsilon_{33}^r + 2\varepsilon_{11}^r$. The correlations of damage extent with the transport properties are illustrated in Figure 6.4 using the UPV damage factor and the open porosity as damage extent variables.

From the aforementioned experimental investigations, it seems that the alteration of transport properties by mechanical loading was better correlated with the volume-based cracking variables like volumetric residual strain, the open porosity, and the UPV damage factor. This alteration was also observed to depend strongly on the crack patterns generated by the compressive loading. The crack patterns with percolated cracks, containing a main crack cluster, alter the transport properties more efficiently than homogeneously distributed cracks. The characterization of the crack network geometry for these specimens is to be given in Section 6.3.3.

6.1.2   Effect of Cracks: Single Crack

Concrete can bear cracks due to either mechanical loading or shrinkage. The cracking patterns are different under these two actions: shrinkage cracks can be more diffusive in the solid matrix, whereas the mechanical cracks are more concentrated; see Figure 6.5. The cracks influence the durability performance of structural concretes mainly by accelerating the mass transfer and exchange between the material and the external environment. This subject can be further split into two subtopics: how one single crack accelerates the mass transfer and how a crack network alters the transport properties of materials. These two aspects are to be treated successively in the following.

A single crack alters the transport process in concrete through the following mechanisms: the fluid flow through the crack, the adsorption of ions from the liquid (flow) in the crack, the dissolution of soluble matters of solid by the liquid flow, and the (possible) chemical reaction between the fluids with the solid phases exposed on the crack surface. The first three processes are physical and the last one is chemical. This section is dedicated to the physical processes.

Fluid Flow Through a Crack

Let us focus first on liquid flow in the cracks. Liquid flow through cracks is one important engineering concern, as the structural concrete is expected to provide liquid-tightness. To estimate the liquid flow rate through concrete cracks with different opening width w, a flow

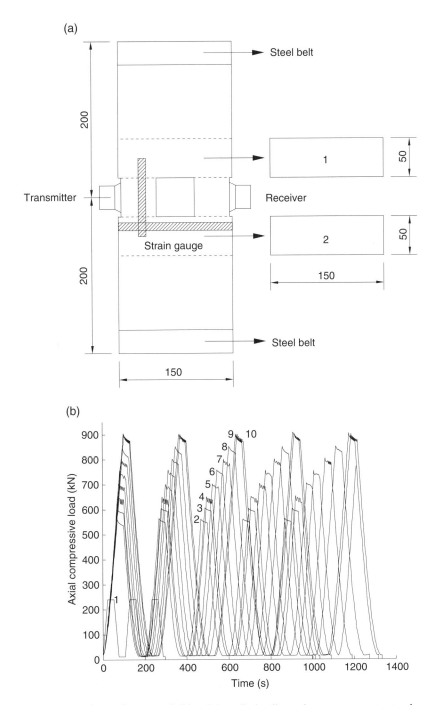

Figure 6.2 (a) Experimental setup and (b) axial cyclic loading scheme on two structural concretes: OPC concrete with $w/c = 0.55$ and $f_c = 50.0$ MPa at 28 days, high-volume fly-ash concrete (HVFC) with $w/c = 0.5$ and $f_c = 44.4$ MPa at 28 days; the achieved damage extents (axial residual strains) are 0 to -702 µε for OPC and 0 to -566 µε for HVFC. *Source:* Zhou *et al.* 2012a, Fig. 2, Fig. 3, Reproduced with permission of Springer.

(a)

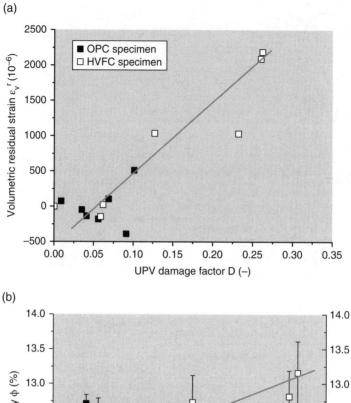

(b)

Figure 6.3 Correlation of UPV damage factor with (a) the axial residual strain and (b) the volumetric residual strain. *Source:* adapted from Zhou *et al.* 2012a, Fig. 5, Fig. 7. Reproduced with permission of Springer.

rate coefficient ξ is introduced into the classical Poiseuille's law to account for the influence of the roughness of the crack wall surface (Tsukamoto and Wörmer, 1991):

$$Q = \xi \frac{1}{12} \frac{w^3}{\mu} \frac{\Delta P}{L} = \frac{1}{12} \frac{w_e^3}{\mu} \frac{\Delta P}{L} \quad \text{with} \quad \xi = \left(\frac{w_e}{w}\right)^3 \qquad (6.3)$$

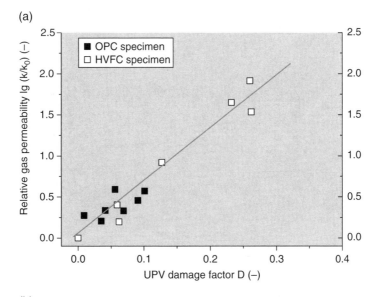

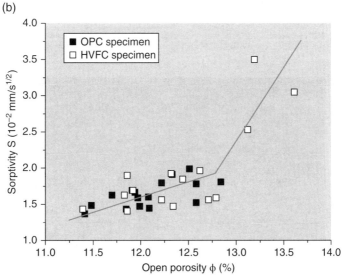

Figure 6.4 Alteration of transport properties in terms of different damage variables: (a) relative gas permeability in terms of UPV damage factor and (b) water sorptivity in terms of the open porosity. *Source:* adapted from Zhou *et al.* 2012a, Fig. 8, Fig. 9. Reproduced with permission of Springer.

where Q (m^2/s) is the liquid flow in the crack, μ (kg/(m s)) is the liquid viscosity, and $\Delta P/L$ (Pa/m) is the pressure gradient. The flow rate coefficient ξ was measured as 0.01–0.15 for different cement-based materials by Tsukamoto and Wörmer (1991). Actually, this coefficient is related to both the crack surface roughness and the crack opening. A recent study from Li *et al.* (2011) helps to deepen the insight on the influence of crack surface roughness. This study created fracture surfaces for different cement-based materials, inserted the fractured

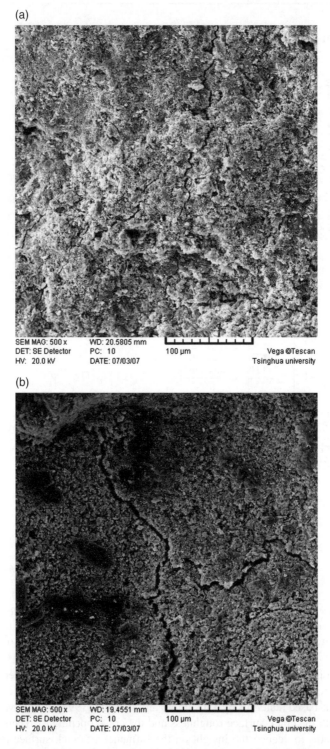

(a)

SEM MAG: 500 x   WD: 20.5805 mm
DET: SE Detector PC: 10   100 µm   Vega ©Tescan
HV: 20.0 kV   DATE: 07/03/07   Tsinghua university

(b)

SEM MAG: 500 x   WD: 19.4551 mm
DET: SE Detector PC: 10   100 µm   Vega ©Tescan
HV: 20.0 kV   DATE: 07/03/07   Tsinghua university

Figure 6.5 Crack patterns in concrete under (a) drying–wetting cycles and (b) mechanical loading. *Source:* Zhou *et al.* 2011, Fig. 1. Reproduced with permission of Elsevier.

specimens into a permeation cell, and measured the water flow through the crack between the fractured surfaces that were mechanically closed. Figure 6.6 shows the surface topography of the fracture surfaces of specimens, and the flow results are given in Table 6.1. In the table the crack opening was obtained through graph analysis from the topography of two opposite fracture surfaces.

The roughness of the fracture surface is noted in terms of the surface fractal dimension, and a higher dimension value corresponds to a larger surface roughness. In Table 6.1, the concrete fracture has the highest surface roughness but also has the largest flow rate, which is somehow against the common sense that high surface roughness reduces the flow rate. Actually, these data were obtained under two fracture surfaces mechanically closed, and high surface roughness leaves more flow channels between the surfaces. This explains the flow rates in terms of the surface roughness. Thus, the surface roughness influences the flow rate via the crack opening: the smaller the crack opening, the more the surface roughness influences the flow rate; under a certain opening limit the flow rate loses its correlation with the average opening but relates to the residual channeling between the rough fracture surfaces.

Note that the coefficient ξ in Table 6.1 refers to the initial flow rate of liquid in the measurement. The liquid flow in cracks was observed to decrease constantly with time, a phenomenon termed "self-healing" of cracks (Edvardsen, 1999). Although it was generally believed that self-healing is related to some chemical reaction between the flowing liquids and the crack surfaces, Li *et al.* (2011) attributed the flow decrease to the nonsteady liquid flow behavior between rough surfaces under a low-pressure gradient.

Adsorption and Dissolution in a Crack

As liquid flows in a crack, mass transfer can occur between the liquid and the fracture surfaces. Two physical phenomena can coexist: adsorption of ions in the liquid solution by the fracture surface and the dissolution of soluble phases on the fracture surface by liquid flow; see Figure 6.7. The two processes can be described through the advection–dispersion equation of mass transport (Bodin *et al.*, 2003):

$$\frac{\partial c_{\mathrm{m}}}{\partial t} + m_{\rightarrow} = D_{\mathrm{L}} \frac{\partial^2 c_{\mathrm{m}}}{\partial x^2} - u \frac{\partial c_{\mathrm{m}}}{\partial x} \tag{6.4}$$

where c_{m} (mol/m^3 or kg/m^3) is the concentration of solute (ion or solvable phase) in the liquid solution, u (m/s) is the flow rate, $m_{\rightarrow}$ (mol/(m^3 s) or kg/(m^3 s)) is the adsorption/dissolution rate of the ion/soluble phase, and D_{L} (m^2/s) is the diffusivity of the solute.

Let us treat the ion adsorption first. Suppose the diffusion contribution in Equation 6.4 can be neglected and an instantaneous adsorption process of first order is assumed. Then the adsorption of fracture surface can be described as[1]

$$\bar{m}_{\rightarrow}^{\mathrm{ads}} = \frac{r_{\mathrm{ads}}^0}{k_{\mathrm{ads}}} \left[1 - \exp\left(-k_{\mathrm{ads}} t\right)\right] \tag{6.5}$$

---

[1]This expression uses the initial condition of zero adsorbed quantity at $t = 0$.

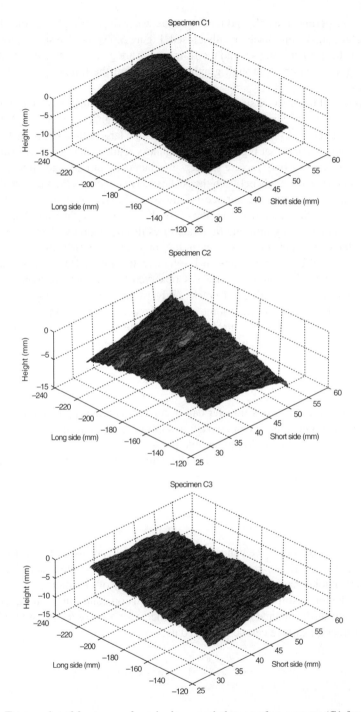

Figure 6.6 Topography of fracture surfaces by large-scale laser surface scanner (C1 for cement paste $w/c = 0.4$, C2 for mortar with fine sands and $w/c = 0.6$, C3 for mortar with coarse sands and $w/c = 0.5$, C4 for mortar with coarse sands and $w/c = 0.6$, C5 for concrete with $w/c = 0.6$, and C6 for concrete with only coarse aggregates and $w/c = 0.6$).

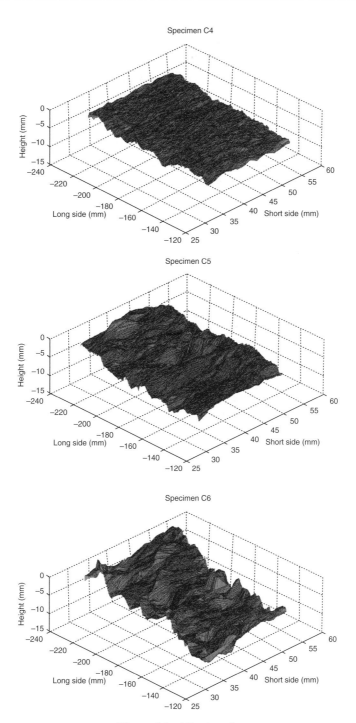

Figure 6.6 (*Continued*)

Table 6.1 Surface roughness and flow rates in cracks

Properties	C1	C2	C3	C4	C5
Surface fractal dimension D	2.0273	2.0703	2.0969	2.1110	2.1270
Average fracture opening w (mm)	0.1394	0.1554	0.2058	0.2182	0.2619
Opening deviation σ_w (mm)	0.0552	0.0695	0.0962	0.0989	0.1710
Initial flow rate q_0 (mL/h)	19.9	46.0	56.5	84.0	153.6
Equivalent hydraulic opening w_e (mm)	0.0123	0.0163	0.0174	0.0199	0.0243
Flow rate coefficient ξ (10^{-3})	0.687	1.154	0.604	0.758	0.798

Source: Li *et al.* 2011, Table 2. Reproduced with permission of Elsevier.

(a)   (b)

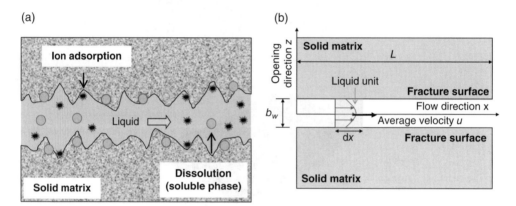

Figure 6.7 (a) Mass transfers between liquid flow and fracture surfaces of crack and (b) convection–dispersion modeling of mass transport. (b) *Source:* Wang and Li 2011a, Fig. 1. Reproduced with permission of *Journal of the Chinese Ceramic Society.*

where $\bar{m}_{\rightarrow}^{ads}$ (kg/m²) is the adsorbed quantity of fracture surface, r_{ads}^0 (kg/(m² s)) is the initial adsorption rate of the surface, k_{ads} (1/s) is the adsorption rate coefficient, and r_{ads}^0/k_{ads} (kg/m²) gives the final/equilibrium adsorption quantity of fracture surface. Note that Equation 6.5 can be used to interpret directly the ion adsorption in a static liquid solution using the balance between the ion quantity in the liquid solution and that on the solid surface. Converting the adsorption rate in Equation 6.5 with respect to liquid volume using the crack opening b_w, the ion transport–adsorption equation is

$$\frac{\partial c_m}{\partial t} + \frac{1}{b_w} r_{ads}^0 \exp\left(-k_{ads} t\right) = -u \frac{\partial c_m}{\partial x} \quad \text{with} \quad m_{\rightarrow}^{ads} = \frac{2}{b_w} \frac{d\bar{m}^{ads}}{dt} \tag{6.6}$$

This equation depicts the adsorption kinetics of ions contained in a flowing liquid between fracture surfaces, with k_{ads} and r_{ads}^0 as basic adsorption parameters. Experimentally, if the ion concentration c_m can be monitored with respect to time and distance in the flow direction, the adsorption parameters k_{ads} and r_{ads}^0 can be regressed numerically.

Then let us consider surface dissolution. Suppose the soluble phase on the fracture surface is portlandite; that is, a crystal of $Ca(OH)_2$. The dissolution equilibrium is governed by the dissolution constant K_{CH}, defined as $[Ca][OH^-]^2$ in the aqueous phase, equal to $10^{-5.18}$ at 25 °C (Thomas et al., 2003). The dissolution rate of portlandite on the fracture surface for a spherical crystal, following Ramachandra and Sharama (1969), is written as

$$\bar{m}_{\rightarrow}^{dis} = k_{dis} \ln \left(\frac{\left[Ca^{2+} \right]\left[OH^- \right]^2}{K_{CH}} \right) \tag{6.7}$$

with k_{dis} (kg/(m² s)) the dissolution rate coefficient. In the following we assume $[Ca^{2+}], [OH^-] = 0$ at $t = 0$ and $[Ca^{2+}] = 2[OH^-]$. Using this assumption, Equation 6.7 can be used to interpret directly the Ca^{2+} leaching rate in static water. As the liquid solution flows in the crack, the transport of Ca^{2+} should include the dissolution contribution from the fracture surfaces:

$$\frac{\partial c\left(Ca^{2+} \right)}{\partial t} + \frac{2}{b_w} k_{dis} \ln \left[\frac{4c^3 \left(Ca^{2+} \right)}{K_{CH}} \right] = -u \frac{\partial c\left(Ca^{2+} \right)}{\partial x} \quad \text{with} \quad m_{\rightarrow}^{dis} = \frac{2}{b_w} \bar{m}_{\rightarrow}^{dis} \tag{6.8}$$

This equation depicts the surface dissolution kinetics of Ca^{2+} from portlandite crystals on the fracture surface, with the dissolution rate coefficient k_{dis} as the basic parameter. Experimentally, if the concentration $[Ca^{2+}]$ can be traced in the crack along the flow and with time, the parameter k_{dis} can be regressed numerically.

Experimental Evidence

Using the models in Equations 6.6 and 6.8, a recent study investigated the chloride adsorption and Ca^{2+} dissolution (leaching) on fracture surfaces of cement-based materials (Wang and Li, 2011a,b). A NaCl solution was used as the flowing liquid for Cl^- adsorption and deionized water for Ca^{2+} leaching. In parallel, the static adsorption of Cl^- and static leaching of Ca^{2+} were measured using the same fracture surfaces. The regressed adsorption and dissolution parameters are given in Table 6.2. The results show that the liquid flow does change substantially the adsorption and dissolution parameters. Compared with the static case, liquid flow greatly decreases the adsorption rate and the adsorption capacity of fracture surfaces, but increases considerably the dissolution rate of portlandite. The contrast is more pronounced for cement

Table 6.2 Adsorption and dissolution parameters of fracture surface of cement pastes in static and flow cases (C1: cement paste with $w/c = 0.4$, C3: mortar with $w/c = 0.5$)

Parameters	Static water	Crack flow
Initial Cl^- adsorption rate r_{ads}^0 (10^{-7} kg/(m² s)) (C1)	10.515	0.988
Initial Cl^- adsorption rate r_{ads}^0 (10^{-7} kg/(m² s)) (C3)	5.260	1.711
Final Cl^- adsorption capacity r_{ads}^0/k_{ads} (10^{-3} kg/m²) (C1)	64.6	14.68
Final Cl^- adsorption capacity r_{ads}^0/k_{ads} (10^{-3} kg/m²) (C3)	41.8	10.55
Leaching rate of portlandite k_{dis} (10^{-9} kg/(m² s)) (C1)	0.0216	0.747
Leaching rate of portlandite k_{dis} (10^{-9} kg/(m² s)) (C3)	0.1227	0.209

paste (C1) than for mortar (C3). The underlying mechanism for the decrease of ion adsorption by water flow is attributed to the decrease of the adsorption layer of ions near the fracture surface: in this layer, multispecies ions exist and the adsorption is stronger for the ions nearer to the solid surface; the flowing liquid takes away the ions with weak adsorption force, decreases the thickness of the adsorption layer, and reduces the adsorption capacity.

6.1.3   Effect of Cracks: Multi-cracks

Mechanical loading and internal shrinkage can generate multi-cracks in the material. These cracks form a network, and the geometry of the network plays a crucial role in the transport and mass exchange, thus altering the related deterioration processes in concrete. This section treats the aspect of multi-cracks, complementary to the single crack behavior in Section 6.1.2. First, some micromechanical results are presented using the effective self-consistent scheme to account for the influence of multi-cracks on the linear physical properties. In this scheme the cracks are taken as inclusions in a solid matrix; see Figure 6.8.

The permeability is taken as the linear physical property to be studied. The formulations from the interaction direct derivative (IDD) method (Zheng and Du, 2001) are retained to describe the influence of cracks. In this method, every single crack, with length $2a$, is modeled as an inclusion in an equivalent porous solid matrix with an ellipse (2D) or ellipsoid (3D) as its atmosphere. The crack density, ρ_2 (2D) or ρ_3 (3D), is

$$\rho_2 = \frac{\sum_i a_i^2}{A}, \qquad \rho_3 = \frac{\sum_i a_i^3}{V} \tag{6.9}$$

Let us study the simplest case of crack inclusion: all cracks have similar atmospheres with the same aspect ratio $\gamma_D = a/(a+b)$. The permeability of the solid matrix is denoted as k_0 and the local permeability of crack as k_i. If the local permeability of cracks is assumed to be much larger than the matrix permeability (i.e., $k_i \gg k_0$), the crack opening can be neglected.[2] In this case, the IDD model provides the following solution for the equivalent permeability k_{IDD} for an isotropic crack distribution (Zhou et al., 2011):[3]

$$\left(\frac{k_{IDD}}{k_m}\right)_{2D} = \frac{1+\dfrac{\pi}{2}\rho_2\left(1-\gamma_D\right)}{1-\dfrac{\pi}{2}\gamma_D\rho_2} \quad \text{and} \quad \left(\frac{k_{IDD}}{k_m}\right)_{3D} = \frac{1+\dfrac{32}{9}\rho_3\left[1-F\left(\gamma_D\right)\right]}{1-\dfrac{32}{9}\rho_3 F\left(\gamma_D\right)} \tag{6.10}$$

with

$$F\left(\gamma_D\right) = \frac{1-g\left(\gamma_D\right)}{2\left(1-\gamma_D^2\right)} \quad \text{and} \quad g\left(\gamma_D\right) = \frac{\gamma_D^2}{\left(\gamma_D^2-1\right)^{1/2}}\arctan\left(\gamma_D^2-1\right)^{1/2} \tag{6.11}$$

---

[2]This assumption holds for cement-based materials: taking the characteristic crack opening as 1 μm, the local permeability is on the order of 10^{-12} m^2, while the usual values of permeability for structural concretes range between 10^{-16} and 10^{-18} m^2. Thus, the local permeability of cracks is four to six orders of magnitude higher than the solid matrix.
[3]The analytical results in the following use the dilute solution of conductivity of a solid containing ellipsoidal inclusions (Shafiro and Kachanov, 2000), and the mathematical details can be found in Zhou et al. (2011).

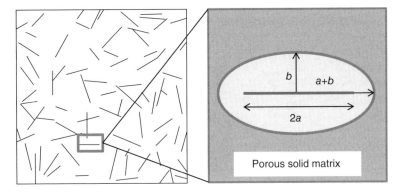

Figure 6.8 Micromechanical modeling of a porous solid matrix with cracks as inclusions.

As the aspect ratio γ_D should respect the global crack density, ρ_2 (2D) or ρ_3 (3D), the aspect ratio can be related to the global crack density through

$$(\gamma_D)_{2D} = \frac{\sqrt{1+(4/\pi\rho_2)}-1}{2\sqrt{1+(4/\pi\rho_2)}} \text{ and } \frac{(\gamma_D)_{3D}}{\left[1-(\gamma_D)_{3D}\right]^3} = \frac{3}{4\pi\rho_3} \qquad (6.12)$$

The results in Equation 6.10 predict a global permeability $(k_{IDD}/k_m)_{2D}$ or $(k_{IDD}/k_m)_{3D}$ increasing gradually with crack density. For the 2D case, $\rho_2 \to \infty$, $\gamma_D \to 0$, but $\rho_2\gamma_D \to 1/\pi$; thus, $(k_{IDD}/k_m)_{2D}$ follows a gradual increase with the crack density. This observation reflects the average nature of the micromechanical model: the crack influence is treated in a homogeneous way and the connectivity or clustering of cracks is not taken into account. Actually, as cracks form a continuous path for fluid flow, the global permeability can tend to infinity and depends no longer on the average-based crack density in Equation 6.9. This phenomenon is called crack percolation. Accordingly, the results in Equation 6.10 can apply to homogeneous crack distribution cases in which the local clustering of cracks is not important. Upon percolation, the strong influence of crack connectivity should be considered.

The results of Equation 6.10 assume $k_c \gg k_0$, and thus crack opening is neglected. This assumption is appropriate for permeability or electrical conductivity of cement-based materials with cracks at the micrometer scale. As the crack gets finer or the permeability of the matrix higher, the crack opening should be considered. Note $2b_c$ the crack opening and γ_c aspect ratio of the crack itself (b_c/a). The following results (after the IDD scheme) are obtained for an isotropic crack distribution:

$$\left(\frac{k_{IDD}}{k_m}\right)_{2D} = \frac{1+\dfrac{\pi}{2}\rho_2(1+\gamma_c)\left[\gamma_c+(1+\gamma_c)(1-\gamma_D)\right]}{1-\dfrac{\pi}{2}\rho_2(1+\gamma_c)\left[-\gamma_c+(1+\gamma_c)\gamma_D\right]} \qquad (6.13)$$

Note that this expression turns to Equation 6.10 as $\gamma_c \to 0$. The global permeability is influenced mutually by the crack opening and the crack density, but the influence of crack opening becomes very limited as $\gamma_c < 0.01$.

(a)   (b)

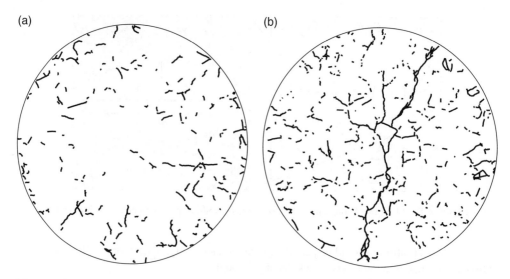

Figure 6.9 Typical crack patterns after cyclic compressive loadings for two structural concretes: (a) OPC specimens with axial residual strain 129 $\mu\varepsilon$ and (b) HVFC specimens with residual strain 74 $\mu\varepsilon$. *Source:* Zhou *et al.* 2012b, Fig. 4. Reproduced with permission of Elsevier.

Experimental Evidence

Investigation of the geometry of crack networks on the transport properties is rare in the literature. Following the experimental investigation on the alteration of transport properties by mechanical loadings in the previous section, the crack geometry was quantified on the specimens sawed out from the damaged concrete cylinders. The typical crack patterns after cyclic loading are presented in Figure 6.9: the cracks were observed with digital microscopy and the cracks with opening larger than 20 μm were captured.

The geometry analysis characterized the crack density, crack opening, crack length distribution, crack orientation, and crack connectivity. The crack connectivity is defined as

$$f_{\text{conv}} = 1 - \sum_{j=1,m} a_j^2 \bigg/ \sum_{k=1,n} a_k^2 \qquad (6.14)$$

where $a_{j=1,m}$ denotes the isolated cracks and $a_{k=1,n}$ is all cracks. Figure 6.10 illustrates the length distributions for the two networks in Figure 6.9, and the length distributions were found to obey a log-normal distribution. The crack connectivity was evaluated for all the concrete specimens and was found to correlate with the crack density; see Figure 6.11. This would imply that the crack connectivity is not an independent variable, but correlated to the crack density, an average quantity, under a specific damage process. Figure 6.12 illustrates the impact of the cracks on the transport properties of the structural concretes in terms of the crack density.

6.2 Multi-fields Problems

The second aspect for structural context pertains to field concepts. Contrary to the material specimens under laboratory investigation, structural material is constantly subject to multi-field conditions during its service life. The basic physical quantities, such as

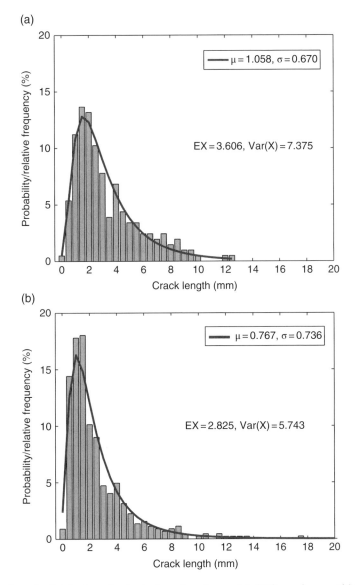

Figure 6.10 Length distributions for typical crack patterns: (a) OPC specimens with axial residual strain 129 με and (b) HVFC specimens with residual strain 74 με. *Source:* Zhou *et al.* 2012b, Fig. 7. Reproduced with permission of Elsevier.

temperature, moisture content, and ion concentration, assume both spatial and temporal distributions in structural elements. Since these quantities are fundamental to the different deterioration processes described so far, this section is dedicated to the field description of the temperature and moisture content, and the possible coupling between the physical fields as well.

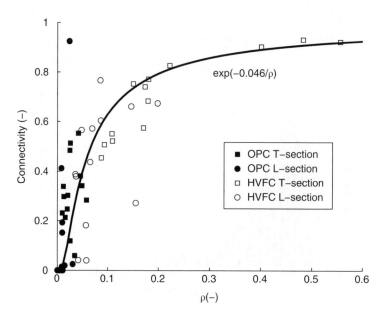

Figure 6.11 Correlation between the crack connectivity and the crack density. *Source:* Zhou *et al.* 2012b, Fig. 14. Reproduced with permission of Elsevier.

6.2.1 Thermal Field

The temperature in structural elements assumes a field concept, being both a spatial and a temporal function. The change of temperature in structural elements is due to heat transfer between the elements and the environment through different transfer modes, including thermal conduction (due to temperature gradient), thermal convection (due to the flow of fluid mass in materials), and thermal radiation (due to the intrinsic generation and absorption of electromagnetic radiation of a thermal body); see Figure 6.13. The heat transfer is described through the heat conservation of a structure or element as

$$C_T \frac{\partial \theta}{\partial t} = -\mathrm{div}(q) + r_{int} \tag{6.15}$$

where θ (°C) is the structural temperature, C_T (J/(m³ °C) or kJ/(m³ °C)) is the heat capacity of concrete materials, q (J/(m² h) or kJ/(m² h)) is the heat flow due to thermal conduction or flow convection – the thermal conduction usually adopts a linear Fourier's law linking the gradient of temperature to the conduction heat flow by

$$q = -K_T \, \mathrm{grad}(\theta) \tag{6.16}$$

with K_T (J/(h m °C) or kJ/(h m °C)) the thermal conductivity of the material – and r_{int} (J/(m³ h) or kJ/(m³ h)) is the internal heat generation source; for concretes the heat from the exothermic reaction of cement hydration should be taken into account at an early age, and this term can be neglected for totally hardened concrete during service conditions.

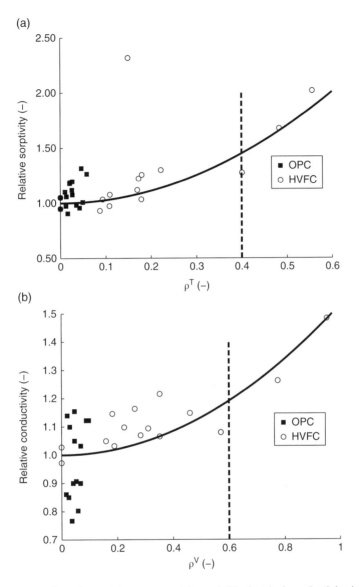

Figure 6.12 Impact of cracks on (a) water sorptivity and (b) electrical conductivity in terms of the crack density. *Source:* Zhou *et al.* 2012b, Fig. 16, Fig. 18. Reproduced with permission of Elsevier.

Neglecting the internal source of heat and expressing the conduction heat flow through Fourier's law give the simplified heat transfer equation

$$C_{\mathrm{T}} \frac{\partial \theta}{\partial t} = -\mathrm{div}\left[K_{\mathrm{T}} \, \mathrm{grad}\left(\theta\right)\right] \tag{6.17}$$

where the thermal conductivity K_{T} can usually be assumed constant. To solve the thermal equation in Equation 6.15 or 6.17, appropriate initial and boundary conditions should be prescribed.

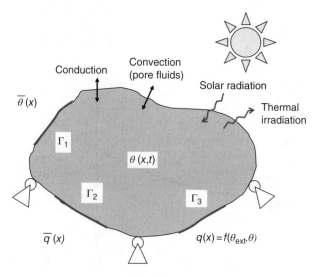

Figure 6.13   Illustration of different modes of thermal transfer in a concrete structure.

Boundary Conditions

While the initial condition can be assumed as an arbitrary field – that is, $q(x,t=0)=q_0(x)$ – the boundary conditions can be prescribed in three forms to take into account the different heat transfer modes on the material–environment interface:

1. *The Dirichlet condition* prescribes the temperature on the boundary Γ_1:

$$\theta(x,t>0)\big|_{\Gamma_1}=\bar{\theta}(x,t) \tag{6.18}$$

   which accounts for the conduction heat flow across the boundary.
2. *The Newmann condition* gives the thermal flow on the boundary Γ_2:

$$q(x,t>0)\big|_{\Gamma_2}=\bar{q}(x,t) \tag{6.19}$$

   which accounts for the heat convection induced by the flow of pore fluids. Also, this boundary condition can represent the heat flow from the absorption of solar radiation:

$$q(x,t>0)\big|_{\Gamma_2}=\lambda_{SA}q_{sw} \tag{6.20}$$

   with λ_{SA} the surface absorptivity of shortwave solar radiation and q_{sw} (J/(m^2 s)) is the shortwave solar radiation.
3. *The exchange condition* provides the thermal flow in terms of the temperature difference on the boundary Γ_3:

$$q(x,t>0)\big|_{\Gamma_3}=f_{ext}(\theta,\theta_{ext};t) \tag{6.21}$$

This boundary condition can depict the convection flow or thermal irradiation across the boundary. For convection flow, a linear form for the exchange function is proposed following the boundary-layer theory for the thermal field near the concrete surface as (Ulm and Coussy, 2001)

$$f\left(\theta,\theta_{\mathrm{ext}}\right)\big|_{\Gamma_3} = \lambda_3\left(\theta_{\mathrm{ext}} - \theta\right) \tag{6.22}$$

where λ_3 (J/(h m^2 K) or kJ/(h m^2 K)) is defined as the exchange coefficient. This condition is also termed the "exchange boundary condition", which is particularly adapted to concrete elements wrapped in formworks, with λ_3 as the thermal conductivity of formworks per unit thickness. This condition can also describe the thermal irradiation flow from the concrete surface:

$$f_{\mathrm{ext}}\left(\theta,\theta_{\mathrm{ext}}\right)\big|_{\Gamma_3} = \varepsilon_{\mathrm{irr}}\sigma_{\mathrm{SB}}\left(T^4 - T_{\mathrm{ext}}^{\;4}\right) \tag{6.23}$$

where $\varepsilon_{\mathrm{irr}}$ is the concrete surface emissivity, σ_{SB} (W/(m^2 K^4)) is the Stefan–Boltzmann constant, and T and T_{ext} (K) are the concrete surface and external environmental temperatures. Table 6.3 gives the usual properties for concrete materials for thermal transfer.

Characteristic Length

As a concrete surface is exposed to a thermal gradient, say $\Delta\theta$, on the boundary, the temperature field evolves due to heat conduction. If heat conduction is the only active transfer mode, the influential depth of thermal transfer, within which the temperature is to be affected by the thermal conduction, can be scaled by a characteristic length $L_{\mathrm{T}}^{\mathrm{c}}$:

$$L_{\mathrm{T}}^{\mathrm{c}} = \sqrt{D_{\mathrm{T}} t_{\mathrm{T}}^{\mathrm{c}}} \tag{6.24}$$

where D_{T} (m^2/s) is the thermal diffusivity, equal to $K_{\mathrm{T}}/C_{\mathrm{T}}$, and $t_{\mathrm{T}}^{\mathrm{c}}$ is a characteristic time for the imposed duration of thermal gradient on the boundary. It is shown that for the region $x \ll L_{\mathrm{T}}^{\mathrm{c}}$, the thermal field is changed substantially by the gradient, while the thermal field can be considered to be undisturbed in the region $x \gg L_{\mathrm{T}}^{\mathrm{c}}$. Taking $D_{\mathrm{T}} = 10^{-6}$ m^2/s in Table 6.3, this characteristic length is equal to 6.3 cm, 30 cm, 169 cm, and 32.2 m respectively for hourly, daily, monthly, and annual thermal events. This length can actually help engineers to determine for structural elements in which region the thermal effect is to be taken into account.

6.2.2 Moisture Field

Concrete is a porous material with pore size ranging from nanometers at the level of C-S-H, to millimeters at the level of air voids. The pores are saturated, or partially saturated, by water of different natures. Usually the pore water that can be dried out by a lower humidity is referred to as *free* water in capillary pores. After hydration, the free water in the capillary pores is in equilibrium with pore gaseous phases, and in turn with the gaseous phase in the external environment. Accordingly, the moisture state in structural concrete can be represented either

Table 6.3 Thermal properties of concrete materials collected from the literature

Property	Value	Source
Heat capacity C_T (kJ/(m^3 °C))	2344–2512	Mehta and Monteiro (2006)
Thermal conductivity K_T (kJ/(h m °C))	7.28–15.44	Mehta and Monteiro (2006)
Thermal diffusivity D_T (10^{-6} m^2 s)	0.81–1.39	Mehta and Monteiro (2006)
Shortwave surface absorptivity λ_{SA}	0.5–0.9	Qin and Hiller (2011)
Shortwave solar radiation (W/m^2)	300–900	Qin and Hiller (2011)
Exchange coefficient, concrete–air λ_3 (kJ/(h m^2 K))	14.4–21.6	Ulm and Coussy (2001)
Exchange coefficient, wood formwork 20 mm λ_3 (kJ/(h m^2 K))	9.2–11.2	Ulm and Coussy (2001)
Exchange coefficient, steel formwork 2 mm λ_3 (kJ/(h m^2 K))	14.4–19.6	Ulm and Coussy (2001)
Concrete surface emissivity ε_{irr} (-)	0.9	Qin and Hiller (2011)
Stefan–Boltzmann constant σ_{SB} (10^{-8} W/(m^2 K^4))	5.67	Mahmoud et al. (2012)

by the saturation degree of the liquid phase in pores or by the relative humidity in equilibrium with this saturation degree.

This moisture equilibrium can be best described by the pore water saturation s_1, in terms of the environmental humidity in equilibrium h:

$$s_1 = s_1(h) \tag{6.25}$$

The liquid saturation s_1 is defined as the volume ratio of liquid in the pore space, and thus ranges from 0 to 1.0. The external humidity h is defined as the partial pressure of water vapor p_v divided by the saturated vapor pressure p_{vs}, ranging also between 0 and 1.0. Actually, this relation is termed the "water retention curve" or the "moisture isotherm" of porous materials, reflecting the fundamental property of water retention capacity of the pore structure of the material. Indeed, this relation can be derived from the ideal gas law and thermodynamic equilibrium at the liquid–gas interface:

$$\rho_1 \frac{RT}{M_v} \ln(h) = p_1 - p_{atm} \tag{6.26}$$

where R (J/(mol K)) is the ideal gas constant, M_v (kg/mol) is the molar mass of vapor, T (K) is the temperature, and ρ_1 (kg/m^3) is the liquid water density. (Equation 6.26 is actually derived from the thermodynamic equilibrium between liquid and gas phases in pores:

$$\frac{dp_1}{\rho_1} = \frac{dp_v}{\rho_v} \tag{6.27}$$

where p_1 and p_v (Pa) are the liquid and vapor pressures. Assuming the vapor phase to be an ideal gas, one obtains

$$p_v M_v = RT \rho_v \tag{6.28}$$

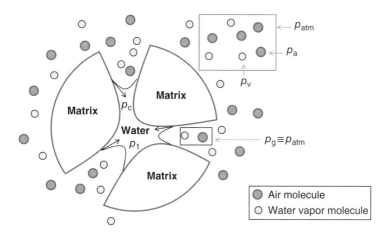

Figure 6.14   Moisture equilibrium between pore phases and external gas phases.

Substituting this expression into Equation 6.27 gives

$$dp_1 = \rho_1 \frac{RT}{M_v} d \ln(p_v) \tag{6.29}$$

Integrating Equation 6.29 with respect to a reference state, air containing saturated vapor under atmospheric pressure p_{atm}, one obtains

$$\int_{p_v}^{p_{vs}} \rho_1 \frac{RT}{M_v} d \ln(p_v) = \int_{p_1}^{p_{atm}} dp_1 \tag{6.30}$$

Using the definition of relative humidity, $h = p_v / p_{vs}$, one obtains (6.26).)

The right side of Equation 6.26 can be expressed further through pore capillary pressure:

$$p_g \equiv p_{atm} : p_1 - p_g = -p_c(s_1) \tag{6.31a}$$

where the gas phases are assumed to be in equilibrium with the external atmosphere; see Figure 6.14. Combining Equations 6.26 and 6.31a gives the definition for the moisture isotherm in Equation 6.25. Note that the relation in Equation 6.31a is also termed the "moisture characteristic curve" for porous materials, measured experimentally through standard methods for porous materials like soils.[4]

---

[4]For concretes, the direct measurement of this curve, p_c–s_1, seems rather problematic and the indirect determination through water vapor adsorption/desorption isotherm has been performed (Mainguy, 1999), and the following expression is obtained:

$$p_c(s_1) = \alpha \left(s_1^{-\beta} - 1 \right)^{1-(1/\beta)} \tag{6.31b}$$

with α and β as experimental coefficients.

The moisture state, represented by s_l or h, in concrete elements also forms a physical field. To quantify this physical field, $s_l(x,t)$ or $h(x,t)$, the transport of moisture in the pore space needs to be described. To this purpose, the concrete is regarded as a porous medium with three phases participating in the transport: water (liquid), vapor (gas), and dry air (gas); see Figure 6.14. These phases transport in pores due to pressure (liquid and gas phases) or concentration gradient (gas phases). Assume also that thermodynamic equilibrium is always kept among the phases during the transport. The mass flux of water vapor J_v (kg/(m² s)) is written as (Philip and De Vries, 1957)

$$J_v = -D_{va}\zeta\nabla\rho_v, \qquad \nabla\rho_v = -\frac{M_v\rho_v}{\rho_l RT}\nabla p_c \tag{6.32}$$

where D_{va} (m²/s) denotes the vapor diffusivity in air and ζ is the resistance of the pore structure to gas diffusion (Millington, 1959). The second part of equation uses the ideal gas assumption for vapor in Equation 6.28. The mass flux of liquid water J_l (kg/(m² s)), is formulated by Darcy's law:

$$J_l = \rho_l \frac{k}{\eta_l} k_{rl}(s_l)\nabla p_c \tag{6.33}$$

where k (m²) is the intrinsic permeability, k_{rl} is the relative liquid permeability, and η_l (kg/(m s)) is the viscosity of liquid. Neglecting the vapor mass compared with liquid mass, the mass conservation of moisture in pores gives

$$\frac{\partial(\rho_l\phi s_l)}{\partial t} = -\mathrm{div}(J_v + J_l) \tag{6.34}$$

Finally, the moisture transport is described through a diffusion-like equation:

$$\frac{\partial s_l}{\partial t} = \mathrm{div}\left[D(s_l)\nabla s_l\right] \tag{6.35}$$

with the moisture diffusivity expressed as

$$D(s_l) = -\frac{1}{\rho_l\phi}\left[\rho_l\frac{k}{\eta_l}k_{rl}(s_l) + D_{va}\zeta\frac{M_v\rho_v}{\rho_l RT}\right]\frac{\partial p_c}{\partial s_l} \tag{6.36}$$

So far, Equations 6.31a, 6.35, and 6.36 give a complete multiphase modeling for moisture transport in concrete with pore saturation s_l as basic variable. Through the moisture isotherm (e.g. Equation 6.25), one can also transform the transport model in terms of relative humidity,

$$C_h(h)\frac{\partial h}{\partial t} = \mathrm{div}\left[D_h(h)\nabla h\right] \quad \text{with} \quad C_h(h) = \frac{\partial s_l}{\partial h}, \quad D_h(h) = D(s_l)\frac{\partial s_l}{\partial h} \tag{6.37}$$

Unlike thermal field description in Equation 6.17, the moisture transport equation, in Equation 6.35 or 6.37, is highly nonlinear with respect to pore saturation s_l or relative humidity h. Figure 6.15 shows the moisture diffusivity and the characteristic curve for two typical structural concretes in literature. (In this application, two supplementary laws are used for the

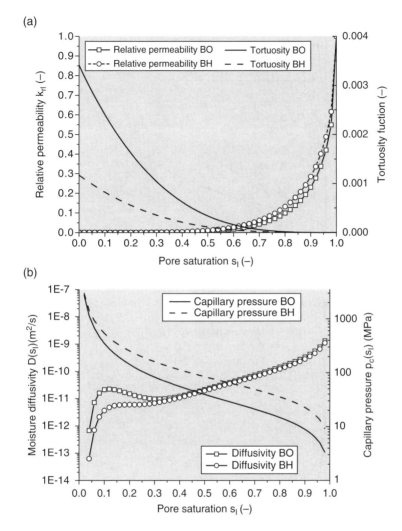

Figure 6.15 (a) Relative permeability and tortuosity and (b) moisture diffusivity and characteristic curves for two typical structural concretes. The parameters are as follows: $R = 8.3147$ J/(K mol), $T = 293.15$ K, $M_v = 0.018$ kg/mol, $\rho_l = 1000$ kg/m^3, $\rho_{vs} = 0.0174$ kg/m^3, $\eta_l = 0.001$ kg/(m s), $D_{va} = 2.45 \times 10^{-5}$ m^2/s; $k_{BO} = 3.0 \times 10^{-21}$ m^2, $k_{BH} = 6.0 \times 10^{-22}$ m^2, $\varphi_{BO} = 0.122$, $\varphi_{BH} = 0.082$, $\alpha_{BO} = 18.62$ MPa, $\beta_{BO} = 2.275$, $\alpha_{BH} = 46.94$ MPa, $\beta_{BH} = 2.06$ (Baroghel-Bouny, 2007).

tortuosity function and the relative permeability. The tortuosity function assumes the following form after Millington (1959):

$$\zeta\left(s_l, \phi\right) = \phi^{2.7}\left(1 - s_l\right)^{10/3} \tag{6.38}$$

and the relative permeability adopts the expression in Equation 2.4.)

The contributions of gas and liquid phases to the moisture diffusivity are well illustrated in Figure 6.15. For low saturation range, $s_l < 0.3$, the transport of gas phases contributes to the

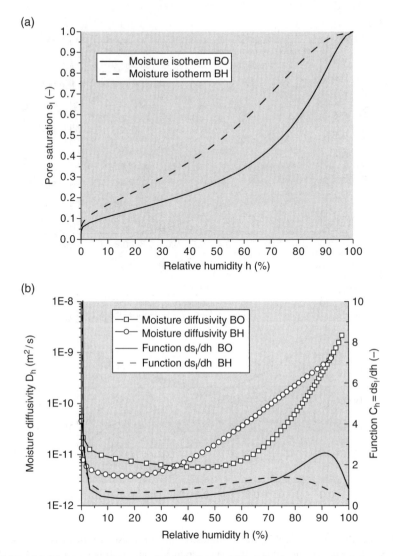

Figure 6.16 (a) Moisture isotherm and (b) moisture diffusivity in terms of relative humidity for BO/BH concretes (same data retained as in Figure 6.15).

whole diffusivity due to fact that the water flow channels are greatly decreased under the low pore saturation; see Figure 6.15b. In other terms, the magnitude of the second term in brackets in Equation 6.36 is important compared with the first term. As $s_1 > 0.3$, liquid flow dominates the global diffusivity. The two concretes show rather different capillary pressures, but the global moisture diffusivity does not differ in magnitude, especially in the range $s_1 > 0.3$.

Figure 6.16 illustrates the moisture isotherm, s_1–h, of the two concretes, and the denser concrete (BH) tends to have higher pore saturation under the same relative humidity. The moisture diffusivity coefficient D_h is presented in terms of relative humidity. Again, the contribution of gas transport in the lower humidity range is well noted.

6.2.3 Multi-field Problems

The previous sections provided the basic tools for field analysis of the temperature and moisture states in structural elements. So far, the impact of the moisture state on the thermal field and the thermal impact on the moisture transport have not been addressed. However, these impacts actually exist for the fields in a structural context, from which the multi-field problems arise. This section attempts to treat the multi-field problem for thermo-hydro couplings (TH couplings) on the basis of the previous sections. To this purpose, the temperature and pore saturation are retained as two basic and independent variables.

The TH coupling comes from mainly two origins: the flow of pore fluids in concrete and the thermal-hydro-dependence of concrete properties. Bearing this aspect in mind, the TH couplings are treated in the following in a straightforward way without providing the mathematical details. The heat conservation in Equation 6.17 is extended to include the flow of pore fluids, using the definition of flow rates in Equations 6.32 and 6.33, as[5]

$$C_T \frac{\partial \theta}{\partial t} = -\text{div}\left(K_T \nabla \theta\right) - \left(\sum_{i=l,g} C_i J_i\right)\nabla \theta \qquad (6.39)$$

where the influence of moisture state s_l is expressed through the following properties and term:

- C_T stands for the heat capacity for concrete, expressed as

$$C_T(s_l) = \rho_s(1-\phi)C_s + \rho_l \phi s_l C_l + \rho_g \phi(1-s_l)C_g \qquad (6.40)$$

  with the subscripts s, l, and g referring to the solid skeleton, the liquid phase, and the gaseous phase respectively.
- K_T stands for the thermal conductivity of concrete, written as

$$K_T(s_l) = K_s\left[1 + \cfrac{\phi}{\cfrac{1-\phi}{3} + \cfrac{K_s}{s_l K_l + (1-s_l)K_g - K_s}}\right] \qquad (6.41)$$

  with K_l, K_g, and K_s being the thermal conductivities of liquid, gas, and solid phase respectively.
- The term $\sum_{i=l,g} C_i J_i$ reflects the convective heat due to the flow of fluids in pores with C_l and C_g the heat capacities for the liquid and gas phases. Considering Equations 6.32 and 6.33, this term is also a function of the pore saturation s_l.

---

[5]This expression can be derived from the energy conservation between the entropy change of material representative elementary volume and the heat transfer due to conduction and convective flow of pore fluids; see Coussy (2010) for details.

The moisture transport equation (Equation 6.35) should be extended to include the thermal influence through the moisture diffusivity:

$$D(\theta,s_1) = -\frac{1}{\rho_1\phi}\left[\rho_1\frac{k}{\eta_1(\theta)}k_{rl}(s_1) + D_{va}(\theta)\zeta\frac{M_v\rho_v}{\rho_1 RT}\right]\frac{\partial p_c}{\partial s_1} \qquad (6.42)$$

where the temperature intervenes explicitly in the vapor part of diffusivity, second term in the brackets; thus, the vapor diffusion is by nature more sensitive to temperature. In addition, the thermal influence on other properties in Equation 6.42 can be described as follows (Mainguy, 1999):

- Liquid viscosity η_1 of pore liquid can change with temperature. As the pore liquid is idealized as water, the thermal dependence of viscosity can be expressed as

$$\eta_1(T) = 2.414\times10^{-5}\exp\left(\frac{570.58}{T-139.85}\right), \quad T = 273.15+\theta \qquad (6.43)$$

Note that the viscosity of water changes from 1.744×10^{-3} kg/(m s) (or Pa s) at 0 °C to 0.650×10^{-3} kg/(m s) (or Pa s) at 40 °C.
- Vapor diffusivity D_{va} (m²/s) depends both on the temperature and the gas pressure:

$$D_{va}(p_g,T) = 0.217\times10^{-4}\frac{p_{atm}}{p_g}\left(\frac{T}{T_0}\right)^{1.88} \qquad (6.44)$$

where T_0 is reference temperature, taken as 273.15 K.
- Vapor density ρ_v can be expressed through the saturated vapor density and relative humidity as

$$\rho_v = \rho_{vs}(T)h(s_1,T), \quad \ln[\rho_{vs}(T)] = 62.013 - \frac{7214.64}{T} - 7.2973\ln T \qquad (6.45)$$

with $h(s_1, T)$ given by Equations 6.30 and 6.31a.
- Characteristic curve $p_c(s_1)$ can be temperature dependent due to the thermal dependence of the surface tension of pore water. The surface tension of water was reported to change from 74.95×10^{-3} N/m at 5 °C to 70.41×10^{-3} N/m at 35 °C (Vargaftik *et al.*, 1983); that is, a moderate change of 6% for a thermal range of 30 °C. This thermal dependence is neglected here.

To gain a quantitative impression of the TH couplings, the two concretes in Figure 6.15 are retained and Figure 6.17 illustrates the hydro-dependence of thermal properties at $T = 293.15$ K ($\theta = 20$ °C) and the thermal dependence of moisture diffusivity. In the analysis, the liquid density ρ_1 and the characteristic curve $p_c(s_1)$ are regarded as constant. From the figure, the influence of pore saturation on the heat capacity is around 20% between $C_T(s_1 = 0)$ and $C_T(s_1 = 1.0)$; and the influence is much smaller for thermal conductivity, with a variation of

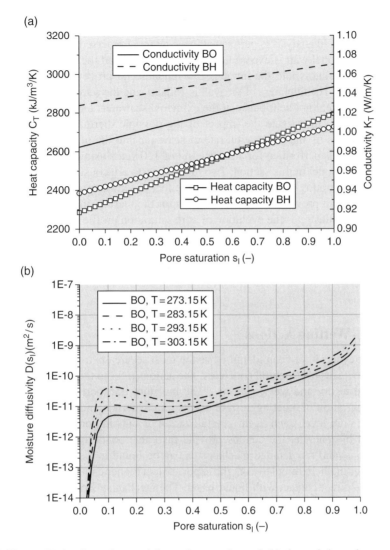

Figure 6.17 (a) Hydro-dependence of thermal properties and (b) thermal dependence of moisture diffusivity of BO/BH concretes. The parameters used are as follows: $C_{sk} = 1000$ J/(kg K), $C_l = 4180$ J/(kg K), $C_v = 1800$ J/(kg K), $C_a = 1000$ J/(kg K); $K_{sk} = 1.12$ W/(m K), $K_l = 0.6$ W/(m K), $K_g = 0.26$ W/(m K); $\rho_{sk} = 2603$ kg/m^3 (BO), 2598 kg/m^3 (BH), $\rho_l = 1000$ kg/m^3, $\rho_a = 1.205$ kg/m^3.

K_T around 5% from $s_l = 0$ to 1.0. The influence of temperature on the moisture diffusivity turns out to be more important: raising the temperature by 10 K leads to an augmentation of moisture diffusivity by 30% and $D(s_l)$ nearly doubles its value from $T = 273.15$ K (0 °C) to $T = 303.15$ K (30 °C). Thus, the thermal impact on moisture transport is more important than the hydro-influence on thermal properties. For engineering applications, the thermal impact on moisture transport should be the first to be considered.

Case Study

The thermo-hydro-field analysis is performed in a concrete immersed tube tunnel in a marine environment. The tunnel wall is exposed directly to seawater and the penetration of seawater through the wall thickness into the intrados is a fundamental process that determines other durability processes. The thickness of the wall is 150 cm, and the seawater penetration occurs with a constant thermal gradient between the extrados (seawater) and intrados (traffic side). This case study is to investigate the impact of the possible thermal gradient between the extrados and intrados on the penetration rate of seawater into the tunnel wall. The thermal-hydro-field analysis is performed for a service life of 120 years using the thermal model and moisture transport model in this section. Two thermal cases are considered: a warming effect of seawater at the extrados by 5 °C and a cooling effect of seawater at the extrados by 10 °C, which are supposed to provide the boundary cases for the thermal impact on seawater penetration. Figure 6.18 illustrates the evolution of pore saturation profile during 120 years due to seawater penetration. From the simulation results, the thermal impact on the saturation profile is notable by a difference of 15 °C imposed on the extrados, and the saturation profile due to the real temperature of seawater on the extrados is expected to be situated between the two boundary profiles.

6.3   Drying–Wetting Actions

Drying–wetting cycles have been identified as the most severe environmental actions for the deterioration processes of structural concretes. The actions refer to the alternative exposure of concrete surface to dry and wet conditions. The moisture transport under these conditions facilitates the migration of other aggressive agents into materials, inducing some detrimental processes at the pore level, such as the crystallization of salts in pores. For the corrosion of steel embedded in concrete, the actions provide favorable conditions for both anodic and cathodic reactions, and the detailed mechanisms can be found in Chapters 1 and 2. Strictly speaking, the wetting here refers to both the exposure to a higher environmental humidity, but not 100%, and a direct contact with liquid water. The mechanisms of moisture transport are fundamentally different for these two cases. In this section, the different moisture transport mechanisms under drying–wetting cycles are first introduced, and then drying–wetting cycles are investigated through some new engineering indices. The objective is to provide simple tools for engineers to estimate the impact of drying–wetting cycles on the mass transport into concrete surface.

6.3.1   Basis for Drying–Wetting Actions

As the concrete surface, with a certain s_1 in pores, is exposed to a humidity h_0 lower than that in equilibrium with s_1 – that is, $h_0 < h^{-1}(s_1)$ – two processes occur simultaneously: the lower pressure of water vapor in the environment attracts the transport of water vapor in pores toward the concrete surface and, in turn, accelerates the evaporation of liquid water in pores into vapor through the liquid–gas interface; meanwhile, the evaporation of liquid tends to increase the curvature of the liquid–gas interface and augment the capillary pressure through surface tension, dragging the liquid water toward the concrete surface. The relative importance of

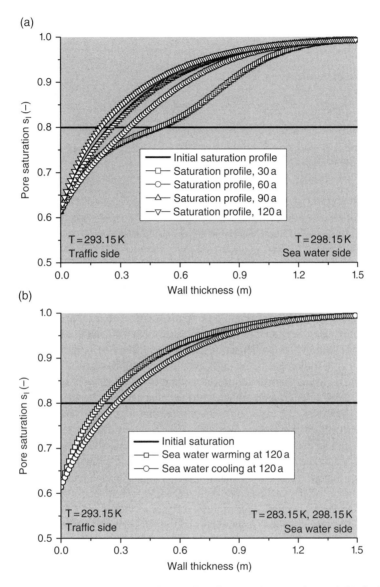

Figure 6.18 (a) Evolution of pore saturation profiles for extrados warming and (b) final saturation profiles for extrados warming and cooling cases at 120 years. The parameters of concrete used are as follows: porosity $\phi = 0.11$, skeleton density $\rho_{sk} = 2663$ kg/m^3, intrinsic permeability $k = 7.2 \times 10^{-21}$m^2, capillary pressure coefficient $\alpha = 29.819$MPa, capillary pressure coefficient $\beta = 2.427$; the relative humidity at intrados is imposed as 70%, corresponding to $s_l = 0.61$ through the moisture isotherm.

these two transport modes depends on the pore structure and the drying extent on the external surface, and some quantitative analysis is available in Mainguy *et al.* (2001). Whatever the relative importance, the water loss by this drying action can be captured by the moisture isotherm in Equation 6.25; that is, $\Delta s_l = s_l(h) - s_l(h_0)$.

(a)   (b)

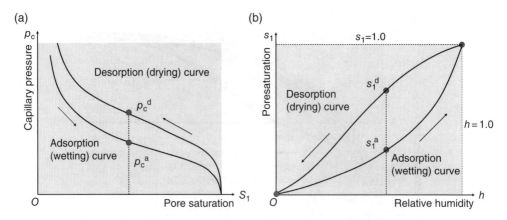

Figure 6.19 Moisture adsorption–desorption hysteresis in terms of (a) characteristic curves and (b) water retention curves.

Focus now on the concrete surface exposed to wetting. Let us limit first the term "wetting" to the exposure to a higher humidity $h_1 > h^{-1}(s_1)$. The two processes in drying are expected to reverse: the transport of water vapor into pore space and the subsequent condensation of water vapor into liquid water; meanwhile, the water flows back to material by the relaxation of capillary pressure due to the condensation. The only difference is that the hysteresis of the water retention curve intervenes in the mass gained in equilibrium; see Figure 6.19. The origin of this hysteresis is usually attributed to the geometry of larger pores connected by smaller pores (Dullien, 1992). For concrete materials, the s_1–h curve during the drying regime is normally situated above the s_1–h curve during wetting; that is, for the same humidity concretes contain more water in drying than in wetting. This actually reflects the hydrophilic nature of concrete: water can be easily stored in the pores, but less easily be driven out by drying.

Then, let us consider the exposure of concrete surface to liquid water. As the surface is in contact with liquid water, the open pores of the material serve as the transport channel for the liquid water into material. Driven by the capillary pressure, the liquid water flows into the open pores. This phenomenon is termed water absorption, a much faster transport process of liquid water into material pores. Figure 6.19 shows the moisture equilibrium inside the material while the water absorption is governed by the flow kinetics of liquid water in the open pores near the surface. Actually, water absorption has been identified as one useful property for the near-surface transport of concrete, and a standard method is available (ASTM, 2013c). Water absorption across the concrete surface I_w (mm³/mm² or mm) is found to scale with the square root of absorption time t (s) as

$$I_w = S_w \sqrt{t} \tag{6.46}$$

where S_w (mm/s$^{1/2}$) is the absorption coefficient or water sorptivity. Actually, water absorption can be divided into short-term and long-term phases with different sorptivity values; see Figure 3.6. The first and faster phase corresponds to the water flow in the connected coarse pores near the surface, and the second and slower phase to the latent water transport in smaller pores. In practice, the sorption rate of the first stage is usually taken as the intrinsic property of concrete surface. Thus, the sorptivity S_w always refers to the first stage of water absorption (i.e., the k_1

value in Figure 3.6) in this chapter. The usual values for the sorptivity of structural concretes can be found in Chapter 8. Note that this sorptivity depends also on the pore saturation s_l: the lower the pore saturation, the higher the sorptivity. The quantitative expression of this dependence is given in Equation 2.5.

In summary, three mechanisms of moisture transport are involved: the drying, the wetting without involving liquid water, and wetting with liquid water. These three cases are to be treated differently in the following.

6.3.2   Drying–Wetting Depth

On the basis of the concepts introduced in Section 6.3.1, we attempt to derive a useful scale length for engineering use: the "influential depth of moisture" under drying–wetting actions. Actually, under the alternative actions of drying and wetting and driven by the aforementioned mechanisms (drying, wetting with or without liquid water contact), moisture transports in and out of the concrete, forming a highly convective zone near the concrete surface. From an engineering point of view, if the depth of this convective zone can be quantified in terms of the drying–wetting actions and concrete properties, it will help greatly in the durability design for concrete elements; for example, specification on the thickness of concrete cover.

To this aim, a one-dimensional moisture transport problem is built in Figure 6.20: a porous material is initially totally saturated by liquid water, and then subject to a drying condition at $x = 0$. The drying humidity is noted as h_0 (<100%). The following mechanisms are taken for the drying process of the saturated pores: as the concrete surface is exposed to the environmental humidity h_0, the liquid water starts to evaporate and a liquid–gas interface is created in the initially saturated pores; then the water vapor transports toward the drying surface by diffusion due to the water vapor pressure gradient between the liquid–gas interface and the drying boundary $x = 0$. By this evaporation–diffusion process a drying front is created, and this front marks the liquid–gas interface inside the pores. The analysis herein is to derive the analytical solution for this drying front X. The mass conservation of water vapor in the range of $0 < x < X$ is written as follows:

$$\frac{\partial c_v}{\partial t} = \text{div}\left[D_v \text{grad}\left(c_v\right)\right] \tag{6.47}$$

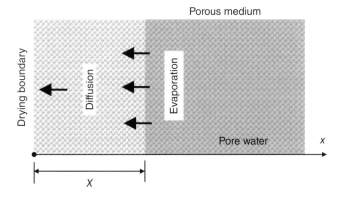

Figure 6.20   Illustration of moisture influential depth.

where c_v (mol/m^3) is the molar concentration of vapor in gas phases and D_v (m^2/s) is the vapor diffusion coefficient in the concrete pores. The vapor concentration c_v can be converted to its partial pressure p_v in gas phases by the Clapeyron equation and further to relative humidity by the standard definition of relative humidity:

$$p_v = RTc_v, \qquad h = \frac{p_v}{p_{vs}} \tag{6.48}$$

where R (J/(mol K)) is the ideal gas constant, T (K) the absolute temperature, and p_{vs} (Pa) the saturated vapor pressure. With these expressions, the mass conservation of water vapor in Equation 6.47 can be expressed in terms of relative humidity:

$$\frac{\partial h}{\partial t} = \mathrm{div}\left[D_v \mathrm{grad}(h)\right] \tag{6.49}$$

The boundary conditions for the evaporation–diffusion process are

$$h(x = X, t) = h_s = 100\%, \qquad h(x = 0, t > 0) = h_0 < 100\% \tag{6.50}$$

Note that the vapor diffusivity in Equation 6.49 is usually dependent on the pore saturation s_1. To precede the analysis, an assumption is taken for the diffusion process of water vapor: evaporation occurs at the liquid–gas interface, increasing locally the vapor concentration as well as the local vapor pressure. If this increase is assumed to be a local one, the vapor diffusivity D_v in Equation 6.49 can be assumed to be a constant, equal to $D_v(s_1^0)$ with $s_1^0 = s_1(h_0)$. Then it is further assumed that the evaporation at the interface is instantaneous compared with the subsequent vapor diffusion from $x = X$ to $x = 0$. Under this assumption, the vapor mass conservation across the liquid–gas interface can be expressed in terms of vapor concentration c_v or relative humidity h:

$$\rho_1\left(\frac{\mathrm{d}X}{\mathrm{d}t}\right) = M_v D_v \left(\frac{\partial c_v}{\partial x}\right)\Big|_{x=X}, \qquad \rho_1\left(\frac{\mathrm{d}X}{\mathrm{d}t}\right) = p_{vs}\frac{M_v}{RT} D_v \left(\frac{\partial c_v}{\partial x}\right)\Big|_{x=X} \tag{6.51}$$

This equation expresses the mass balance between the water vapor flow across the interface due to evaporation, on the right side of Equation 6.51, and the advancement of the liquid–gas interface, on the left side. Accordingly, the evaporation–diffusion process is described by Equations 6.49, 6.50, and 6.51, actually forming a moving-boundary problem. The solution of this problem can be found in Li *et al.* (2009), and the drying front X is expressed as

$$X = 2\lambda_D \sqrt{D_v t} \tag{6.52}$$

with λ_D as constant related to concrete properties and drying actions. It is the solution of the following equation:

$$\lambda_D \mathrm{erf}(\lambda_D)\exp(\lambda_D^2) = \frac{1}{\sqrt{\pi}}\frac{p_{vs}}{\rho_1}\frac{M_v}{RT}(h_s - h_0) \tag{6.53}$$

where erf is the mathematical error function. Since the main variable in Equation 6.53 is the imposed humidity gradient $(h_s - h_0)$ the constant λ_D is regarded as a parameter related to the drying action. The drying front X is the main result of this evaporation–diffusion model. On the basis of this concept, the influential depth can be derived considering both drying and wetting actions. Note that the wetting action herein refers to the direct contact with liquid water.

For a drying–wetting cycle, note t_d the drying duration, t_w the wetting duration, and τ the time ratio between drying and wetting durations; that is, $\tau = t_d/t_w$. The aforementioned evaporation–diffusion model is used to describe the moisture transport during the drying, while the water absorption in Equation 6.46 is employed to express the water intake during wetting. Consider first an equilibrium situation that the moisture loss during drying duration I_d is equal to the water intake during wetting duration I_w. Thus, after each cycle, the pore water is totally restored. Using the preceding equations, this equilibrium is expressed as

$$I_d = I_w : \left(1 - s_l^0\right) X\left(t_d\right) = S_w \sqrt{t_w} \tag{6.54}$$

Using Equation 6.52, an equilibrium time ratio τ_s can be deduced:

$$\tau_s = \frac{t_d}{t_w} = \left[\frac{S_w}{2\lambda_D \sqrt{D_v}\left(1 - s_l^0\right)}\right] \tag{6.55}$$

Note that the equilibrium time ratio depends on both the intrinsic properties of concrete, D_v and S_w, and the action parameters, s_l^0 and λ_D. Using the equilibrium time ratio, the drying–wetting actions can be classified into three categories:

1. *Drying-dominated drying–wetting action.* The time ratio of drying–wetting cycle $\tau > \tau_s$; that is, the moisture loss is more than the water intake after one drying–wetting cycle. In consequence, the drying front goes deeper and deeper into the material and the drying front scales the influence depth of the moisture transport under drying–wetting cycles. The mass loss of water mass can be expressed as

$$\Delta I = 2\lambda_D \sqrt{D_v t_d}\left(1 - s_l^0\right) - S_w \sqrt{t_w} = \left(1 - s_l^0\right) dX \tag{6.56}$$

And after n cycles the influential depth X_n can reach

$$X_n = ndX = n\left(2\lambda_D \sqrt{D_v t_d} - \frac{S_w \sqrt{t_w}}{1 - s_l^0}\right) \tag{6.57}$$

2. *Equilibrium drying–wetting action.* The time ratio of drying–wetting cycle $\tau = \tau_s$; that is, the water loss and water intake is in equilibrium after each cycle. In this case, the influential depth of moisture transport is stabilized and equal to the drying front after each drying phase t_d:

$$X_n \equiv 2\lambda_D \sqrt{D_v t_d} \tag{6.58}$$

3. *Wetting-dominated drying–wetting action.* The time ratio of drying–wetting cycle $\tau < \tau_s$; that is, the water loss is less than the water intake after one cycle. Since the concrete is assumed to be saturated initially, the water intake during the wetting phase is only to restore the water mass loss during the drying. Thus, the influential depth of moisture transport is also stabilized and can be evaluated through Equation 6.58.

Further Analysis

Through the evaporation–diffusion model and the simple water absorption model, the influential depth of moisture transport under drying–wetting actions can be evaluated through Equations 6.57 and 6.58 following different drying–wetting patterns: drying-dominated, equilibrium and wetting-dominated drying–wetting cycles. The key parameter is the equilibrium time ratio in Equation 6.55. This expression can provide to engineers the first estimation of the depth of the convective zone near the surface of concrete elements. To obtain some quantitative results on the influential depth, one needs some supplementary laws for the parameters involved in Equation 6.55: the saturated vapor pressure p_{vs} from Equation 6.45, the diffusion coefficient of vapor in concrete $D_v = D_{va}\zeta$ (cm²/s) with the vapor diffusivity in air D_{va} (cm²/s) from Equation 6.44 and the tortuosity function ζ from Equation 6.38, and the moisture isotherm from Equation 6.31b. Figure 6.21 gives the constant λ_D in terms of the environmental

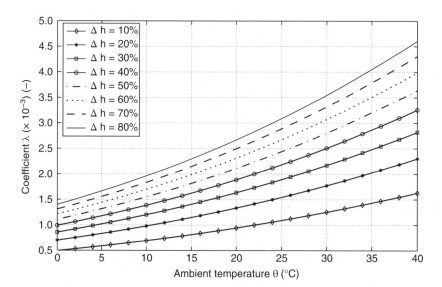

Figure 6.21 Influential constant λ_D in terms of drying gradient and ambient temperature. The parameters used are as follows: the perfect gas constant $R = 8.3147$ J/(K mol), the reference temperature $T_0 = 293.15$ K, the saturated vapor pressure at 20 °C $p_{vs} = 2.337$ kPa, the atmospheric pressure $p_{atm} = 101.325$ kPa, the water density at 20 °C $\rho_l = 1000$ kg/m³, the molar mass of vapor $M_v = 0.018$ kg/mol, concrete internal humidity $h_s = 100\%$, the drying humidity $h_0 = 20 - 90\%$, the capillary pressure coefficients $\alpha = 18.624$ MPa, $\beta = 2.275$. *Source:* Li *et al.* 2009, Fig. 1a, Reproduced with permission of Elsevier.

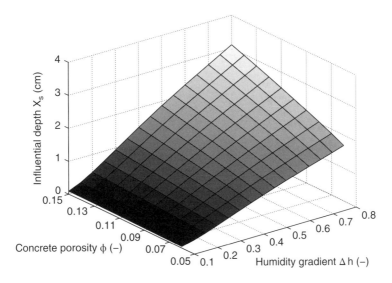

Figure 6.22 Stabilized drying front X_s in terms of the concrete porosity and the drying gradient with the same set of data in Figure 6.21. *Source:* Li *et al.* 2009, Fig. 4a, Reproduced with permission of Elsevier.

temperature and the drying humidity gradient. Figure 6.22 gives the stabilized influential depth in terms of the drying gradient for structural concretes with different porosities.

6.3.3 Moisture Transport under Drying–Wetting Actions

In Section 6.3.2 the moisture transport process is greatly simplified for the sake of an analytical analysis on the influential depth. To be more realistic, the moisture transport should take into account all the pore phases. Here, a multiphase model of moisture transport is introduced on the basis of mass transport theory of porous media, and this model is further adapted to include the drying–wetting actions discussed so far.

As discussed in Section 6.3.2, the moisture characteristic curve can adopt different paths for adsorption (wetting) and desorption (drying), and a different mechanism is involved as the concrete surface is in direct contact with liquid water. Let us define three moisture conditions: "drying condition" for concrete exposed to a drying humidity $h_0 < h(s_1)$, "humidifying condition" for concrete exposed to a higher humidity $h_0 > h(s_1)$, and "wetting condition" for concrete in direct contact with liquid water. The multiphase model, described in Equations 6.31a, 6.35, and 6.36 can be extended to account for these three conditions through different diffusivities. For drying and for humidifying conditions, the moisture diffusivity can use the expression in Equation 6.36 by adopting the respective characteristic curves in desorption and adsorption:

$$D_{dry,hum}\left(s_1\right) = -\frac{1}{\rho_1\phi}\left[\rho_1\frac{k}{\eta_1}k_{rl}\left(s_1\right) + D_{va}\zeta\frac{M_v\rho_v}{\rho_1RT}\right]\frac{\partial\left(p_c\right)_{des,ads}}{\partial s_1} \qquad (6.59)$$

Table 6.4 Parameters and properties for the moisture transport under drying–wetting action

Parameter/property	Value
Concrete w/c ratio	0.40
Cubic strength at 28 days (MPa)	57.1
Capillary porosity by gravimetry ϕ	0.141
Intrinsic permeability of concrete k (m/s)	3.0×10^{-21}
Dynamic viscosity η_1 (kg/(m s))	0.001
Diffusivity of vapor in air D_{va} (m²/s)	2.45×10^{-5}
Characteristic curve coefficient α (MPa) adsorption, desorption	6.02, 33.3
Characteristic curve coefficient β, adsorption, desorption	3.26, 2.77
Wall thickness (mm)	500

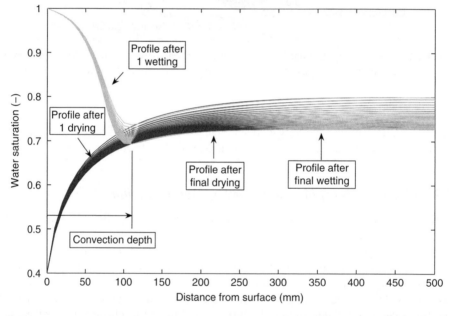

Figure 6.23 Moisture profiles for the concrete wall in the disposal facility: annual drying–wetting cycles with 30 days' wetting. *Source:* Pang *et al.* 2013, Fig. 12.5, Reproduced with permission of Springer.

For the wetting condition, the moisture diffusivity is converted from the sorptivity following Lockington *et al.* (1999) as

$$D_{wet}\left(s_1\right) = D_{wet}^0 \exp\left(n_w s_1\right) \tag{6.60}$$

with D_{wet}^0 the moisture diffusivity at $s_1 = 0$, n_w a model coefficient, taking a value of 6.0 for concrete materials (Leech *et al.*, 2003). The updated model consists of Equations 6.31a, 6.35, 6.59, and 6.60 with different moisture diffusivities for the three exposure conditions.

Case Study

A concrete wall in a near-surface disposal facility is investigated for the moisture transport. The structural concrete is characterized for the transport properties, given in Table 6.4. The concrete wall is not exposed to the atmosphere but buried in near-surface soil. Seasonal surface precipitation can infiltrate into the soil and the wall can be in contact with liquid water. Considering the low permeability of the soil in the disposal site, the liquid water flow is largely limited and the concrete surface is less likely to absorbed free water to a significant extent. Thus, the concrete surface is supposed to be exposed to alternate drying and humidifying conditions with an initial pore saturation of 0.8. The drying–humidifying cycle is taken as an annual event, and the humidifying duration is taken as 30 days per year. Figure 6.23 gives the modeling results for the moisture influential depth after 30 years of exposure.

From the results, the convection depth reaches 110 mm with a wetting period of 30 days. Deeper into the concrete wall, the water saturation profile follows a drying pattern with 30 days' wetting, indicating 30 days' wetting is not enough to compensate the water loss during the drying period. For durability consideration, the reinforcement steel bars should be protected against the risk of corrosion, otherwise a hydrophobic treatment of the wall surface in contact with soil can be considered.

Part Three

Durability Design of Concrete Structures

7

Durability Design: Approaches and Methods

As the first chapter in Part 3, this chapter gives a comprehensive framework for durability design of concrete structures. Fundamental concepts for durability design of concrete structures include the structural performance deterioration, durability limit states (DLSs), and service life. On the basis of these concepts a durability design can be achieved through prescription-based or performance-based approaches. The former provides directly the design requirements for materials and structural details, while the latter bases the design on the specified DLS and performance deterioration. The use of models in durability design is discussed in depth, with a comprehensive review of the strengths and weaknesses of the available durability models. Moreover, the uncertainty of modeling is especially addressed through deterministic and probabilistic design methods. Life-cycle engineering is recalled to illustrate the general context of durability design and life-cycle cost analysis (LCCA) is introduced as a quantitative tool for durability design in a life-cycle context. As a crucial part of durability design under life-cycle engineering, the maintenance design is treated for its principles and methods with an application case.

7.1 Fundamentals

The basic idea of durability design is illustrated in Figure 7.1. In the figure a certain structural performance f deteriorates with time from its initial value f_0. This performance, without loss of generality, is idealized as a decreasing curve with time. The acceptable level of this performance for structural users is denoted by a horizontal line on the figure, and this level line intersects with the decreasing curve at time t_s. The fundamental concepts for durability hence include the performance deterioration law $f(t)$, the DLS f_{DLS}, and the service life t_s. Whichever approaches and methods are used for durability design, one will always cope with these three fundamental concepts, which are to be detailed in the following.

Durability Design of Concrete Structures: Phenomena, Modeling, and Practice, First Edition. Kefei Li.
© 2016 John Wiley & Sons Singapore Pte. Ltd. Published 2016 by John Wiley & Sons Singapore Pte. Ltd.

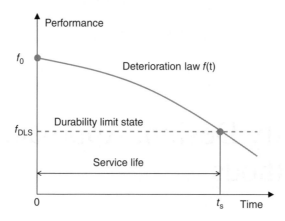

Figure 7.1 Conceptual illustration of durability design for concrete structures.

7.1.1 Performance Deterioration

Performance deterioration refers to the whole structure, structural elements, or the structural materials, depicting to what extent the durability processes affect the performance level. Compared with traditional structural engineering, performance has a broader meaning here: it includes the mechanical resistance of the structure, the serviceability conditions of the structural elements and structural materials, and esthetic aspects as well.

The durability deterioration processes are extensively investigated regarding the phenomena, mechanism, and modeling in the chapters in Part 1, and aspects related to the transfer to structural contexts were investigated in Part 2. Thus, the deterioration law of performance is assumed to be known for given service conditions.[1] Conceptually, the deterioration law can be expressed in terms of the service life t, service conditions $\mathbf{s}$, and the material properties $\mathbf{m}$:

$$f = f\left(\mathbf{s}, \mathbf{m}; t\right) \tag{7.1}$$

Note that the deterioration pattern induced by durability processes can be local or general. The local pattern refers to deterioration inducing only local damage to structural elements and the performance at the element or structural level is not necessarily affected. In this case, the performance deterioration in Equation 7.1 should better be defined for local damage. Steel corrosion induced by chloride ingress, freeze–thaw damage, and salts crystallization damage can have local deterioration patterns, while other durability processes, such as carbonation-induced steel corrosion, usually induce a general deterioration pattern; see Table 7.1.

7.1.2 Durability Limit States

Nowadays the design of concrete structures resorts to several limit states: the service limit states (SLSs) and ultimate limit states (ULSs). The SLS and ULS are respectively related to the acceptable levels for structural responses under service loadings and ultimate loadings

---

[1] The service conditions include both mechanical loadings and environmental actions.

Table 7.1 Durability performance and deterioration patterns

Durability processes	Performance deterioration	Deterioration pattern
Carbonation-induced steel corrosion	Steel corrosion	General
	Concrete cover cracking	Local
	Concrete–steel bond failure	General
	Element rigidity/resistance	General/local
Chloride-induced steel corrosion	Steel corrosion	Local
	Concrete cover cracking	Local
	Concrete–steel bond failure	Local
	Element rigidity/resistance	Local
Concrete deterioration by freeze–thaw actions/salts crystallization	Surface spalling	Local or general
	Element rigidity/resistance	Local or general
Concrete deterioration by leaching	Surface leaching	Local or general
	Element rigidity/resistance	Local or general

on structure (*fib*, 1999). The SLS refers usually to the deformation, cracking extent, sensitivity to vibration, or stress level of elements under service conditions, while the ULS refers to the stability failure of structures under extreme loading cases. The DLS corresponds to the acceptable level of the intended performance, subject to long-term deterioration processes. The DLS defines the extent to which the target performance is allowed to deteriorate at the end of the service life. Thus, the DLS, by definition, is related to both the deteriorated performance and the service life. Together with the SLS and ULS, these limit states provide the boundaries for design of concrete structures with respect to a target service life.

But, at which level do we address this DLS: structural materials, structural elements, or structure as a whole? The level at which the DLS is prescribed determines also at which level the durability design is to be performed. This level is actually decided by the deterioration patterns: if the deterioration is a local one, the DLS is limited to the affected materials; if the deterioration extends to elements or even the whole structure, the DLS corresponds to the affected element or structural performance. Thus, unlike SLS and ULS, the DLS is not always prescribed at the structure/element level. Usually, the durability processes undergo a long time span and the deterioration manifests during the service life, so the DLS is naturally close to the SLS. Thus, in most design cases the DLS can be regarded as a part of the SLS requirements for structure design; for example, the durability considerations are included in 'Serviceability' in the AASHTO code (AASHTO, 2004). This will facilitate the implementation of durability design into the standard design procedure, especially as the safety margin is to be specified for performance deterioration. Actually, the design with respect to the DLS can adopt the same reliability level as prescribed for the SLS.

However, the DLS can be nearer to the ULS than the SLS in some cases; for example, the corrosion of prestressed tendons. These tendons are made from high-strength steel wires and work under high stress level, usually above $0.8f_t$, during service life. In post-tensioned prestressed concrete structures, these tendons are primary elements, carrying the external loadings and constituting the structural rigidity. According to the crevice corrosion mechanisms, steel corrosion can be greatly enhanced and accelerated under a high stress level. To worsen the situation, these tendons are usually capsulated in impermeable ducts, making

Table 7.2   DLSs for different deterioration processes

Deterioration process	DLS	Design case
Carbonation-induced steel corrosion	(C1a) Specified carbonation depth	Principal elements High-stress wires Elements with high maintenance cost
	(C1b) Specified corrosion extent	Replaceable elements Elements with low maintenance costs
Chloride-induced steel corrosion	(C2a) Specified chloride ingress extent	Principal elements High-stress wires Elements with high maintenance cost
	(C2b) Specified corrosion extent	Replaceable elements Elements with low maintenance costs
Concrete deterioration by freeze–thaw	(C3a) Critical saturation state	Elements with spalling-free or cracking-free requirement
	(C3b) Specified spalling extent	Elements without spalling-free requirement, possibly superposed with other DLSs
Concrete deterioration by leaching	(C4a) Specified CH content for concrete	Elements with internal pH values required to remain above prescribed values
	(C4b) Specified leaching depth	Elements without pH requirements, possibly superposed with other DLSs
Concrete deterioration by salt crystallization	(C5a) Critical supersaturation degree	Elements with crack-free requirements
	(C5b) Specified spalling extent	Elements without crack-free requirements, possibly superposed with other DLSs

direct inspection impossible. All these factors lead to the situation that these tendons, once corroded, can induce rapidly the structural failure by rupture of steel wires in tendons. The collapse of eight walking bridges in the UK from 1967 to 1992 were reported and the investigation attributed the cause of failure to insufficient protection of prestressed tendons (BCS, 2002). Thus, the corrosion of high-stress steel wires and tendons should be considered more as a ULS than an SLS in the design, meaning a much higher safety margin should be adopted for the prescribed DLS.

On the basis of the mechanisms of deterioration processes in Chapters 1–5, the possible DLSs are proposed in Table 7.2 with the application for design cases. For each deterioration process, two DLSs are conceptually proposed with the first one, (Cxa), corresponding to an earlier stage (initiation) of deterioration, and the second one, (Cxb), to a certain extent of deterioration. Adopting the DLS of deterioration initiation provides a larger safety margin for durability performance of a structure than adopting the DLS of deterioration of specified extent. Meanwhile, the choice between (Cxa) and (Cxb) also involves cost considerations for the whole service life. This aspect is to be detailed later in Section 7.3.

Take chloride-induced steel corrosion for example. The (C2a) state corresponds to the accumulation of chloride ions to the critical level to initiate steel corrosion, and the (C2b) state corresponds to the corrosion developing to a specified extent after corrosion initiation. The designer decides to what extent this deterioration process is allowed to develop, involving structural safety and cost considerations. In practice, the (C2a) state is often adopted for the principal elements and the elements for which the maintenance works are difficult to perform, implying that allowing the corrosion to develop to some extent generates high maintenance costs. From the same consideration, some corrosion extent can be allowed for replaceable elements or elements that can be retrofitted with relatively low costs. The same concept applies to the other deterioration processes in Table 7.2, and the physical meaning of the (Cxa) and (Cxb) states can be found in the "design basis" sections in Sections 1.4, 2.4, 3.4, 4.4, and 5.4.

As one element is subject simultaneously to several deterioration processes, or to one deterioration process with very different exposure conditions, durability design should assign a DLS to each deterioration process or each exposure condition. Take the concrete element in Figure 5.1 as an example: the element is half-buried in salt-bearing soil and half-exposed to the atmosphere, salt crystallization affects the element to about 0.5 m above ground level, and the higher part is subject to carbonation. In this case two DLSs should be considered for this element: (C1a) for the carbonation part and (C5a) for the crystallization part can be used for high maintenance/replacement cost, and (C1b) and (C5b) for low cost. The situation will be more complicated as several processes act simultaneously on the same part of the reinforced concrete (RC) element. For example, those underground elements in contact with chloride-rich flowing water. These elements are exposed simultaneously to leaching by flowing water and chloride ingress. For this case, a conservative design is to superpose the requirements for concrete cover to resist respectively chloride ingress and leaching; for example, prescribe the cover thickness as the addition of the thickness satisfying (C2a,b) and the thickness satisfying (C4a,b). Usually, the required thickness by leaching is regarded as a sacrificial thickness in design. The less conservative design necessitates considering the interaction between leaching and chloride ingress in the same range of concrete cover.

7.1.3   Service Life

The service life of concrete structures can be defined as the duration in which the structural performances remain above the expected levels, usually prescribed by the corresponding DLSs. The life span illustrated in Figure 7.1 can be regarded as the design life for structures under design or as an acquired life span for existing structures. Nowadays, the "service life" has become a standard terminology, but bears different names under different contexts. For design purposes this term is often called "design service life" or "design working life"; for assessment of existing structures the term becomes "actual service life" or "service life." According to ISO 2394 (ISO, 1998), this term is defined as the "*assumed* period for which a structure or a structural element is to be used for its intended purpose without major repair being necessary." The standard ISO 13823 (ISO, 2008) gives a similar definition: the "*actual* period of time during which a structure or any of its components satisfy the design performance requirements without unforeseen major repair." Evidently, ISO 2394 refers to structural design and ISO 13823 pertains to structural assessment.

Apart from the different contexts, a major difference exists in the two definitions and merits further discussion: how the service life is related to the maintenance or repair works. According to ISO 2394, the service life should be regarded as terminated once a major maintenance occurs. This definition is more adapted to concrete structures mildly exposed to environment actions, such as building structures. A major repair means the main structure is to be retrofitted and the corresponding function of structure ceases. However, this definition is less adapted to structures heavily exposed to environmental actions and service loadings. For roads and bridges, the asphalt pavement can best endure a life span of 20 years due to the combined actions of traffic loads and natural ageing of the pavement. Thus, during the expected life span of roads and bridges, usually more than 50 years, major maintenance occurs surely to replace the pavement. This applies actually to all complex structures with elements having a life span much shorter than the expected life of the whole structure. In contrast, the definition from ISO 13823 changes the limitation on major maintenance to "without unforeseen major repair"; that is, the maintenance actions, including major repairs, can be included into the service life of concrete structures as long as these actions have been planned at the design phase. Accordingly, the service life of concrete structures was alternatively defined as "the duration of time span for structures during which all intended functions are satisfied under expected loading and environmental actions, and expected maintenance actions" (CNS, 2008).

Service Life: Technical or Economical?

So far, the term "service life" is only addressed from the technical aspect, determined by the performance deterioration law and the DLS; see Figure 7.1. But, in the design procedure, the service life (or design working life) is a specified value and one of the most important design targets. In this sense, the specification of service lives in the design phase surpasses the technical aspect, involving also the economic aspect, the legal aspect, and possibly the ecological aspect. The technical aspect defines the valid duration during which all intended requirements can be technically realized with or without human intervention (by maintenance works); the economic aspect treats the benefit and cost of a concrete structure and provides the economic cycle of the structure as a part of the economic body; the legal aspect considers the social constraints on the operation of structures; the ecological aspect will emphasize the ecological impact of concrete structures in the local region. In design practice, the owner's requirement on service life is often on an economic basis. The final specification of service life in design is a result of a synthetic analysis of the aforementioned aspects. Certainly, during the design of a particular structure the value of design life is often adopted from design codes and standards, and will not go as far as the above analysis. But, the specification of different service lives in design codes does need to consider the aforementioned aspects. Table 7.3 recapitulates the specification for design service lives for concrete structures in different sectors from some active codes and standards.

Example (specification of design service life for concrete structures in ports)
The background of this specification is to introduce the expected service life as an explicit design target into the design codes for concrete structures in marine ports in China. The research started in 2007 and ended in 2011, from which rational value ranges for service life were proposed for concrete structures and elements in ports and served as basis for relevant design codes. To this purpose, the economic life, the legal constraints, and the technical working life from in-situ investigations were synthetized. The economic analysis showed that the coastal regions of China maintain a steady marine transportation needs in long term, and recommended the service life to be situated after the initiation and expansion phases but before the declined phase of the economic cycle of ports. The investment return periods were calculated for several major ports on the Chinese coast, including Yan-tian Port (Shenzhen),

Table 7.3 Specification of service lives from design codes and standards

Structure	Classification	Service life (years)	Source
Building	Temporary	1–5	ISO (1998), CEN 2002)
	Industrial buildings	30	CCES (2005)
	Residential buildings	50	ISO (1998), CEN (2002)
	Monumental buildings	>100	ISO (1998), CEN (2002)
Bridge	Small to medium bridges, viaducts over normal urban roads	30–50	CCES (2005)
	Highway bridges	75	CSA (2000)
	Large bridges, large-scale viaduct, urban light rail transit, viaduct over express roads	100	CCES (2005)
	Highway and railway bridges	120	BSI (2003)
Port	Open piled jetties, mooring and berthing, ro-ro link spans	15–30	BSI (2000)
	Reclamation, shore protection, breakwaters, gravity quay walls	50	BSI (2000)
	National or regional port, flood defense, coastal management infrastructure	100	BSI (2000)
Tunnel	Road tunnels	120	HWA (1999)
	Road tunnels	150	FHWA (2009)
Nuclear energy	Structures for nuclear power plants	30–40	NEA (2000)
	Radwaste containers	>300	IAEA (1999)

Shanghai Port (Shanghai), Tian-jin Port (Tianjin), and Ying-kou Port (Ying-kou), on the basis of annual incomes during 2005–2007, the basic discount rates, and the return on total assets. The return period was evaluated as 27 years for Shanghai Port. Thus, the minimum economic life of ports was proposed as 30 years. The technical life of concrete structures in ports in Chinese coastal regions was based on the investigation of 59 ports constructed during 1956–2008. Considering the evolution of design codes and concrete technology, the technical life of concrete structures in ports was evaluated as 30–50 years, depending on the structure types; for example, high-pile wharfs usually have shorter service lives, while gravity quay walls have longer lives. Considering the economic life, the technical life as well as other constraints, the basic value for design service life of concrete structures in marine ports was proposed as 50 years.

Service Life: for Whole Structure or Elements?

Another crucial aspect of service life is the level of its specification: for the structure as a whole or for a constitutive element. So far, there is no difference in the term "service life" on this point. From the two cited definitions from ISO 2394 and ISO 13823, the differences between the specification levels are to some extent blurred; literally, the term refers to both the structure and the elements. This point must be clarified for the durability design and the life-cycle engineering discussed later in this chapter. A rational understanding would be as follows: a concrete structure is composed of multiple structural elements, and in the design phase a service life is first specified and attributed to the whole structure; then taking this service life as a global design target, the constitutive elements are designed in such a way that this global service life can be ensured. Certainly, the design service lives will also be assigned to the constitutive elements so that the design of an elements has its respective target. This assignment of design service lives to elements, on the basis of the whole service life, may be rather straightforward for a simple structure but can be less evident for a complex one. Normally, the principal elements and sheltered elements that are not maintainable must be designed to have the same service life as the whole structure. Owing to technical constraints some elements may not reach the service life of the whole structure, and these elements should be designed as replaceable or maintainable. Conceptually, the service life of the whole structure is to approach the minimum of service lives of constitutive elements without maintenance, and the service life of the whole structure can approach the maximum of the service lives of elements with proper maintenance planning. Taking the service life, either at the whole structure level or at the element level, as a design target, a sufficient safety margin should be adopted for this target. Following the general reliability levels established in terms of the cost of safety measures and the importance of failure consequence, a set of reliability indices has been proposed by ISO 2394 (Table E.2), giving a first specification for the safety margin of the service life as a design target. This aspect is to be detailed in the next section.

7.2   Approaches and Methods

7.2.1   Objectives

The basic task of durability design is to assure the service life of the structure through the design procedure. Traditionally, the design procedure of concrete structures can be divided into conceptual design, preliminary design, and detailed design. The conceptual design makes fundamental decisions related to the form and the nature of the structure; for example, functionality, esthetics, environmental/ecological compatibility, and economy. The design in this phase is rather qualitative but has a decisive impact on the subsequent phases. The preliminary design involves the dimensioning of the structure at a general level and the most preferred alternative is determined, including a check for the SLS, ULS for loadings, and DLS for environmental actions. In the detailed design that follows, the design on all structural details is performed, including the detailed drawings, reinforcement and formwork drawings, as well as the specification for types and qualities of building materials. Subsequently, the structure goes into the construction phase and service phase (life). Since concrete materials are hardened in-place during the construction phase, the durability quality depends greatly on the concrete technology used in the construction. Thus, it is crucial to ensure that concretes in situ achieve the expected quality levels, usually specified in the design phases. During the service phase of the structure, maintenance is important to keep the performance levels above the expected or regulated levels, including the expected DLS.

To ensure the service life for the structure, these phases, from conceptual design to the service phase, are equally important. Thus, the durability design adopts a broader sense, enveloping design, construction, and maintenance during the whole life cycle of the concrete structure. Following this broader sense, the objectives and contents of durability

Table 7.4   Contents for durability design of concrete structures in different phases

Phase	Design content/task
Conceptual design	Determination of service life and environmental actions
Preliminary design	Choice of structural alternatives with lower intensity of exposure, better access to inspection and maintenance, and rational life-cycle cost Determination of the intensity of environmental actions Assignment of service lives to elements Durability design strategy for elements (replaceable or permanent)
Detailed design	Specification for constitutive materials (concrete, steel) and quality Requirements on the quality of concrete cover to steel bars Crack control of elements Waterproof and drainage Additional protection measures for elements Maintenance planning for different element groups
Construction	Avoidance of cracking at early age (by curing scheme) Quality control of in-situ concrete
Maintenance	Inspection and monitoring of durability performance Maintenance actions

design in different phases of the life cycle can be detailed as follows, and are summarized in Table 7.4:

- *Conceptual design phase.* The fundamentals of durability design are determined, such as service life and the environmental actions.
- *Preliminary design phase.* The study should favor the structural alternatives with lower intensity of environmental actions, better access to inspection and maintenance, and rational life-cycle cost. The design service lives should be assigned to structural elements on the basis of the service life of the whole structure. The elements are determined to be designed as replaceable or permanent elements. The type and intensity of environmental actions should be determined for elements.
- *Detailed design phase.* Detailed durability design is performed at the element level. On the basis of the design lives, the designer should specify, for replaceable elements and permanent elements, the constitutive materials of elements, the required quality for bulk concrete and steel, the quality requirements for concrete cover to steel bars, the crack control of concrete elements, and the waterproof and drainage considerations. For permanent elements with a long design life or elements under severe environmental actions, additional protection measures should be specified. A basic maintenance scheme should be set up for the different groups of elements.
- *Construction phase.* Sufficient curing should be effectuated on concrete elements, and shrinkage cracking at an early age should be avoided. The quality of the in-situ structural concrete should be demonstrated to have reached the expected quality level.
- *Service phase.* The durability performance of structural elements should be inspected and, if necessary, monitored. The maintenance should be executed following the established scheme, and can further be updated by inspection/monitoring results to keep the performance above the expected levels (DLS).

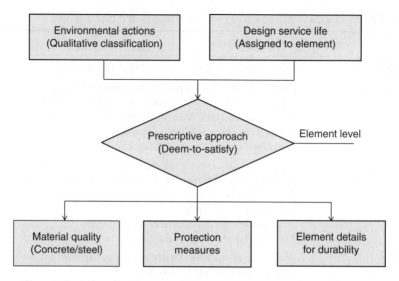

Figure 7.2   Flow chart for prescriptive approach at structural element level.

7.2.2   Global Approaches

Traditionally, two basic approaches have been proposed for durability design of concrete structures: a prescriptive approach and a performance-based approach. The prescriptive approach, also called the "deem-to-satisfy" approach, provides the material proportioning and composition, structural details, and protection measures with respect to the environmental actions, and eventually the expected service life. For concrete materials these requirements include the w/c (w/b) ratio, cement type and content, chloride content, air content (for freeze–thaw environment), and mineral admixtures. Since durability has for a long time been regarded as a material-level problem, most design codes and standards rely on these requirements to ensure the service life of structure. The flowing chart for prescriptive approach is given in Figure 7.2 for concrete elements. The basis of this approach is the acquired experiences on the durability performance of existing concrete structures. Because experience is usually insufficient to support quantitative requirements, most of the requirements for durability are formulated in a qualitative and empirical way. The strength of this approach is its flexibility to account for experience and the easiness for use in design works. The weakness of this approach is also evident: from limited experience it is always insufficient to support all requirements so that some requirements have to be extrapolated from the available data by reasoning, and even less could these prescriptions be demonstrated to ensure the expected service life. Moreover, the experience-based prescriptions can also potentially hinder the use of novel materials in durability design.

In contrast, the performance-based approach performs the design in such a way that the structure can satisfy its intended functions, formulating the requirements for durability design on an application-dependent basis; that is, with respect to the environmental actions and the expected service life. The flow chart is given in Figure 7.3. This approach does not specify the constitutive materials of structural elements or the process of these materials but gives the requirements on the structural responses with respect to the service actions.

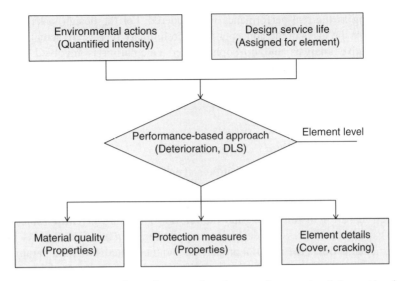

Figure 7.3   Flow chart for performance-based approach at structural element level.

The performance-based approach is widely used in the conventional design procedure of concrete structures. Whether taking into account the actual response of the structure or not is the major difference between the prescriptive and performance-based approaches. Take carbonation-induced steel corrosion, for example. The prescriptive approach puts requirements on concrete proportioning, such as maximum w/b ratio, cement content, and mineral admixture content, to assure a good concrete mixing and placing, and a minimum strength class and sufficient concrete cover thickness to assure the protection of reinforcement steel bars. The performance-based approach will choose the deterioration performance in Table 7.1 and the corresponding DLS in Table 7.2, and then establish the relation between the environmental CO_2 actions and the resulting service life of structural elements. Accordingly, the deterioration law is the central issue in the performance-based approach. The strength of the performance-based approach is its pertinence to the durability responses, and the service life can be achieved in a more scientific and reliable way. The challenge of this approach is that the deterioration law of durability performance needs support from both scientific research and long-term in-situ observations. Moreover, the uncertainty associated with the deterioration law should be correctly accounted for in the design process.

Since the prescriptive approach involves qualitative analysis and the performance-based approach needs to quantify the deterioration law, the prescriptive approach is sometimes referred to as "qualitative" and the performance approach as "quantitative." This division is not accurate and is misleading, since the difference between two approaches lies in considering the actual response of structure, not in the nature, qualitative or quantitative, of durability requirements. For example, if on the project site concrete structures exist that achieved or could be proved to achieve the target service life, then the durability design of new structures can be achieved through rational analysis of the durability measures taken in these existing structures. This procedure, though qualitative, is actually performance based since the response of the structure is considered. The methods used in the different phases of durability design can be different, and these methods all belong to the performance-based approach as long as the real performance is considered. For the design tasks in Table 7.4, a large number of the tasks in the conceptual and preliminary phases are actually effectuated through qualitative analysis and no sophisticated calculation is involved. Quantitative analysis is used more for tasks in the detailed design phase. So, the qualitative and quantitative methods are both useful for durability design, and are rather complementary in the whole design procedure.

7.2.3   Model-based Methods

In the performance-based approach, scientific and engineering models of durability are often needed to evaluate the deterioration law for performance, and these methods are termed model-based methods. The available models for performance deterioration were comprehensively treated in Part 1 of this book, and the relevant DLSs are given in Table 7.2. These models are expected to express the deterioration performance $f(t)$ in Equation 7.1 in terms of the environmental actions, the material properties, and the related structural parameters.

The choice of models for durability design is the central issue for the model-based method. A model is by nature a simplified vision of the facts, and the model is considered to be usable for durability design following two basic criteria. First, the model should be physically correct, considering the main influential factors from environmental actions and material properties. In this sense, the mechanism-based scientific models are always preferred to the empirical models from simple data regression. The disadvantage of an empirical model lies in the fact that its validity depends solely on the scope of data from which the model is regressed. As the design case is out of the regression scope, the prediction from the empirical model has to be made through extrapolation, leading to the design being on the unsafe side. Second, the uncertainty of the model should be quantified and mastered. Two kinds of uncertainly coexist in durability models: uncertainty arising from randomness of the physical parameters involved is called "inherent uncertainty" (e.g., the fluctuation of atmospheric temperature/humidity); uncertainty from model defects is termed "artificial uncertainty" (e.g., some crucial mechanisms misunderstood or neglected in the model). For inherent uncertainty the statistical properties of design variables should be taken into account, while the artificial uncertainty can only be decreased by better knowledge of the deterioration processes. Table 7.5 gives the first analysis on the strength and weakness of the available durability models presented so far.

Depending on how the uncertainty of models is considered, the model-based method, following the flow chart in Figure 7.3, can adopt deterministic or probabilistic formats. To illustrate the two formats, the deteriorated performance in Equation 7.1 is further expressed as

$$f\left(\mathbf{s},\mathbf{m};t\right) = R\left(\mathbf{m};t\right) - P\left(\mathbf{s};t\right) \tag{7.2}$$

where R stands conceptually for the resistance to the performance deterioration, expressed in terms of the material properties $\mathbf{m}$ and time t; P idealizes the structural effects of environmental actions $\mathbf{s}$, also expressed in terms of t. The dependence of P on time t reflects mainly the evolving nature of the deterioration by environmental actions, including also the evolution of environmental actions themselves (e.g., the change of CO_2 concentration in the atmosphere for carbonation-induced steel corrosion). Both the resistance R and the effect P are assumed

Table 7.5 Strength and weakness analysis of available models for durability deterioration

Deterioration	Performance	Model	Model type	Uncertainty		Model reliability
				Inherent	Artificial	
Carbonation-induced steel corrosion	Carbonation depth	C-1	Scientific	n.a.	Low	++
		C-2	Empirical	Mastered	Moderate	+++
	Corrosion initiation	C-3	Empirical	n.a.	Moderate	++
	Corrosion current	C-4	Empirical	n.a.	High	+
Chloride-induced steel corrosion	Chloride ingress	Cl-1	Scientific	n.a.	Low	++
		Cl-2	Empirical	Mastered	Moderate	+++
	Corrosion initiation	Cl-3	Empirical	n.a.	Moderate	++
	Corrosion current	Cl-4	Empirical	n.a.	High	+
Freeze–thaw damage	Critical pore saturation	FT-1	Empirical	n.a.	High	+
	Pore stress	FT-2	Scientific	n.a.	High	+
Leaching	Leaching depth	L-1	Scientific	n.a.	High	+
		L-2	Scientific	n.a.	Moderate	++
Salt crystallization damage	Critical supersaturation	CT-1	Scientific	n.a.	Low	++
	Pore stress	CT-2	Scientific	n.a.	High	+

+++: very reliable; ++: reasonably reliable; +: poorly reliable.

to be described quantitatively by relevant models. In the deterministic format, the durability design is to assure (for an expected service life t_{SL})

$$f\left(\mathbf{s},\mathbf{m};t_{SL}\right) \geq 0 : R\left(\mathbf{m};t_{SL}\right) \geq P\left(\mathbf{s};t_{SL}\right) \tag{7.3}$$

But this equation does not consider the model uncertainty. In other terms, the design results from $R\left(\mathbf{m};t_{SL}\right) = P\left(\mathbf{s};t_{SL}\right)$ provides a guarantee probability of only 50% for Equation 7.3. Thus, a safety margin should be introduced into the design equation to consider the uncertainty of the models. To this purpose, partial factors can be adopted, and Equation 7.3 becomes

$$\frac{1}{\gamma_R} R\left(\frac{m_i}{\gamma_i^m};t_{SL}\right) \geq \gamma_P P\left(\gamma_j^s s_j;t_{SL}\right) \quad \text{with} \quad \gamma_R,\gamma_i^m,\gamma_P,\gamma_j^s > 1.0 \tag{7.4}$$

where γ_R and γ_P are respectively the partial factors for resistance and loading effect; γ_i^m and γ_j^s are partial factors for material properties m_i and environmental action s_j. To gain a safety margin for design results, all the partial factors adopt values greater than 1.0. The determination of the partial factors involves both subjective and objective aspects. The objective aspect refers to the consideration of the statistical properties of the material properties **m** and environmental actions **s**. Actually, the partial factors can be calibrated for an expected reliability level through structural reliability theory if the statistical properties are known for **m** and **s** (*fib*, 1999). The subjective aspect relates to the artificial uncertainty of the models used in design. The factors γ_R and γ_P can reflect the importance of elements and the failure consequence; that is, the values can be increased for the principal and permanent elements to obtain larger safety margins if the designer feels it necessary.

The probabilistic format considers the randomness of parameters, through their statistical properties, in an explicit way. The design equation (Equation 7.3) holds for the probabilistic format, and the design variables, **m** and **s**, are all random variables with statistical properties. Here, a further assumption is made regarding the variables: the time dependence of the statistical properties of **m** and **s** is neglected, and the time t becomes a simple variable in the design equation (Equation 7.3). In probabilistic format, the failure probability p_f is calculated for the design function:

$$p_f(t) = P\big[f(\mathbf{s},\mathbf{m};t) < 0\big] = P\big[R(\mathbf{m};t) < P(\mathbf{s};t)\big] \qquad (7.5)$$

and the service life t_{SL} corresponds to a prescribed value for the target failure probability p_{fs}:

$$p_f(t_{SL}) = P\big[f(\mathbf{s},\mathbf{m};t_{SL}) < 0\big] = p_{fs} \qquad (7.6)$$

The specified value for failure probability p_{fs} is basically decided by the social perception of structure failure considering its consequences, the expense, and the ethical considerations. Since most DLSs are rather close to SLS requirements, the durability design can adopt the values of failure probability similar to those prescribed for the SLS design. Taking an annual failure probability of 0.13% for irreversible processes in the SLS and the corresponding reliability index $\beta_1 = 3.0$ (*fib*, 1999), the failure probabilities $p_{fs} = 6.7\%$ and 12.2% can be recommended for the reference period of 50 years and 100 years.

7.3 Life Cycle Consideration

In the previous sections, the durability design of concrete structures is detailed for its objectives, contents, and fundamentals, as well as for its approaches and methods. Basically, the choice of DLS and the associated safety margin assigned to the DLS, in terms of partial factors or failure probability, will determine the acquired durability of the concrete structures during their service life. However, these fundamental indices, such as DLS or failure probability associated with the DLS, cannot be determined by durability design itself, and resort must be made to some top planning at the whole structure level for the whole span of the service life. Actually, the choice of these indices comes from two considerations: maintaining the structural performance above an expected level (DLS for durability performance), and maintaining a reasonable financial balance among the different life-cycle phases of conception, design, construction, service, and eventually demolition. The choice of a conservative

DLS and high safety margin at the design phase usually leads to higher construction costs but lower maintenance costs during service. On the contrary, adopting a less conservative DLS and low safety margin reduces the construction costs but augments the maintenance costs. The concepts of life-cycle engineering can help to optimize the choice of these indices for durability design.

7.3.1 Fundamentals for Life-cycle Engineering

Life-cycle engineering refers to the technical activities and methods used to ensure the functionality of products or systems with consideration to the environmental, ecological, financial, and cultural impacts during the full life cycle. This concept stems from the industrial sectors in 1960s which optimized the production procedure of industrial units by minimizing the total cost during the full life cycle. Today, this concept has extended to other engineering fields, including civil engineering, and some tools are available to implement this concept into the design process, such as life-cycle costing (LCC) (Woodward, 1997), the integrated life-cycle design method (Sarja, 2002), and life-cycle management (PIANC, 1998). For concrete structures, the full life cycle consists of the conception (design) phase, construction phase, service phase, and demolition or reuse phase. The requirement of life-cycle engineering is to ensure the structural performance through the full life cycle under the financial, ecological, social, and cultural constraints. One important concept, brought by life-cycle engineering, is the structure should no longer be regarded as an element of civil engineering but more broadly as a part of a financial, ecological, or social and cultural unit within its surroundings.

Again, life-cycle engineering is quite conceptual and its requirements can be very dependent on the application cases. For concrete structures, the fundamentals of life-cycle engineering can be interpreted from two aspects: the required performance of structure, and the constraints. The required performance includes the structural functions to which the structure is designed for, and the quality level with respect to which these functions are satisfied. The constraints refer to the financial, environmental, ecological, social and cultural issues related to the construction and the use of the structure. Moreover, these two aspects should be put in the context of full life cycle of the structure. Figure 7.4 illustrates the conceptual contents of life-cycle engineering applied to concrete structures.

For durability design of concrete structures, the concepts of life-cycle engineering can help to determine the fundamental indices (e.g., DLS and the associated safety margin), considering both the performance and the constraints. For example, a conservative DLS and larger safety margins are preferred for concrete structures with social and cultural importance. For conventional structures, these two indices can be optimized through financial costing methods considering the costs and benefits generated during the full life cycle. The assignment of service lives to structural elements can also resort to the constraints; for example, the financial constraint can help to decide in the choice between short-lived but replaceable elements, and long-lived but expensive elements.

7.3.2 Life-cycle Cost Analysis

The most widely used tool in life-cycle engineering is the costing method, LCCA or LCC. The method uses the financial cost to optimize the design alternatives or options, transferring

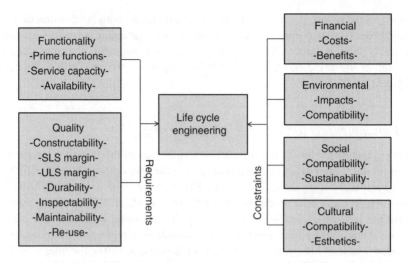

Figure 7.4 Concepts for life-cycle engineering for concrete structures.

the life-cycle engineering concepts into a cost-based optimization problem. The basic approach is to establish the total cost of the structure for its whole service life t_{SL}:

$$C_{T} = C_{const} + \sum_{t=0}^{t_{SL}} \frac{C_{m}+C_{u}+C_{in}}{\left(1+r_{disc}\right)^{t}} + \frac{C_{dem}}{\left(1+r_{disc}\right)^{t_{SL}}} \qquad (7.7)$$

where C_{T} is the full life-cycle cost, C_{const} is the construction cost, C_{m} is the maintenance cost, C_{u} is the user cost during maintenance, C_{in} is the indirect cost during service, and C_{dem} is the demolition cost. The term r_{disc} is the discount rate for cash flow, a financial factor used to evaluate the present value of future cash flow.

The cost in the construction phase C_{const} includes the construction and installation costs, costs of equipment and utilities, as well as the financial cost of loans and interests, and the starting time of cost analysis is placed implicitly at the end of construction. The cost in the service phase consists of maintenance cost C_{m}, the user cost C_{u} caused by the maintenance, and other indirect costs C_{in} (e.g., from adverse environmental impacts). The cost in the demolition phase C_{dem} refers to technical costs of structure removal and the costs related to the recovery of the project site following ecological or environmental requirements. Note that these future costs need to be converted into their present values through a discount rate r_{disc}. The discount rate in the ith year from the reference time (e.g., end of construction) can be expressed through (Cady, 1983)

$$r_{disc,i} = \frac{r_{soc,i} - f_{i}}{1+f_{i}} \qquad (7.8)$$

where $r_{soc,i}$ is the social discount rate converting the future cost/benefit of an investment to its present value from the perspective of the whole social economic growth, f_{i} is the annual changing rate of the producer price index. Financially, the discount rate $r_{disc,i}$ reflects to what extent the future value is discount as present value. However, it can also be taken as a subjective indicator for the partition of financial responsibility between the present generation and future generations.[2] Actually, if this rate is high then future intervention has little effect on the total cost in Equation 7.7; that is, the present decision has more impact and the connection between the generations can be weak. On the contrary, future intervention will have larger effects on total cost if the rate is low; that is, the future decision has more impact and the connection among the generations will be strong.

The central idea of LCCA is to use the full life-cycle cost C_{T} in Equation 7.7 as the global target function and the design alternatives are selected by optimizing the cost C_{T}. It can have two major applications in design: selection of alternatives in the phase of preliminary design, and maintenance planning in the phase of detailed design; see Table 7.4. Using total cost C_{T} as an optimizing target, the selection of design alternatives A_i can be expressed as

$$\min_{A_i}\left[C_{\mathrm{T}}\left(A_i\right)\right]=\min_{A_i}\left[C_{\mathrm{c}}\left(A_i\right)+\sum_{t=0}^{t_{\mathrm{SL}}}\frac{C_{\mathrm{m}}\left(A_i\right)+C_{\mathrm{u}}\left(A_i\right)+C_{\mathrm{in}}\left(A_i\right)}{\left(1+r_{\mathrm{disc}}\right)^{t}}+\frac{C_{\mathrm{dem}}\left(A_i\right)}{\left(1+r_{\mathrm{disc}}\right)^{t_{\mathrm{SL}}}}\right] \quad (7.9)$$

with the costs for different alternatives determined. In the detailed design phase, the design alternative has been determined and the design focuses on the materials, section detailing, and protection of elements. The maintenance planning in this phase consists of determining the protection measures and establishing the maintenance scheme for structural elements during the service life. Assuming C_{c} and C_{dem} in Equation 7.7 are not affected by maintenance works, the optimized maintenance scheme M_{opt} can be obtained through optimizing the total costs of C_{m}, C_{u}, and C_{in}:

$$M_i \rightarrow M_{\mathrm{opt}} : \min_{M_i}\left[\sum_{t=0}^{t_{\mathrm{SL}}}\frac{C_{\mathrm{m}}\left(M_i\right)+C_{\mathrm{u}}\left(M_i\right)+C_{\mathrm{in}}\left(M_i\right)}{\left(1+r_{\mathrm{disc}}\right)^{t}}\right] \quad (7.10)$$

The application of LCCA merits more discussion. As a useful tool to implement life-cycle engineering, the LCCA method has its strong point and weak point, both stemming from the postulation of total cost optimization in Equation 7.7. The strong point of LCCA is its capacity to solve the selection of design alternatives and maintenance planning quantitatively through optimizing the relevant costs. The concepts behind Equations 7.9 and 7.10 are rather straight-forward, and can be adapted to different applications. However, the weak point lies also in the fact that the selection procedure is converted into a merely cost-related problem, and some costs, like indirect cost C_{in}, are difficult to quantify in an unbiased way. In short, LCCA can be helpful only if the relevant costs are determined correctly.

7.3.3 Maintenance Design

Maintenance design or planning is one major task in the detailed design phase, concerning both replaceable elements and permanent elements. For the replaceable elements, the maintenance is a mandatory action for the element replacement. For the permanent elements, the

---

[2] Personal communication with Professor Bruce Ellingwood, 2014.

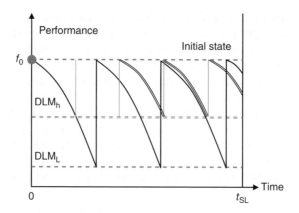

Figure 7.5   Illustration for DLM and intervention frequency for maintenance.

maintenance is necessary to maintain their performance level above the expected DLS. For concrete structures with very long design service lives (e.g., more than 100 years), it occurs that the uncertainty associated with the durability design can by no means be reduced to a satisfactory level, or the safety margin cannot be demonstrated in the design phase, due to the inherent uncertainty or artificial uncertainty. Within this context, maintenance during service becomes indispensable to reduce this uncertainty and helps the structure to achieve the service life. In this sense, maintenance is not an auxiliary action to structural design but an efficient tool to reduce the uncertainty left in the design phase and to help achieve the expected service life.

Maintenance design consists of establishing a life-cycle planning of maintenance actions with respect to the in-situ evolution of the deteriorated performance. Maintenance planning should take into account three basic elements: at which deterioration state the maintenance actions are to be undertaken (deterioration level for maintenance), what techniques are to be used for maintenance (maintenance techniques), and how the maintenance actions are to be organized at the whole structure level (maintenance scheme). These three aspects are detailed in the following.

Deterioration level for maintenance (DLM). The concept of deterioration level is rather near the concept of DLS: both prescribe the acceptable level for the performance deterioration with DLS used in the design phase and DLM for maintenance planning. For a given deterioration law, a lower DLM allows the deterioration to develop to a more advanced stage before the intervention, while a higher DLM limits the deterioration to an earlier stage before maintenance. Usually, the lower DLM leads to fewer interventions during the service life and a higher DLM induces more interventions; see Figure 7.5. But the DLM corresponding to optimized total maintenance cost remains to be evaluated. In the probabilistic format of durability design, the DLM can be represented by the failure probability associated with a certain DLS.

Usually, three DLMs can be considered for maintenance planning: preventive level, necessary level, and mandatory level. The preventive level corresponds to an early initiation stage of deterioration processes (e.g., (C1a)–(C5a) in Table 7.2). The intervention at this level necessitates usually maintenance works with low cost and has only a slight or no influence on the structural functions. However, sophisticated equipment is necessary to detect the early

Table 7.6 Maintenance techniques for concrete and reinforcement steel according to the state-of-the-art of techniques

Objective	Technique	Protection duration (years)	Cost
Concrete	Surface silane impregnation	15–20	+
	Surface coating	10–15	+
	Surface polyuria layer	15–20	+
	Permeable crystallization treatment	—	+
	Electrochemical salt extraction	—	+
	Cracking sealing	—	+
	Reconstruction of concrete cover	—	++
Steel	Surface protection by epoxy resin	—	++
	Replacement of corroded steel bars	—	++
	Cathodic protection (artificial anode)	20	+++
	Cathodic protection (imposed current)	50	++++

+: lowest cost; ++++: highest cost.

stages of deterioration, generating additional costs for maintenance. For steel corrosion the DLM is recommended as a failure probability of 2% associated with (C1a) and (C2a) in Table 7.2. The necessary level corresponds to the deterioration developed to an allowed extent, (e.g., (C1b)–(C5b) in Table 7.2. The mandatory level corresponds to such a deterioration stage that intervention should be performed otherwise the relevant regulations on safety will be violated. The corresponding deterioration is much more advanced than the DLS prescribed in Table 7.2, and normally the mechanical resistance of elements has been affected. It is a rather late stage for intervention, in response to the structural deterioration in a passive manner.

Maintenance techniques. The maintenance techniques are the support for the maintenance planning, aiming to recover the deteriorated performance to its initial level. Table 7.6 summarizes the maintenance techniques to concrete and reinforcement steel according to the state-of-the-art in construction techniques. The later the intervention is performed, the more sophisticated are the techniques involved in maintenance works. Table 7.6 also provides indicative protection durations for some techniques. These values are based on the state of the art of the maintenance techniques, and closely related to the workmanship. Note that maintenance and repair are active technical fields, and novel techniques and materials are updated rapidly. This is reason that maintenance planning should always be treated with a certain flexibility, especially for structures with very long design lives, because the planning can only be based on the techniques available in the design phase. One cannot predict to what extent the maintenance techniques can evolve during the service life of structures.

Maintenance scheme. For sophisticated structures, the maintenance planning will be confronted with the situation where different element groups have quite different deterioration rates. The practical problem is how to organize the maintenance activities for the different element groups. A natural strategy is to treat the different element groups separately and perform the intervention for each group once its deterioration reaches the expected DLM, in what is termed an unsynchronized scheme, as shown in Figure 7.6a. The disadvantage of this strategy is that the interventions for different elements are dispersed during the service life and will induce more user costs by more total intervention times. Another strategy is to

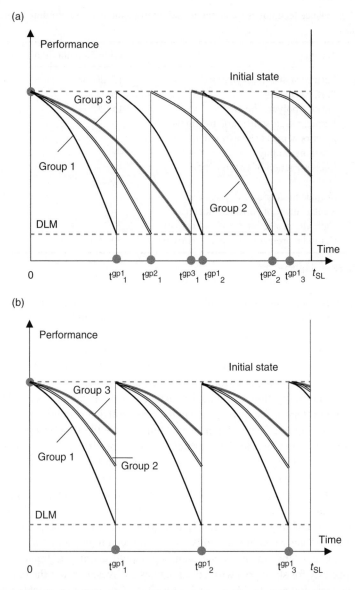

Figure 7.6 (a) Unsynchronized scheme and (b) synchronized scheme for maintenance planning.

synchronize the maintenance for all element groups, and the intervention time is determined by the most deteriorated group; see Figure 7.6b. The advantage of this strategy is the reduced total intervention times but the less deteriorated groups are subject to maintenance before their expected DLM. Certainly, other schemes can also be established to optimize the total maintenance cost.

On the basis of the aforementioned three basic elements, the maintenance planning can be established. Note the available DLM as $R_D = DLM_i$ ($i = 1, m$), the corresponding maintenance techniques as $R_{\mathrm{MT}} = \mathrm{MT}_i^j$ $(i = 1, m; \; j = 1, n)$, and the possible maintenance schemes as $R_{\mathrm{MS}} = \mathrm{MS}_k$ $(k = 1, l)$. The maintenance planning can be performed through optimizing the maintenance-related costs in Equation 7.10:

$$M_{\mathrm{opt}} \in R_{\mathrm{D}} \times R_{\mathrm{MT}} \times R_{\mathrm{MS}} : \sum_{t=0}^{t_{\mathrm{SL}}} \frac{C_{\mathrm{m}}\left(M_{\mathrm{opt}}\right) + C_{\mathrm{u}}\left(M_{\mathrm{opt}}\right) + C_{\mathrm{in}}\left(M_{\mathrm{opt}}\right)}{\left(1+r\right)^t} \to \min \qquad (7.11)$$

The maintenance planning is illustrated for a marine wharf having a design service life of 100 years following the method in Equation 7.11.

Study Case: Container Wharf of High-driven Piles

Structure of Container Wharf
The port of the container wharf was built in 2002 as the extension part of an international hub port, located on the southeastern coast of China. The wharf measures 1250 m long by 54.5 m wide, including four modern container berths with annual throughput capacity of 1.8 million TEU (twenty-foot equivalent units). The structure of the wharf is a RC beam–slab system supported by high-driven piles. The expected service life of the wharf is 100 years. A vertical view of the concrete structures is illustrated in Figure 7.7, and the description of the elements is given in Table 7.7. The structure is exposed to a northern subtropical monsoon climate: the annual average temperature is 15.7 °C, the annual average relative humidity is 79%, and the average annual rainfall is 1124 mm. On the basis of hydrological data, the exposure zones of RC elements are classified as immersed zone, tidal zone, splash zone, and atmospheric zone; see Figure 7.7. The case study aims to establish cost-effective maintenance planning for the RC elements in the wharf.

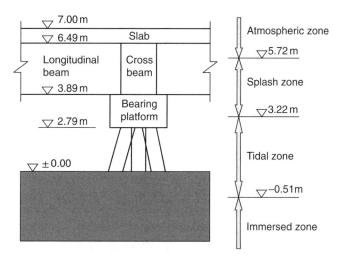

Figure 7.7 Illustration of RC elements in different exposure zones in the container wharf. *Source:* Yang *et al.* 2013, Fig. 1. Reproduced with permission of Springer.

Table 7.7 RC elements in the container wharf

Element	Dimensions (m)	Surface exposed to splash/tidal zones (m²)	Number of elements
Pile	0.6 × 0.6 × 50	3.96	2685
Bearing platform	2.7 × 2 × 1.1	15.74	2046
Longitudinal beam (rail)	7 × 1.1 × 2.6	44.10	356
Longitudinal beam (middle)	7 × 0.8 × 1.8	30.80	892
Longitudinal beam (front)	7 × 0.5 × 1.09	18.76	178
Longitudinal beam (rear)	7 × 0.7 × 1.8	30.10	178
Cross-beam	54.5 × 1.0 × 2.1	283.4	179

Source: Yang *et al.* 2013, Table 1. Reproduced with permission of Springer.

Performance Deterioration of Reinforced Concrete Elements

The control deterioration process for the RC elements is chloride-induced corrosion of reinforcement steel. The DLS is chosen as the steel corrosion initiation. The Cl-2 model in Equation 2.19 is retained for the chloride ingress and the performance function for deterioration is written

$$G = C(x,t) - C_{crit} = \left(C_0 + (C_S - C_0) \left\{ 1 - \text{erf} \left[\frac{x_d}{2\sqrt{D_{Cl}^0 (t_0/t_s)^n t}} \right] \right\} \right) - C_{crit} \leq 0 \quad (7.12)$$

In this equation, $C(x,t)$ (%binder mass) is the chloride content in concrete, C_{crit} (%binder mass) is the critical chloride content to initiate the corrosion, x_d (mm) is the concrete cover thickness, t (years) is the service age; C_0 (%binder mass) is the initial chloride content in concrete, C_S (%binder mass) is the surface chloride content, D_{Cl}^0 (mm²/year) is the chloride diffusion coefficient of concrete at age of t_0, n is the ageing exponent of chloride diffusion coefficient, and t_s is the ageing duration of chloride diffusion coefficient, taken as 30 years. Equation 7.12 is used to calculate the failure probability of RC elements under chloride ingress through Monte Carlo simulations, and the parameters and their statistical properties are given in Table 7.8.

Maintenance Levels and Techniques

Three DLMs are defined in terms of the failure probability with respect to the performance function (Equation 7.12). The corresponding maintenance techniques are summarized for different intervention levels in Table 7.9, in which the cost is normalized by the price of 1 m³ of commercial concrete of C30 grade.

Maintenance Planning by Life-cycle Cost Analysis

The cost model is simplified, and only the operation costs directly related to the maintenance actions and the user costs are considered in the maintenance cost C_M:

$$C_M = \sum_{i=1}^{m} \frac{C_{mi}}{(1 + r_{disc})^i} + \sum_{i=1}^{m} \frac{C_{ui}}{(1 + r_{disc})^i} \quad (7.13)$$

Table 7.8   Statistical characteristics of parameters in deterioration model

Parameter (unit)	Distribution	Mean value μ	Standard deviation
Initial chloride content C_0 (%binder)	Lognormal	0.1	0.01
Surface chloride content C_s (% binder)	Lognormal	3.9	0.39
Concrete cover x_d (mm)	Lognormal	46 (68) [75][a]	0.1μ
Critical chloride content C_{crit} (%binder)	Lognormal	0.50	0.05
Chloride diffusion coefficient at 28 days D_{Cl}^0 (mm²/year)	Lognormal	63 (158) [158][a]	0.1μ
Ageing exponent n	Lognormal	0.42	0.084

Source: adapted from Yang *et al.* 2013, Table 2, Table 3. Reproduced with permission of Springer
[a] The three values represent respectively the driven piles (cross and longitudinal beams) and [bearing platforms].

Table 7.9   Failure probability and operation costs for different intervention levels

DLM	Techniques	Operation cost	Service disturbance
Preventive, $p_f = 2\%$ (Level-1)	Chloride extraction	0.91	No disturbance
Necessary, $p_f = 5\%$ (Level-2)	Chloride extraction Surface treatment	1.06	Limited disturbance, berth service maintained
Mandatory, $p_f = 20\%$ (Level-3)	Cover reconstruction Steel Supplementation	1.82	Stoppage of berth service for 90 days

Source: Yang *et al.* 2013, Table 4. Reproduced with permission of Springer.

and a stepwise function is adopted for the discount rate: $r_{disc,i} = 2.0\%$ (0–30 years), 2.7% (31–60 years), 3.3% (61–90 years), and 4.0% (>90 years). The daily loss due to stoppage of berthing is evaluated as 2424, calculated from the annual lease price of the port divided by 365 and the commercial price of 1 m³ of structural concrete of C30 grade. The user cost is calculated from this daily loss in the maintenance durations. Three maintenance schemes are studied: (1) the unsynchronized scheme, performing maintenance for each element group (piles, beams, and bearing platforms) once the expected DLM for this group is reached; (2) the synchronized scheme, performing the maintenance at the same instants once the most deteriorated elements (beams) attain the DLM; (3) coating-synchronized scheme, applying coating on the beam elements and delaying the intervention time for beams so that the three groups of elements can reach their DLMs ($p_f = 2\%$, 5%, 20%) at nearly the same instant. The coating cost is taken as 0.04, and the protection duration of surface coating is conservatively taken as

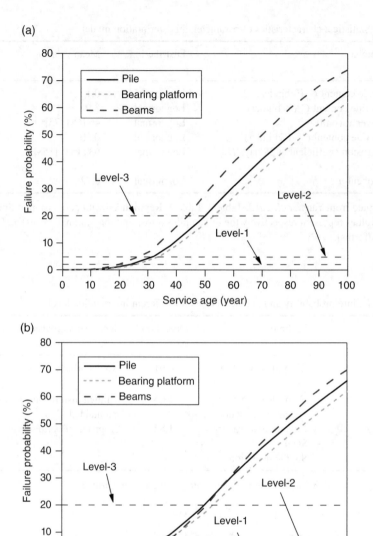

Figure 7.8 Failure probability of corrosion initiation for three groups of elements without coating (a) and with coating (b). *Source:* Yang *et al.* 2013, Fig. 2, Fig. 5. Reproduced with permission of Springer.

7 years. Figure 7.8 shows the failure probability with respect to Equation 7.12 for three groups of elements without and with coating protection on beams. The evolution of failure probability for the three maintenance schemes is provided in Figure 7.9 and the total maintenance costs are given in Table 7.10.

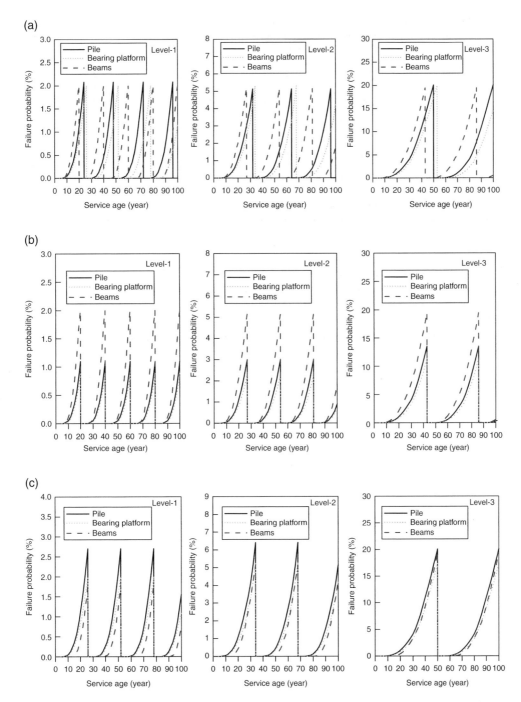

Figure 7.9 Evolution of failure probability for (a) unsynchronized scheme, (b) synchronized scheme, and (c) coating-synchronized scheme. *Source:* Yang *et al*. 2013, Fig. 3, Fig. 4, Fig. 6. Reproduced with permission of Springer.

Table 7.10 Maintenance costs of different maintenance schemes (Yang *et al.* 2013)

Maintenance scheme	DLM	Interventions (elements)	Maintenance cost C_M (10^3)
Unsynchronized	Preventive	4 (beam, pile), 3 (platform)	208.1
	Necessary	3 (beam, pile), 2 (platform)	168.6
	Mandatory	2 (beam), 1 (pile, platform)	371.8
Synchronized	Preventive	4 (beam, pile, platform)	242.0
	Necessary	3 (beam, pile, platform)	181.9
	Mandatory	2 (beam, pile, platform)	245.4
Coating-synchronized	Preventive	3 (beam, pile, platform)	169.1
	Necessary	2 (beam, pile, platform)	140.9
	Mandatory	1 (beam, pile, platform)	172.6

Case Summary

Three observations can be obtained from the results in Table 7.10: (1) the DLM at the necessary level is always the most cost-effective case because the preventive DLM induces more frequent interventions and the "mandatory" DLM generates very significant user costs due to berth service interruption; (2) the coating-synchronized scheme, regardless of DLMs, gives the most optimized organization for maintenance, and the coating delays the deterioration of beams so that the maintenance can be performed for three element groups at nearly the same time; (3) the coating-synchronized scheme with a "necessary" DLM achieves the most optimized maintenance cost, and thus is recommended for the container wharf.

8

Durability Design: Properties and Indicators

This chapter focuses on the material properties related to the durability of concrete structures. The deterioration of structural performance depends on both the environmental actions and the material properties. Accordingly, the material properties are central parameters for the performance deterioration laws, including the chemical properties (e.g., CH, C-S-H contents), electrochemical properties (e.g., corrosion resistance of steel), physical properties (e.g., diffusivity and permeability of concrete), and mechanical properties (e.g., fracture strength). Some properties, such as concrete strength, can be readily measured in the laboratory, while the measurement of other properties is more difficult. In some cases, the pertinent properties can only be deduced indirectly; for example, the ion and moisture diffusivities in unsaturated concrete. This is the very reason to extend the term "durability properties" to "durability indicators," including any material characteristic reflecting the material resistance to deterioration. These characteristics can be performance test results in the laboratory using artificial environmental actions.

This chapter first reviews the basic properties related to the deterioration processes, followed by some fundamental relationships among the different properties of concrete. Then, the characterization methods for these properties are introduced, together with the performance tests for certain deterioration processes. The uncertainty, repeatability, and reproducibility associated with these tests are also analyzed. Finally, the durability indicators are investigated for the most probable sets of material properties and characteristics with respect to each deterioration process, and some recent results in this regard are reviewed.

8.1 Basic Properties for Durability

The material properties presented are limited to the deterioration processes from Chapters 1–5: carbonation and carbonation-induced corrosion, chloride ingress and chloride-induced corrosion, freeze–thaw damage, leaching, and the salt crystallization process. The properties involved are introduced in the following for the chemical properties, the microstructure and related properties, the transport properties and the mechanical properties. The chemical properties refer to the chemical and mineral composition in concrete and the electrochemical properties of reinforcement steel; the microstructure and related properties include the pore structure of concrete, the pore-structure-related properties of concrete, and the characteristics of the concrete–steel interface; the transport properties involve the permeability of fluids and diffusivity of moisture and ions in concrete as a porous medium.

8.1.1 Chemical Properties

CH and C-S-H Contents

The portlandite (CH) and calcium silicate hydrates (C-S-H) are the main hydration products of cement and secondary cementitious materials (SCM). These hydrates determine the chemistry of pore solution and maintain the alkaline environment in concrete. Since the pH value of pore solution is crucial for the electrochemical stability of reinforcement steel, the CH and C-S-H contents are crucial to the carbonation-induced and chloride-induced corrosions of reinforcement steel. Meanwhile, the CH and C-S-H contents are pertinent properties of concrete against leaching. Normally, CH content of hardened cement pastes (HCPs) depends on the w/c ratio and the cement composition. The HCP made from Portland cement has a CH content of about 20%, and this value can be less than 10% for binder incorporating a large fraction of SCM (Taylor et al., 2013). Concretes having a CH content above 20% (mass ratio of binder) are considered to have good durability potential against the deteriorations mentioned earlier (AFGC, 2007).

Calcium Equilibrium

The calcium equilibrium between the liquid solution and the solid phases is fundamental for the chemical stability for concrete since this equilibrium assures the stability of C-S-H, and thus the mechanical resistance of structural concretes. This equilibrium can be simply written as

$$m_{Ca}^s = m_{Ca}^s \left(c_{Ca} \right) \tag{8.1}$$

with m_{Ca}^s and c_{Ca} being solid calcium content and Ca^{2+} concentration in the aqueous phase. This equilibrium is particularly useful for the leaching process, during which the aqueous Ca^{2+} is transported to the external environment from concrete pores, promoting the dissolution of CH and C-S-H dissolution into pore solution. The detailed mechanism can be found in Chapter 4.

Corrosion Resistance of Steel

The corrosion resistance of reinforcement steel depends on both the steel composition and the chemistry of the pore solution. The resistance to corrosion initiation can be described by the Pourbaix diagram of iron in aqueous solution with or without chloride ions; see Figures 1.11 and 2.11. As an engineering approach, the resistance of steel against carbonation-induced corrosion is related to the pH value of the pore solution surrounding the steel, and this property is further simplified as the carbonation depth of concrete; that is, the corrosion is assumed to be initiated as the carbonation occurs in the pore solution surrounding the steel – see Section 1.3. The resistance of steel corrosion against chloride-containing electrolytic solution is expressed as the concentration ratio between chloride (Cl⁻) and hydroxide (OH⁻); see Section 2.3. For a pore solution with a certain pH value, the maximum chloride concentration resisted by steel passivation is called the critical chloride concentration (content), a property enveloping steel composition and pore chemistry for corrosion resistance.

8.1.2 Microstructure and Related Properties

Porosity

Porosity is defined as the void space in concrete occupied by fluid phases of gas and liquid. The concrete pores include the structural pores in cement hydrates (gel pores), the interstitial space left by the filling of hydration products (capillary pores), and the trapped or intentionally entrained air during mixing (air voids). The porosity is related to both the mechanical and transport properties. The pores can be connected, isolated, or occulted in solid phases, and the interconnected pores (e.g., capillary pores) are assumed to be determinant for transport properties. Normally the porosity of structural concretes is below 20%, and nowadays the porosity of compact concretes can decrease below 6%. Figure 8.1 collects recent literature data on the porosity of structural concretes in terms of their compressive strength. The porosity of concrete can be controlled by the water-to-binder ratio (*w/b*) in mixtures, and the porosity of structural

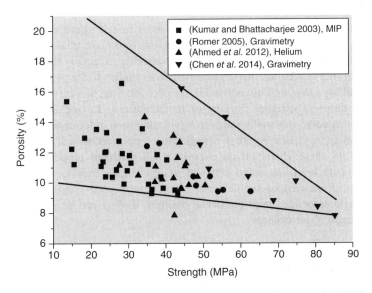

Figure 8.1 Literature data on the concrete porosity in terms of compressive strength. "MIP": measurement by mercury intrusion porosimetry; "Gravimetry": porosity evaluation by weight loss through drying; "Helium": measurement by helium pycnometry.

concrete can contain air voids up to 3%. A high fluidity of the concrete mixture tends to reduce the air void content in the hardened concrete. Usually, a concrete with porosity below 10% can be assumed to be compact, providing high mechanical resistance and low transport rates.

Pore Structure

The pore structure of concrete refers to the geometry of the pore network, including at least the total porosity, pore size distribution, and pore connectivity. Nowadays, it is recognized that the concrete pores assume a multiscale nature, ranging from C-S-H interlayer pores at the nanometer level, capillary pores at nanometer–micrometer level, and air voids at micrometer–millimeter scale. The relative portion of these pores of different sizes and their connectivity determine the transport path for external agents. Since the real geometry of the pore network is very complex, the characterization of pore size distribution always needs a pre-assumption on the pore geometry. Another geometry parameter for pore structure is the tortuosity of connected pore path, defined as the ratio between real percolating distance between two points and their straight-line distance. The tortuosity is also difficult to quantify at the pore level but can be deduced from some macroscopic transport properties; for example, by electrical resistance measurement through Archie's law (Archie, 1942). A particular aspect of pore structure related to freeze–thaw resistance is the average spacing factor of air voids in concrete. Following the mechanisms of pore freezing in Chapter 3, this spacing factor can be regarded as a pertinent indicator for freeze–thaw resistance of concrete; that is, concrete can be assumed exempted from freeze–thaw damage with a spacing factor less than 250 μm.

Characteristic Curve for Phase Changes

These characteristic curves of phase change in pores refer to pore drying, freezing, and crystallization. Owing to the interface energy arising from these phase changes, the phase change processes are actually related to the pore structure, especially the pore size distribution. For pore drying or wetting, this characteristic curve refers to the pore saturation in terms of external humidity $s_l(h)$ (see Equation 6.25); for pore freezing, this curve pertains to the pore ice content in terms of external freezing or temperature $s_C(T)$ (see Figure 3.7); for pore crystallization this curve is actually the pore liquid saturation in terms of salt supersaturation in pore solution $s_L(z/z_{sat})$; see Figure 5.7. Following a poromechanical description (Coussy, 2010; Chapter 9), these characteristic curves are all pertinent properties to evaluate the pressure at the pore level associated with the phase changes; that is, $p_c = p_c(s_{l,C,L})$. These characteristic relations envelop the detailed information of pore structure, interface energy, and contact angle associated with the phase changes, so they can be regarded as pertinent properties for these phase changes.

Properties of Concrete–Steel Interface

The interface between concrete and steel plays a central role in the corrosion process of steel and the subsequent mechanical damage of concrete cover by corrosion products. The interface actually refers to the steel surface and cement paste, and the resulting wall effect leads to a larger local *w/b* ratio and thus higher porosity and larger CH content compared with the bulk cement paste and concrete materials (Horne *et al.*, 2007). Larger porosity tends to facilitate the corrosion initiation by aggressive agents like chlorides transported to the steel surface, while the higher CH content seems to counteract this effect by maintaining the pH level of the pore solution. It was also shown that defects at larger scale (on the order of a centimeter) have clearly adverse effects on steel corrosion (Sandberg, 1998). These defects of the interface are the results of concrete segregation and bleeding, and promote the steel corrosion by macro-cell corrosion mechanisms. Accordingly, the compactness or porosity (voids) of the concrete–steel interface can be a pertinent property for the steel corrosion processes in RC elements.

8.1.3   Transport Properties

Water Permeability

Water is the most important transport medium for deterioration processes in concrete, and also the indispensable reactant for corrosion, freezing, and crystallization. The water permeability depicts the rate of water transport under a pressure gradient through the pores of concrete, regarded as one of the key properties related to durability. The water permeability can be defined, following Darcy's law, from the relationship between water flow rate J_D (m/s) and the gradient of water pressure p_w (Pa) or water head h_w (m):

$$J_D = -\frac{k}{\eta_w}\,\mathrm{grad}\left(p_w\right) \text{ or } J_D = -k_S^w\mathrm{grad}\left(h_w\right) \tag{8.2}$$

with k_w (m²) the intrinsic permeability of concrete, k_S^w (m/s) the saturated permeability of concrete for water, and η_w (kg/(m s) or Pa s) the dynamic viscosity of water; see Equation 6.43.

The water permeability of concrete is sensitive to the pore structure, especially the connected capillary pores, and the saturated water permeability was recorded between 10^{-16} and 10^{-12} m/s for structural concretes (El-Dieb and Hooton, 1995). In practice, concretes can be regarded as impermeable materials and used for storage and sealing purposes if the permeability is low enough, e.g. reaching the order of 10^{-11} m/s. Figure 8.2 summarizes the literature data on water permeability measured by different authors in terms of concrete strength and porosity. It seems that the water permeability correlates better to porosity than to strength. The water permeability of concrete is controlled by both the *w/b* ratio and the maximum size of aggregates: the former determines the pore structure in bulk cement pastes and the latter conditions the interface transition zone between the cement paste and aggregates (Mehta and Monteiro, 2006). The water permeability of concrete is also highly dependent on the pore saturation, and the relative water permeability is defined as the ratio between the water permeability at a certain pore saturation s_l and the saturated water permeability. Owing to the wide use of SCM and superplasticizer in concrete technology, concrete materials can be very compact today and adopt low values of permeability, which poses a practical problem for the characterization of water permeability in the laboratory.

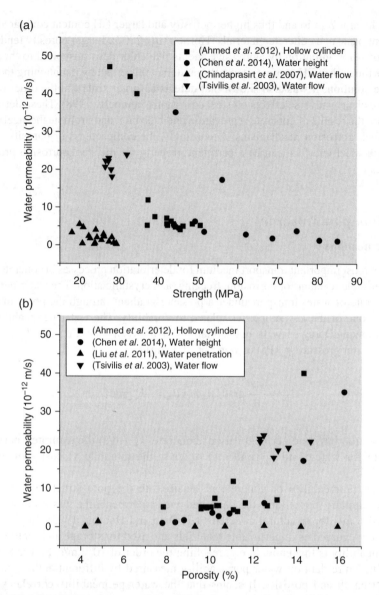

Figure 8.2 Literature data on the water permeability of concretes in terms of (a) compressive strength and (b) porosity. "Hollow cylinder," "water height," "water penetration," and "water flow" correspond to the experimental setups from different sources.

Gas Permeability

Like water permeability, gas permeability depicts the flow rate of gas through concrete pores under a pressure gradient. Gas flow is necessary for concrete carbonation (CO_2) and cathodic reaction of steel corrosion (O_2). The gas permeability can be deduced from Darcy's law considering the compressibility of gas and is written as (Dullien, 1992)

$$J_D\left(P_{out}\right) = -\frac{k_g}{\mu_g}\frac{P_{out}+P_{in}}{2P_{out}}\,\text{grad}\left(p_g\right) \tag{8.3}$$

with P_{in} and P_{out} (Pa) the inlet and outlet pressures of gas, k_g (m²) the gas permeability, and μ_g (Pa s) the dynamic viscosity of gas. Owing to boundary slip during gas flow through the pores, this gas permeability is found to be pressure dependent (Klinkenberg, 1941):

$$k_g = k_g^0\left(1+\frac{b_K}{P_m}\right)\text{ with }P_m = \frac{P_{in}+P_{out}}{2} \tag{8.4}$$

where b_K (Pa) is the Klinkenberg coefficient and k_g^0 is the intrinsic permeability for gas flow corresponding to P_m approaching infinity.

The gas permeability, as a property, refers usually to the permeability measured for all the connected pores of concrete; that is, the pore saturation equal to zero.[1] Theoretically the intrinsic permeability depends only on the pore structure of concrete; thus, the gas intrinsic permeability k_g^0 in Equation 8.4 should be equal to the water intrinsic permeability k_w in Equation 8.2 in total saturation. However, the literature data showed that $k_g^0 \gg k_w$, and k_g^0 is two or three orders higher than k_w (Baroghel-Bouny, 2007). The difference was attributed to the different flow modes of gas and liquids through concrete pores, or the active portion of pores is different for gas and liquid flows. Note that the gas permeability, once considering gas viscosity in Equation 8.3, becomes independent of the type of gas. The gas used for permeability measurement for concrete can be nitrogen, oxygen, or dry air, and nitrogen is preferred owing to its smaller molecule size and inert property with regard to concrete pore phases. Reported gas permeability k_g^0 ranges from 10^{-19} to 10^{-16} m², and concretes with gas permeability lower than 10^{-17} m² can be considered as having high compactness. Figure 8.3 summarizes the literature data on gas permeability of concretes, measured by the CemBureau method (Kollek, 1989) in the dry state, in terms of strength and porosity. The influence factors of concrete materials on k_g^0 are essentially the same as those for water permeability.

Gas Diffusivity

The gas phase can also transport in concrete pores under a concentration gradient, and the gas diffusivity can be expressed through Fick's law:

$$J_F^g = -D_g\left(s_1\right)\text{grad}\left(c_g\right) \tag{8.5}$$

---

[1] The definition of zero pore saturation is a theoretical state since liquid water in the nanopores of concrete cannot be dried out without disturbing the original microstructure. Usually, a rule of thumb in the laboratory is to dry the concrete specimens at low temperature (50–60 °C) until constant weight (mass change rate less than 0.1% during 1 week) with the stabilized state taken as the totally dry state of concrete pores.

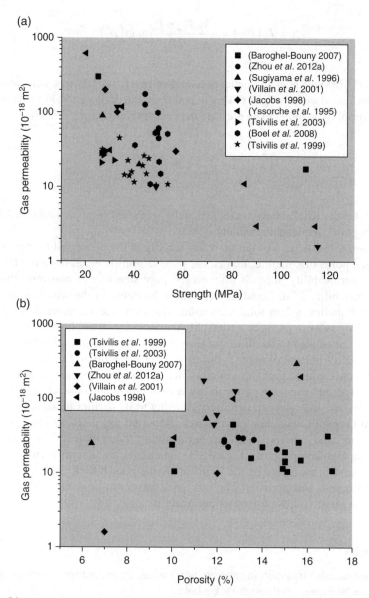

Figure 8.3 Literature data on the gas permeability of concretes in terms of (a) compressive strength and (b) porosity. The data were all obtained by the CemBureau method.

with c_g the gas concentration in the gaseous mixture. Apart from the pore structure, the gas diffusivity also depends on the pore saturation, since the liquid-occupied pore space cannot provide a path for gas diffusion. By nature the gas diffusivity is also temperature dependent due to the fact that gas molecules move faster at higher temperature. The diffusivity of CO_2 and O_2 in concrete can be estimated through models from the literature (see Equation 1.3 or 1.4) as on the order of 10^{-8}–10^{-7} m²/s in totally dried pores of concrete.

Moisture Diffusivity

Moisture refers to water in both the gas and liquid phases. Both phases transport simultaneously through concrete pores, obeying the thermodynamic equilibrium between the two phases. One global way to treat moisture transport without referring to the different water phases is to write Fick's law for moisture flow J_F^m (m/s) directly in terms of the pore saturation gradient:

$$J_F^m = -D_m(s_1)\,\mathrm{grad}(s_1) \tag{8.6}$$

with D_m (m²/s) the moisture diffusivity. The moisture diffusivity can be expressed further in terms of the moisture characteristic curves, $s_1 = s_1(h)$ or $p_c = p_c(s_1)$, through multiphase modeling of moisture transport in concrete materials; see Equation 6.36. The reported values of moisture diffusivity are in the range 10^{-15}–10^{-9} m²/s, and a substantial difference was noted for the drying and wetting processes (Leech *et al.*, 2003). The moisture diffusivity depends also on the pore saturation, and its value at total saturation, $s_1 = 1.0$, can be taken as a reference value.

Ion Diffusivity

The ions transport through the concrete pores via the aqueous pore solution. The ion diffusivity is defined as the coefficient relating the gradient of the ion concentration c_i to the diffusion flow of this ion J_F^i. The diffusivity can be described by Fick's law:

$$J_F^i = -D_i(s_1)\,\mathrm{grad}(c_i) \tag{8.7}$$

with D_i (m²/s) the diffusivity of ions in concrete. This diffusivity depends on the pore structure and the pore saturation of concrete. Noting D_i^0 is the diffusivity in the saturated state, and the following expression is proposed (Saetta and Scotta, 1993):

$$D_i(s_1) = D_i^0(\phi) f_{\mathrm{pore}}(s_1) \tag{8.8}$$

with the function $f_{\mathrm{pore}}(s_1)$ being the influence of pore saturation on the ion diffusivity; see Equation 2.3. The saturated diffusivity of chloride ions in concrete is normally between 10^{-13} and 10^{-10} m²/s, and a concrete having a chloride diffusivity on the order of 10^{-12} m²/s is usually considered to have a high resistance to chloride ingress. The literature data on chloride diffusivity of concrete are summarized in Figure 8.4 in terms of compressive strength and porosity.

Water Sorptivity

Water can also transport into concrete through a surface absorption. This process is more related to liquid water flow in the connected pores open to the concrete surface. Although water absorption is also due to pore capillary pressure, the same reason for moisture diffusion in pores, the absorption rate is far more important than the diffusion process. This phenomenon

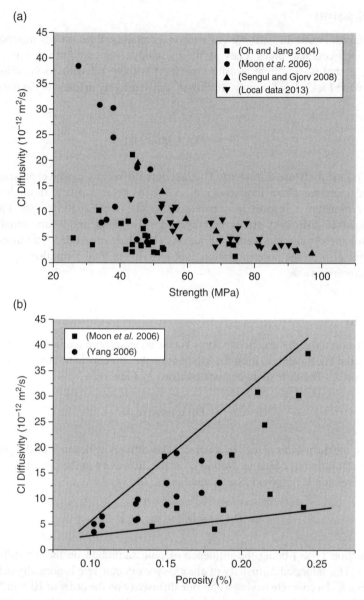

Figure 8.4 Literature data on the chloride diffusivity of concretes in terms of (a) compressive strength and (b) porosity. The data were all obtained by NT Build492 (RCM) method. The source (Local data 2013) refers to local laboratory tests on high-performance concretes of C45 grade used in a marine environment with different binders and for different curing ages.

occurs only as a dry concrete surface comes into contact with liquid water, and the absorption rate actually reflects the water intake capacity and kinetics of the concrete surface. This absorption rate is regarded as a pertinent property to durability because surface drying–wetting is considered to be the most severe action for multiple deterioration processes. In practice, the water sorptivity S_w is defined as the ratio between the absorbed water mass (volume) on unit area and the square root of absorption time, adopting the dimension $[M/L^2T^{1/2}]$ ($[L/T^{1/2}]$); see Equation 6.46. Figure 8.5 summarizes the literature data on water sorptivity, measured by different authors in laboratory for initially dried concretes, in terms of compressive strength and porosity of concrete. Note that the data in the figure refer to the water sorptivity in the first hours of water absorption, and the range 0.005–0.035 mm/s$^{1/2}$ (1.5–10.3 mm/day$^{1/2}$) is noted for the collected data. Relatively good correlation is observed for sorptivity–strength and sorptivity–porosity.

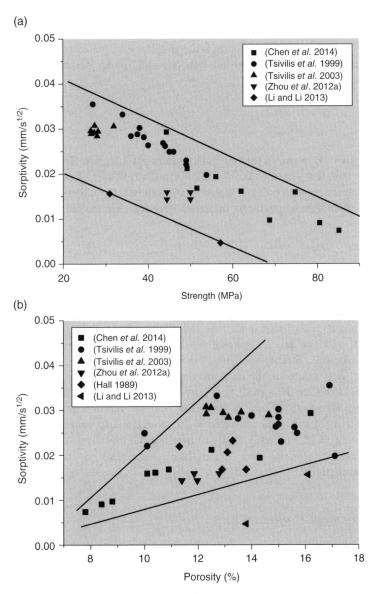

Figure 8.5 Literature data on the water sorptivity of concretes in terms of (a) compressive strength and (b) porosity. The data correspond to the short-term sorption of initially dried concretes in the laboratory.

Ion Sorptivity

Aggressive ions, like chloride, can be bound onto the solid phases of the concrete matrix; thus, the sorption of ions by solid phases can substantially retard the ingress of external ions. The effective absorbent of ions is C-S-H due to their very large specific area, on the order of 100–700 m^2/g. The ion sorption isotherm is defined as the quantity of adsorbed ion by unit mass of solid phases (cement) (moles per kilogram) in terms of the concentration of aqueous ions in pore solution (moles per cubic meter). For chloride sorption by C-S-H, different mechanisms are involved: the formation of Friedel's salt at low concentration, and ion exchange with the OH^- at the C-S-H surface at higher concentration; see Figure 2.6. Different laws are available for the chloride sorption isotherms. Besides the mechanism-based sorption model in Equation 2.6, the linear sorption isotherm is the simplest one: the adsorbed ions c_b (moles per kilogram concrete) is proportional to the aqueous ions (moles per cubic meter) through a sorptivity constant:

$$c_b = r_{Cl} c_{Cl} \qquad\qquad (8.9)$$

where the sorptivity r_{Cl} assumes the unit cubic meters per kilogram. The chloride sorptivity is found to be sensitive to the chemical composition, especially to the C_3A content in cement binder, and high Al content in cement provides high chloride sorptivity for hardened concrete. The typical values for linear chloride sorptivity for concretes were found in the range 10^{-5}–10^{-4} m^3/kg; see Figure 2.6.

Electrical Resistivity

The pore solution of concrete can conduct electrical current while the electrical conductivity of solid phases is very low. As a property, the electrical resistivity of concrete reflects the pore solution and its percolation in the material. Electrical resistivity also plays an important role in the steel corrosion process, controlling the amplitude of corrosion current in the resistivity-controlled range. Strictly speaking, the electrical resistivity is not *per se* a transport property, but closely related to the transport properties. The electrical resistivity of concrete can also be used to define the tortuosity of pore structure through Archie's law. The electrical resistivity depends on both the pore structure and pore saturation. The conventional values of electrical resistivity of structural concretes, in the saturated state, range from 50 to 1000 Ω m, and incorporating SCM, like slag, tends to increase the electrical resistivity substantially. Figure 8.6 summarizes the literature data on the electrical resistivity of concretes in terms of compressive strength.

8.1.4   Mechanical Properties

Compressive Strength

No direct relation can be established for the transport properties and the compressive strength due to the very different mechanisms between the transport processes in concrete pores and the mechanical failure under axial compression. However, the compressive strength of concretes

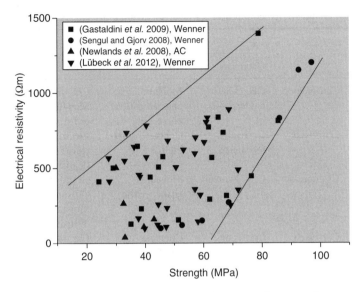

Figure 8.6 Literature data on the electrical resistivity of concretes in terms of compressive strength: the concretes were water saturated. "Wenner" indicates measurement using the four-electrode Wenner method; "AC" indicates electrical resistance measurement through alternating current.

has always been taken as an important indicator for concrete quality due to its wide use in engineering practice and its simple and standard test method. So, the axial compressive strength of concrete is regarded as one fundamental indicator for durability.

Fracture Resistance

Although the durability processes begin with mass transfer and phase changes, the performance deterioration of concrete structures or elements arises from mechanical damage. Thus, the mechanical properties of structural concretes are also fundamental to durability performance. The fracture resistance, or cracking strength, is defined as the resistance of concrete to the propagation of cracking from initial defects. This property is pertinent to the concrete damage from the corrosion-induced cracking, freeze–thaw, and salt crystallization at pore level.

8.1.5 Fundamental Relationships

The physical properties involved in the deterioration processes are listed in Table 8.1. Note that the microstructure, especially the pore structure, is fundamental to all the transport processes. The information of pore structure is implicitly included in the transport properties of moisture, ions and gas. Nowadays, advanced computational methods, like homogenization techniques from micromechanics, allow for predicting the transport properties from the micro-structure of concretes. Here, only some fundamental relations among the properties are recalled, to show that these transport properties, stemming from the same microstructure, can be interdependent. These relations also serve as a basis for specification of durability indica-tors through substitute properties.

Table 8.1   Recapitulation of material properties involved in durability processes

Process		Properties			
		Chemical	Microstructure	Transport	Mechanical
Carbonation-induced corrosion	Carbonation	CH, C-S-H contents	Moisture characteristic curve	CO_2 diffusivity Moisture diffusivity	—
	Corrosion	Corrosion resistance	Concrete–steel interface	Electrical resistivity O_2 diffusivity	Fracture resistance
Chloride-induced corrosion	Cl^- ingress	C-S-H content	Moisture characteristic curve	Cl^- diffusivity Cl^- sorptivity Moisture diffusivity	—
	Corrosion	Corrosion resistance (critical Cl^- content)	Concrete–steel interface	Electrical resistivity O_2 diffusivity	Fracture resistance
Freeze–thaw damage		—	Average spacing factor Freezing characteristic curve	Water sorptivity	Fracture resistance
Leaching		CH, C-S-H content Calcium equilibrium	—	Water permeability Ca^{2+} diffusivity	—
Salt crystallization damage		—	Crystallization characteristic curve	Water sorptivity	Fracture resistance

Archie's Law

Archie's law originally addressed the relationship between the electrical conductivity of the pore solution S_0 and that of the saturated porous material S_p through a formation factor F (Archie, 1942):

$$F(\phi) = \frac{S_0}{S_p} \tag{8.10}$$

The formation factor F here accounts for both the porosity and the tortuosity of the pore structure. A logical reasoning is that Archie's law applies also to the diffusivity of ions; that is

$$F(\phi) = \frac{D_0}{D_p} \tag{8.11}$$

If the formation factors in Equations 8.10 and 8.11 coincide with each other, the diffusivity and resistivity of concrete can be predicted one from the other. Figure 8.7 shows the calculation results on 116 concrete specimens of five types of concretes with $w/b = 0.25$ to 0.50. Rather reasonable agreement is observed between F factors from the conductivity ratio and diffusivity ratios, providing a basis for the following equation:

$$\frac{D_0}{D_p} = \frac{S_0}{S_p} \tag{8.12}$$

Accordingly, the ion diffusivity can be reasonably predicted from Equation 8.12 with measurement on S_p (relatively easy) and S_0 and D_0 as known parameters.

Katz–Thompson Relationship

This relationship links the intrinsic permeability to the formation factor from Archie's law and a characteristic diameter d_c, of pore structure:

$$k = \frac{d_c^2}{C_{KT}F} = \frac{d_c^2}{C_{KT}} \frac{S_p}{S_0} \tag{8.13}$$

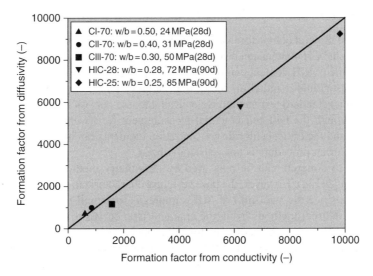

Figure 8.7 Formation factors from conductivity and diffusivity estimation: average results from CI-70 (36 specimens), CII-70 (36 specimens), CIII-70 (36 specimens), HIC-28 (5 specimens), and HIC-25 (3 specimens); the relative error of conductivity measurement is 6.9–8.8%, and the relative error of chloride diffusivity measurement is 19–28%.

with C_{KT} a constant related to pore structure, taken as 226 in Katz and Thompson (1987). This equation has proved quite successful for sedimentary rocks, and links explicitly the pore structure characteristics, d_c^2/C_{KT}, and conductivity (diffusivity) with permeability. The characteristic diameter d_c can adopt the critical pore size from a MIP experiment. However, its application to concrete seems questionable according to the available results. Halamickova *et al.* (1995), using a formation factor from chloride diffusivity and $C_{KT} = 180$, found that this equation is more adapted to pore structure with connected capillary pores (concretes with $w/c = 0.5$). Anyway, this relationship remains a fundamental one for concrete, and its validity remains to be explored for different types of structural concretes.

8.2   Characterization of Durability-related Properties

For the aforementioned durability-related properties, some can be measured through well-defined experimental procedures with quantified experimental uncertainty, while the characterization for other properties is far from achieved. For these properties, the durability performance tests come on to the scene: instead of intrinsic properties, the performance of structural concrete under durability actions (e.g., artificial environmental actions in the laboratory) is measured and used to characterize the durability resistance of concrete. These tests can also be useful for engineering judgment if the underlying mechanism of deterioration process is not clarified. The characterization of durability properties is reviewed in Sections 8.2.1 and 8.2.2 and summarized in Table 8.2. The durability performance tests are treated in Section 8.2.3.

8.2.1   Characterization of Chemical and Microstructural Properties

The chemical and mineral composition of hardened concrete can be measured following standard procedures: the CH content can be quantified using TGA and chemical analysis methods (AFGC, 2007); the C-S-H content can be measured through indirect methods such as nano-indentation (Constantinides and Ulm, 2004); the calcium equilibrium is established from literature data and no standard test method is yet available; the corrosion resistance of steel can be quantified through the half-cell potential measurement for its electrochemical potential (ASTM, 2009); and the critical chloride concentration (content) can be measured in the laboratory, but the relevant procedure is not yet standardized.

The porosity of concrete can be measured by gravimetry methods following standard methods (ASTM, 2013a). However, the pore structure characterization for concrete remains a tricky problem since the size limit of MIP samples is too small to be representative for concrete. Some indirect methods using the characteristic curves to deduce the pore structure can be used. The average spacing factor in hardened concretes can be quantified following the standard method (ASTM, 2012a). The characteristic curves associated with the phase changes in pores follow mainly research procedures: the moisture characteristic curve can be quantified from the water vapor sorption isotherm (Baroghel-Bouny, 2007); the freezing characteristic curve can be measured through dielectric capacitive methods (Fabbri and Fen-Chong, 2013); and the crystallization characteristic curve has no experimental procedure so far.

Table 8.2 Characterization of properties for durability (micro-level and macro-level)

Level	Properties		Characterization method	Single operator (multilaboratory) precision	Source
Micro-level	Pore structure	Porosity	Gravimetry	n.a.	ASTM (2013a)
		Pore size	Mercury intrusion porosimetry	n.a.	—
		Spacing factor of air voids	Optical analysis	3.7–12.4%	ASTM (2012a)
		Moisture characteristic curve	Water vapor isotherm	n.a.	—
	Chemical	CH content	TGA or chemical analysis	n.a.	—
		C-S-H content	Nano-indentation	n.a.	Constantinides and Ulm (2004)
		Steel corrosion resistance	Half-cell potential	~10%	ASTM (2009)
		Critical chloride content	Chemical analysis after corrosion	26–41%	Local data
Macro-level	Transport	Water permeability	Water intrusion under pressure	n.a.	BSI (2009)
		Gas permeability	CemBureau method	n.a.	RILEM (1999)
			Torrent method	24–27%	Torrent et al. (2013)
		Water sorptivity	Liquid water absorption	6%	ASTM (2013c)
		Gas diffusivity	Gas diffusion cell	n.a.	Sercombe et al. (2007)
		Chloride diffusivity	Chloride migration test	15.2% (23.6%)	Chlortest (2005)
			Immersion test	20% (28%)	Chlortest (2005)
				20.2% (56.6%)	ASTM (2011b)
			Steady migration test	22% (76%)	Chlortest (2005)
				n.a.	Tang and Nilsson (1993)
	Electrical	Chloride sorptivity	Sorption tests		
		Electrical resistivity	Direct current	9.2%	ASTM (2012d)
			Alternative current test	10.5% (25.1%)	Chlortest (2005)
			Wenner method	17.9%	Local data
	Mechanical	Compressive strength	Axial compression	6–10.6% (14%)	ASTM (2014)
		Fracture resistance	Three-point bending	16%	ASTM (2012c)

n.a., not available.

8.2.2   Characterization of Transport and Mechanical Properties

Water permeability can be measured by water flow under imposed pressure, and the standard method can be referred to BS EN 12390-8:2009 (BSI 2009), but the practical difficulty is not negligible as the concrete is very compact and the water permeation across the specimen is actually not possible. In this case, water permeability can resort to the measurement of water penetration depth under external pressure and the water permeability can be estimated from the water penetration depth. Certainly the water permeability thus measured is not under steady flow regime, and thus should be termed "apparent water permeability." Otherwise, the water permeability can be evaluated, with the help of moisture transport modeling, from the water vapor sorption isotherm measurement (Baroghel-Bouny, 2007). Note that in-situ measurement devices have been developed for water permeability measurement on a concrete surface.

Gas permeability can be measured through the steady gas flow under imposed pressure gradients, and this method, originally a CemBureau recommendation (Kollek, 1989), was further standardized with a pretreatment of concrete specimens to initial pore saturation of 75% and nitrogen as infiltrating gas (RILEM, 1999). Thus, the measurement from this procedure does not correspond to the permeability of all pores in concrete. This procedure can measure the intrinsic permeability of concrete between 10^{-19} and 10^{-16} m^2. The relative error of the device itself is rather small, within 5% for repeatability tests on a single specimen, and the relative dispersion among different specimens depends on the homogeneity of the concrete: dispersion as large as 35% was recorded for two specimens of the same concrete. Some devices have been developed for the in-situ measurement of air permeability of concrete surfaces (Torrent, 1992; Yang et al., 2014). The relative error of in-situ measurement is more important due to the more significant randomness of concrete surfaces exposed to air: 24–27% was observed for the logarithmic deviation of the measurement in a recent exercise using a TORRENT device (Torrent et al., 2013). The gas diffusivity can be measured through a gas diffusion cell, and the measurement remains a research procedure (Sercombe et al., 2007). The moisture diffusivity can be measured using a gravimetry method under a humidity gradient, which also remains a research procedure (Baroghel-Bouny, 2007).

The measurement of chloride diffusivity was subjected to extensive investigation and the procedures have been standardized in recent decades. Following the available procedures, the diffusivity can be measured in the steady or unsteady regime, and can be obtained from pure diffusion or accelerated migration tests. The NT Build 492 method is an accelerated migration test in the unsteady regime (Nordtest, 1999) and the NT Build 443 method is a diffusion test in the unsteady regime (Nordtest, 1995), whereas the INSA method is a migration test in the steady regime (Chlortest, 2005). The results from these different tests need interpretation before comparison, and some theoretical relations have been proposed for the values of diffusivity from migration and diffusion tests (Tang, 1999). Nowadays, the method used most is the rapid chloride migration (RCM) test of the NT Build 492 method, and its relative dispersion in the same group of specimens was recorded as 20%. An in-situ device has been set up to measure the steady migration diffusivity on concrete surface (Basheer et al., 2005). The chloride sorptivity of concrete can be measured by the solution equilibrium method, and remains a research procedure (Tang and Nilsson, 1993).

Water sorptivity can be measured on a pretreated concrete specimen, with one surface placed into liquid water and the other exposed to air. The water absorption is measured by the mass change of the specimen. This procedure has been standardized in ASTM C1585 (ASTM,

2013c) with a specified pretreatment of concrete: specimens are dried at 60 °C and relative humidity of 80% for 3 days, then put into equilibrium with ambient temperature of 23 °C for 15 days. The liquid water absorption is recorded by the gravimetry method for 7–9 days, and the initial and secondary absorption rates are deduced respectively from the 1–6 h and 1–7 days' time ranges. Thus, the water sorptivity measured in this procedure does not correspond to all pores in concrete but to a portion thereof. The stability of this test is judged good, and the single operator error is 6%. Since this test needs very little labor effort, it can be considered as a good alternative for the water permeability measurement in laboratory. Note that some in-situ devices can also measure the water sorptivity of concrete surface at a low water pressure (Yang *et al.*, 2014).

Electrical resistivity can be measured using either the direct current method (ASTM, 2012d) or alternative current method (Chlortest, 2005). The alternative current method assures a minimum electrical current of 40 mA, uses concrete specimens of 10 cm in diameter and 5 cm in thickness, and saturates the specimens by vacuum before measurement. The advantage of this method is that the device can be easily set up in the laboratory and the same specimen can be measured repeatedly. Literature data show the repetitive and reproducibility errors are respectively 10.5% and 25.1%. Through Archie's law, the electrical resistivity can yield a good estimation for diffusivity properties; see Figure 8.7. An in-situ device of electrical resistivity measurement is also available; for example, the four-electrode Wenner-type device (Sengul and Gjorv, 2008). The in-situ measurement of electrical resistivity is conducted on the concrete surface and shows much larger dispersion compared with laboratory measurement due to the heterogeneity of in-situ concrete, the randomness of pore saturation at the concrete surface, and the in-situ temperature. Moreover, the in-situ device measures the electrical resistivity of the concrete surface only to a rather limited depth, on the order of centimeters, so the values are different from the electrical resistivity of bulk concrete. Since the electrical resistivity is sensitive to pore saturation, the in-situ resistivity device is also used for pore saturation characterization; for example, for in-situ gas permeability measurement (Torrent, 1992).

The fracture resistance of concrete can be measured following the standard method (ASTM, 2012c), splitting beam specimens under a three-point bending test. And the fracture toughness is interpreted through the force–displacement of the bending tests. The relative error of the measurement was quantified as 16% for fiber-reinforced concrete specimens, and a larger error is expected for plain concrete without fibers. Though the fracture toughness is pertinent to material damage of durability deterioration processes, it is not yet a conventional method in the laboratory for plain concrete due to the high brittleness. In this case the tensile strength, especially from three-point bending tests, can be retained as the substitute experiment.

8.2.3   Durability Performance Tests

The durability performance tests are complementary to the characterization of properties. As the pertinent properties for durability are difficult to quantify or the underlying deterioration mechanisms are not available, engineers tend to test the materials or elements under similar environmental actions in the laboratory and judge their resistance by the performance observed in the test. These tests usually use accelerated techniques by imposing much higher intensity of actions to obtain measurable behaviors within an acceptable experimental duration. These tests are globally termed as "durability performance tests."

To what extent these tests can represent the durability resistance in natural environments is still debated among scientists and engineers. Some, mostly engineers, tend to set up some "similarity" between the performance tests and the natural deterioration, while others, mostly scientists, insist that the accelerated tests distort the natural deterioration process and can only be used for comparative study. It is not intended to arbitrate on the opposing ideas on performance tests here. A reasonable judgement is that performance is probably somewhere in between: if the material state and the deterioration mechanism between the accelerated test and natural case are the same and only the action intensity differs, then a similarity between the two can be obtained with enough experimental results; otherwise, the similarity is less probable. The established performance tests are reviewed in the following and summarized in Table 8.3.

Table 8.3 Performance tests for durability of structural concretes

Target performance	Test method	Result	Single operator (multilaboratory) precision	Source/method
Carbonation	Accelerated carbonation test	Carbonation depth	n.a.	DuraCrete (1999)
		ACC resistance	5.3–20%	DuraCrete (1999)
Freeze–thaw	Rapid freeze–thaw test	DF factor by dynamic modulus	20.1% (56.9%)	ASTM (2003), Procedure A
		Length change	n.a.	ASTM (2003), Procedure A
	Frost resistance test	Dilatation	5.5% (7.5%)	RILEM (2001), IDC method
		Relative ultrasonic speed	7.3% (7.6%)	RILEM (2001), IDC-Procedure A
		Relative dynamic modulus	n.a.	RILEM (2001), IDC-Procedure B
	Salt scaling	Scaling mass	10.4% (17.5%)	RILEM (1996), CDF method
Sulfate resistance	Sulfate immersion under drying-wetting	KS factor by mass loss	n.a.	CNS (2009)
Leaching resistance	Accelerated leaching by ammonium nitrate solution	Residual strength	n.a.	Garde et al. (1996)
		Degraded depth	n.a.	Segura et al. (2013)
		Degraded pore structure	n.a.	Segura et al. (2013)
Combined actions	Corrosion under micro cracking	Correlation between cracking and corrosion	n.a.	François and Arliguie (1999)
	Sulfate attack under loads and freeze–thaw actions	Change of dynamic modulus, flexural strength change, mass loss	n.a.	Nehdi and Bassuoni (2008)

n.a., not available.

Accelerated Carbonation Test

This test was set up to study concrete resistance to carbonation under higher CO_2 concentration (20%), specified relative humidity (60%), and specified temperature (20 °C) for a given period (28 days) on specimens of 28 days' age cured and treated in standard conditions (DuraCrete, 1999). This procedure has been adopted as a standard test for carbonation resistance characterization of concrete (CNS, 2009). The accelerated carbonation (ACC) resistance R_{ACC}^{-1} is noted in terms of the carbonation depth in $(mm^2/a)/kg\ CO_2/m^3$, and the repeatability error is noted as 20%. The relevance of accelerated resistance to natural carbonation (NAC) resistance R_{NAC}^{-1} is considered to be high since the carbonation mechanism is basically the same between the two procedures, and the main difference lies in the concrete hardening age, the humidity, and the temperature. Some long-term observations seem to confirm this judgement (see Figure 1.7), and a correlation has been proposed for the two resistances with the dispersion quantified (*fib*, 2006); see Equation 1.15.

Rapid Freeze–Thaw Test

The rapid freeze–thaw test imposes freezing and thawing cycles on concrete specimens, and after a certain number of cycles the properties alteration is quantified to represent the concrete resistance to the freeze–thaw actions. Several procedures are available. The ASTM C-666 test (ASTM, 2003) uses the temperature range −18 to +4 °C and 2–5 h for one freeze–thaw cycle, measures the alteration of dynamic elastic modulus and the specimen mass after 300 cycles; and a durability factor DF is defined as the ratio, in percent, between the elastic modulus after 300 cycles and the initial value. The ASTM C-672 test (ASTM, 2012a) builds a salt solution pond (4% $CaCl_2$ of 6 mm) on the specimen top, freezes the specimen to −18 C within 16–18 h, and thaws at 23 °C for 6–8 h, using visual inspection to evaluate the damage extent. The RILEM TC-117 CDF method (RILEM, 1996) dips one surface of concrete specimens into a 3% NaCl solution of 5 mm, imposes the freezing and thawing rates of 10 °C/h with each cycle of 24 h, and characterizes the concrete damage extent by the scaling mass. The RILEM TC-176 IDC method (RILEM, 2001) exposes one surface of the concrete specimen to water (salt water). The temperature range is −20 to +20 °C with 24 h as a cycle, and the ultrasonic velocity and mass loss are recorded during 56 cycles (possibly extended to 112 cycles), and the damage extent of the concrete specimens is characterized by the length dilatation and relative ultrasonic velocity (procedure A) or by the relative dynamic modulus (procedure B).

All tests involve long experiment duration and heavy labor in specimen preparation and testing. The errors were reported as 20.1% (repeatability) and 56.9% (reproducibility) for DF factor from the ASTM method, 0.4% (repeatability) and 17.5% (reproducibility) for mass scaling from the RILEM TC-117 CDF method, and 7.3% (repeatability) and 7.6% (reproducibility) for relative ultrasonic speed from the RILEM TC-176 IDC method. The rapid freeze–thaw tests use concrete specimens of 28 days' age at contact with the water, so the concrete under test is rather young and the pore saturation is rather high. These two factors can, to a large extent, distort the representativity of the accelerated freeze–thaw results for the freeze–thaw resistance of in-situ structural concrete, which normally has a much greater age at contact with water and a lower pore saturation. Accordingly, these tests can be used for comparison of the freeze–thaw resistance of different concrete mixtures, and the interpretation of freeze–thaw resistance of in-situ concrete from these test results remains to be explored.

Sulfate Resistance Test

The research during recent decades on sulfate attack on cement-based materials provides a vast variety of test methods for both external and internal sulfate attack (RILEM, 2013). One of the most widely used tests for sulfate crystallization damage is the drying–wetting method (Almeida, 1991), now established as a standard method (CNS, 2009). This method uses cubic concrete specimens of 100 mm at age of 28 days, subjected to drying–wetting cycles with each cycle lasting 24 h (soaked in 5% Na_2SO_4 solution for 15 h, air dried for 1 h, oven dried at 80 °C for 6 h and cooled to room temperature for 2 h). The mass loss and compressive strength decrease are recorded for a specified number of cycles (e.g., 150 cycles). The mass change ratio KS, in percent, is noted as the test result. Under this procedure, it is expected that concrete damage is mainly due to pore crystallization of sulfate salts, though the formation of other products from sulfate reaction, such as ettringite, cannot be excluded. The KS index provides the resistance of concrete to a particular, certainly extremely severe, cycle of crystallization. Again, the correlation between the KS index and the crystallization resistance in the natural environment is to be established.

Accelerated Leaching Test

Various procedures are available for concrete leaching tests. These tests usually adopt a dynamic scheme (i.e., the leaching fluid is changed regularly), and several techniques are adopted to accelerate the leaching, including using acid solutions, adding pressure to liquids, or imposing electrical fields; see RILEM (2013: chapter 9). The leaching behaviors of concrete can be quantified through the analysis of leachate or the property change of leached concrete, but no property or performance index has been proposed for accelerated leaching tests yet. It was shown that the acceleration factor of leaching tests using deionized water is around 50–100 compared with natural leaching by pore solution, this factor can rise to 500 under an electrical field of 1300 V/m (Gérard et al., 2002), and it is estimated as 300 when using a 6 M NH_4NO_3 solution compared with leaching in deionized water (Heukamp et al., 2001). For engineering use, the accelerated tests are expected to provide pertinent indicators relevant either to properties or to performance.

Accelerated Test under Combined Actions

Some more sophisticated accelerated tests have been set up recently to investigate the performance of concretes under combined actions; for example, freezing and chloride ingress (Saito et al., 1994), chloride ingress and mechanical loadings (Desmettre and Charron, 2013), and sulfate reactions under mechanical loads and freeze–thaw actions (Nehdi and Bassuoni, 2008). These investigations cast new light on concrete performance under the combined effects of actions, while the interpretation becomes more sophisticated. Note that the durability performance investigated on carbonated concrete makes practical sense, since other processes always occur along with carbonation in the structural elements exposed to the atmosphere. More fundamental research and in-field validation are needed to calibrate these results for realistic evaluation of durability performance of structural elements in service.

8.3 Durability Indicators for Design

8.3.1 Nature of Durability Indicators

As mentioned before, durability indicators include both the properties and the performance characteristics related to a certain durability process. Suppose that the deterioration law in Equation 7.1 can be expressed in terms of durability indicators. Thus, the prescription of durability indicators for a certain durability process involves the solution of these indicators through the deterioration law with respect to the DLS, the expected service life, and the environmental actions, expressed as

$$f\left(\mathbf{s}, \mathbf{m} ; t_{SL}\right) - f_{DLS} \geq 0 \tag{8.14}$$

with $\mathbf{m}$ the durability indicators in the deterioration model and $\mathbf{s}$ the environmental actions. Generally speaking, the durability indicators refer to the material properties, micro-level ($\mathbf{m}_{micro}$) or macro-level ($\mathbf{m}_{macro}$), the performance test results ($\mathbf{m}_{pt}$), or even the concrete proportioning parameters.

To illustrate the concept of durability indicators, we take the chloride ingress with corrosion initiation as DLS as an example. For this process, the durability model in Equation 2.19 is retained for analysis:

$$\left(C_{crit} - C_0\right) - \left(C_s - C_0\right)\left[1 - \mathrm{erf}\left(\frac{x_d}{2\sqrt{D_{Cl}^{app} t}}\right)\right] \geq 0 \tag{8.15}$$

with C_{crit}, C_0, and C_s the critical, initial, and surface contents of chloride, x_d the thickness of concrete cover, and D_{Cl}^{app} the apparent diffusion coefficient of chloride in concrete. Actually, the pertinent durability indicator in Equation 8.15 is the combination of parameters in the error function, and this combination, DI_{Cl}, for a design case $\left(C_{crit,S,0}^d ; t_{SL}\right)$, is to satisfy

$$\mathrm{DI}_{Cl} = \frac{x_d}{2\sqrt{D_{Cl}^{app} t_{SL}}} \geq \mathrm{erf}^{-1}\left(\frac{C_S^d - C_{crit}^d}{C_S^d - C_0^d}\right) \tag{8.16}$$

and the appropriate choice of x_d and D_{Cl}^{app} should satisfy Equation 8.16. The indicators (x_d, D_{Cl}^{app}), thus defined, form the parameter sets for durability design and constitute the sufficient conditions for design according to Equation 8.15.

However, chloride transport in the natural environment involves far more mechanisms than diffusion (see Section 2.2). As one uses Equation 8.15 to envelop all these mechanisms, the parameter D_{Cl}^{app} becomes rather a performance parameter than a simple property; that is, an apparent characteristic resulting from the interaction between the environmental actions and the material microstructure. Figure 8.8 illustrates the nature of the parameter D_{Cl}^{app}. Actually, the parameter D_{Cl}^{app} relates to the environmental actions, including temperature and humidity, and the material characteristics of the concrete cover, such as cracking and loading states. Models at different levels can help to determine the apparent chloride diffusivity: the micro–macro models can predict the macroscopic transport properties from the microstructure and

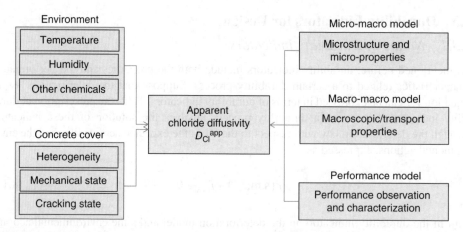

Figure 8.8 Illustration of the relation of the apparent chloride diffusivity to environmental actions and concrete cover materials (left) and to the properties at different levels (right).

micro-properties of concrete components; the macro–macro models provide the correlation among the chloride diffusion coefficients characterized by different tests or the correlation with different properties; the performance models relate directly the apparent diffusivity to the environmental and material conditions through in-situ observations or performance tests.

Two aspects break the nature of the sufficient condition of parameter D_{Cl}^{app} in Equations 8.15 and 8.16: (1) no matter how the models of prediction (micro–macro, macro–macro, or performance models) are elaborated, the environmental actions in a specified project can never be described completely; (2) the models themselves can never be adapted to all structural concretes with different chemical compositions and microstructures. Note that the performance models do simulate the real environmental actions, but due to the limited experimental duration the performance test results should always be extrapolated, to a much longer temporal scale, to be predictive. This raises the temporal similarity of these performance tests. These problems with the parameter D_{Cl}^{app} are rather representative for durability indicators in other processes.

Accordingly, the nature of durability indicators turns from one of sufficient conditions, to assure the design equations, to one of necessary conditions, to provide a most probable set of material characteristics satisfying the design equations. Hence, we define **DI** as the most probable parameter set; that is

$$\mathbf{DI} = \left\{ \mathbf{m}_{micro} \otimes \mathbf{m}_{macro} \otimes \mathbf{m}_{pt} \mid \left(\mathbf{m}_{micro}, \mathbf{m}_{macro}, \mathbf{m}_{pt} \right) \rightarrow f\left(\mathbf{s}; \mathbf{m}; t_{SL} \right) - f_{DLS} \geq 0 \right\} \quad (8.17)$$

with $\mathbf{m}_{micro}$, $\mathbf{m}_{macro}$, and $\mathbf{m}_{pt}$ standing for micro-properties, macro-properties, and performance test characterization respectively for a certain deterioration process.

8.3.2   Durability Indicators for Deterioration

On the basis of the foregoing analysis, the probable sets of durability indicators (DIs) can be formulated, and Table 8.4 summarizes the most probable durability indicators for different durability processes. These DI groups are results of correlation analysis between the respective

Table 8.4 Probable durability indicators for different deterioration processes

Level	Properties/characteristics		Carbonation-induced corrosion	Chloride-induced corrosion	Freeze–thaw	Leaching	Salt crystallization
Micro-level Properties	Microstructure	Porosity	++	+	+	+	+
		Pore size distribution	+				++
		Air voids spacing			++		++
		Characteristic curve		+	++ (ice)		++ (crystal)
	Chemical	CH content	++	++		++	
		C-S-H content				++	
		Steel corrosion resistance	++	++			
		Critical chloride content		++			
Macro-level properties	Transport	Water permeability	+	+	+	++	+
		Gas permeability	+	+			
		Water sorptivity		+	++		++
		Gas diffusivity	++ (CO$_2$, O$_2$)	++ (O$_2$)			
		Ion diffusivity		++ (Cl$^-$)		++ (Ca^{2+})	
		Ion sorptivity		++ (Cl$^-$)			
	Electrical	Electrical resistivity	++	++			
	Mechanical	Compressive strength	+	+	+	+	+
		Fracture resistance	++	++	++	++	++
Performance test		Performance resistance	++ (ACC)	+	+	+	+
Structural properties	Concrete cover	Thickness	++	++			
		Cracking	+	+			
		Surface protection	+	+	+	+	+

++: strong relevance; +: moderate relevance.

mechanisms and the material properties (characteristics). No attempt is undertaken here to propose quantified values for the different DI groups, and the limit values for relevant durability indicators can be referred to the design applications in Chapter 9 and the practice of design codes in Chapter 10.

In Table 8.4, the probable indicators for one deterioration process are made different according to their relevance to the deterioration mechanisms. For the strongly related indicators, direct characterization is recommended; for example, chloride diffusivity for chloride-induced corrosion. For indicators with moderate relevance, substitute properties can be used as the direct characterization is not available. The credibility and the relevance of performance tests depend strongly on the similarity of environmental actions, the physical state of specimens, and the experimental duration. Accelerated carbonation tests have been extensively used to characterize the carbonation resistance of structural concretes, and the measured resistance is more regarded as a property nowadays.

8.3.3   Durability Indicators: State of the Art

On the basis of the probable durability indicators in Table 8.4, the durability indicators can be structured following the respective mechanisms and modeling. The specification of values for the adopted durability indicators involves, moreover, two important aspects: (1) the reliability and dispersion of the characterization methods for the DI; and (2) the global durability margin expected for the structural elements. The former refers to the experimental basis for specification of durability indicators, and the latter to the design strategy for structural durability. Owing to the lack of reliable characterization methods and modeling, the performance tests are preferred as laboratory characterization methods, which is the case of carbonation. Thus, the specification of DI values depends strongly on the laboratory characterization methods and their reliability. Nowadays, DI has been widely adopted in durability design works and durability specification in codes: the durability design for projects with long service life normally involves model-based evaluation, and certain durability-related characteristics of concrete are targeted as design parameters; for design codes the requirements on durability are also specified through DI-like parameters. Two systematic studies are reviewed here for durability indicators.

Durability Indicators from Association Française de Génie Civil

Systematic durability indicators have been investigated by AFGC (2007) with the purpose to help the performance-based durability design in the framework of EuroCode (CEN, 2000). The durability indicators are proposed at several levels: basic properties (compressive strength at 28 days/90 days), general indicators for durability (porosity, chloride diffusivity, water/gas permeability, CH content), indicators specific to durability process (e.g., reactive silica release rate, active alkaline content, and swelling strain of specimens for alkali–silica reaction), substitute indicators for general indicators (e.g., electrical resistance, CO_2 diffusivity, capillary sorption), and complementary indicators (e.g., C-S-H content, shrinkage, tensile strength, cracking). Three general indicators are especially valued in this study: the porosity, water permeability, and gas permeability. The specification of these three properties, as general indicators, is given in Table 8.5. More comprehensive specification of general properties and substitute properties with respect to environmental actions and service lives can be found in AFGC (2007).

Table 8.5 Specification of general durability indicators for different carbonation environments –adapted from AFGC (2007: tables 9, 11, 13)

Durability indicator	Environmental classification (concrete cover 30 mm)				Characterization method
	X0/XC1	XC2	XC3	XC4	
Porosity P_w (%)	<14 (12)	<14 (12)	<12 (9)	<12 (9)	Drying under 105 °C
Water permeability K_l (10^{-18} m²)	—	—	—	<0.1 (0.01)	Direct flow method
Gas permeability K_g (10^{-18} m²)	— (<100)	— (<100)	<100 (30)	— (<30)	Inlet pressure 0.2 MPa, specimen dried at 105 °C

X0 and XC1 refer to dry (relative humidity <65%) and permanently immersed conditions respectively, XC2 to humid condition (relative humidity >80%), XC3 to moderately humid condition (relative humidity 65–80%), and XC4 to the drying–wetting conditions without chlorides.
Value outside parentheses for service life 50–100 years, and values in parentheses for service life 100–120 years.

Table 8.6 Suggested values for chloride conductivity (mS/cm) for monumental structures ($t_{SL} = 100$ years, concrete cover 50 mm)

Environmental class according to EN206-1	Binder		
	CEMI:FA (70:30)	CEM:GGBS (50:50)	CEMI:SF (90:10)
XS1 (airborne salt)	2.50	2.80	0.80
XS2a (submerged)	2.15	2.30	0.50
XS2b (submerged, abrasion), XS3a (tidal, splashing)	1.10	1.35	0.35
XS3b (tidal, splashing + abrasion)	0.90	1.05	0.25

Source: adapted from Nganga *et al.* 2013, Table 1. Reproduced with permission of Elsevier.

Durability Indicators from South Africa

The study proposes durability indices based on three properties tests: oxygen permeability test, chloride conductivity test, and water sorptivity test (Nganga *et al.*, 2013). The values of durability indices are determined on model-based methods for carbonation-induced and chloride-induced corrosion environments, similar to the classification of EN 206-1 (CEN, 2000). The approach has been applied to the Gauteng Freeway Improvement Project (GFIP), a large-scale infrastructure improvement project. The oxygen permeability index (OPI), the negative logarithm of oxygen permeability, is considered to have good correlation with the carbonation resistance of concrete and is used as input parameter to estimate the carbonation depth. For laboratory concrete, the typical limit values are: OPI >10, water sorptivity <6 mm/h$^{1/2}$, and chloride conductivity <0.75 mS/cm. For the structural concrete in GFIP with an expected service life of 100 years in a carbonation environment, the durability indices are prescribed as: OPI >9.7 and water sorptivity <10 mm/h$^{1/2}$ for a concrete cover of 40 mm. Table 8.6 summarizes the proposed values for chloride conductivity for monumental structures in a marine environment.

9

Durability Design: Applications

This chapter applies the durability design concepts and approaches to real engineering cases. Two cases with long design service life are presented in this chapter: concrete structures in a sea link project on the southeastern coast of China with design service life of 120 years, and high-integrity containers (HICs) used for radioactive waste disposal with expected service life of 300 years. Both cases necessitate a global design strategy and model-based design for target deterioration processes. The concrete structures in the sea link project are exposed in a marine environment, and the design strategy involves durability requirements on both material and structural levels. With chloride ingress identified as the control process, a model-based method is used to determine the design value for the chloride diffusivity and the corresponding quality control parameters as well. The HICs serve in a near-surface geological environment, expected to assure the containment function and mechanical stability during the disposal life. The model-based method is used to solve the combined process of radionuclide transport and external leaching, the control process for container design, and to help the choice of key design parameters. Finally, some further aspects related to long service life design are addressed, in the context of life-cycle engineering, referring to the design uncertainty, the quality control, and the durability redesign options.

9.1 Sea Link Project for 120 Years

9.1.1 Project Introduction

The sea link project of Hong Kong–Zhuhai–Macau (HZM) consists of sea bridges of 28.8 km (four navigable spans), two artificial islands, and an immersed tube tunnel of 6.8 km with a total investment near US$12 billion. The project intends to connect mainland China to Hong

Durability Design of Concrete Structures: Phenomena, Modeling, and Practice, First Edition. Kefei Li.
© 2016 John Wiley & Sons Singapore Pte. Ltd. Published 2016 by John Wiley & Sons Singapore Pte. Ltd.

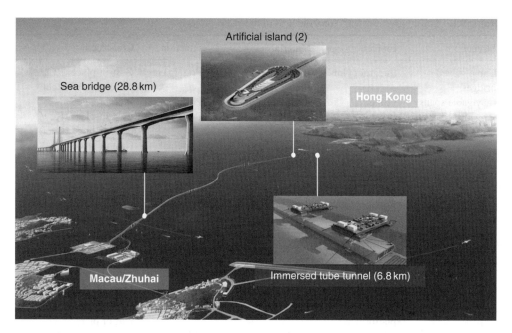

Artificial island (2)

Sea bridge (28.8 km)

Hong Kong

Macau/Zhuhai

Immersed tube tunnel (6.8 km)

Figure 9.1 Plan view of Hong Kong–Zhuhai–Macau sea link project. *Source:* courtesy of Rui Chai.

Kong city, and to facilitate communication between the two important economic bodies of China: Hong Kong city and Guangdong Province. The plan view of this project is given in Figure 9.1 and the constitutive elements for the main structures of the HZM project are given in Table 9.1. The design working life of the project is 120 years (HZMBA, 2010). The preliminary study of the project began from 2008 to 2010, the detailed study phase started from 2010, and construction works are expected to end in 2017.

The HZM sea link project is situated in the southern subtropical marine monsoon region of China. The annual average temperature is between 22.3 and 23.1 °C with 28.4–28.8 °C in July and 14.8–15.9 °C in January. Historically, the highest and lowest recorded temperatures are 38.9 °C and −1.8 °C (Macau site). The annual average humidity is between 77% and 80%, with large seasonal variation, and the seasonal humidity can reach 100% (spring and summer) and drop to 10% (winter). The predominant wind directions are east and southeast. The annual wind speeds are recorded as 3.1 m/s (Zhuhai site), 3.6 m/s (Macau site), and 6.6 m/s (Hong Kong site). The maximum wind speeds are 44.6 m/s, 58.6 m/s, and 71.9 m/s for Zhuhai, Macau, and Hong Kong observation stations respectively. The hydrology data show that the chloride ions (Cl^-) content in seawater at the project site is in the range 10 700–17 020 mg/L and the content of sulfate ions (SO_4^{2-}) is in the range 1140–2260 mg/L. The pH values of seawater samples are between 6.65 and 8.63. The salinity of seawater in the eastern part of the project site (Hong Kong side) is higher than in the western part (Zhuhai–Macau side). For a given site, the seawater salinity increases with the water depth. The recorded highest salinity is 32.9 for the eastern part and 25.4 for the western part (bottom, on seabed), and the lowest salinities recorded are 8.1 and 10.4 for the eastern and western regions respectively. The tide

Table 9.1 Constitutive elements for structures of the HZM project

Structure	Element	Material	Function	Construction
Cable-stayed bridges (navigable spans)	Integral pylon	RC (base)/steel (upper part)	Principal	Cast-in-place (RC)/ prefabricated (steel)
	Cable	High-strength steel	Principal	Prefabricated
	Box girder	RC (deck)/steel (box girder)	Principal	Cast-in-place (RC)/ prefabricated (steel)
	Pier (auxiliary)	RC	Principal	Cast in place
	Bearing platform	RC	Principal	Cast in place
	Pile	Steel tube/ Concrete	Principal	Cast in place (concrete)/ driven pile (steel tube)
Steel box-girder bridges (non-navigable)	Box girder	Steel (in sea)	Principal	Prefabricated
		PC (approach to artificial islands)	Principal	Cast-in-place (PC)
	Pier	RC	Principal	Cast-in-place (upper)/ prefabrication (base)
	Bearing platform	RC	Principal	Prefabrication (with base part of pier)
	Pile	Steel tube/concrete	Principal	Cast-in-place (concrete)/ driven pile (steel tube)
Immersed tube tunnel	Tube segment	RC (in sea)	Principal	Prefabricated
		RC (approach to artificial islands)	Principal	Cast-in-place
	Anti-shock pier	RC	Secondary	Prefabricated
	Segment joint	Rubber	Principal	Prefabricated
	Pile	Compacted sand	Principal	Bored hole, cast-in-place
Artificial islands	Retaining wall	Concrete	Principal	Cast-in-place
	Breakwater	Concrete	Principal	Prefabrication
	Facilities	RC/steel	Secondary	Cast in place (RC)/ prefabrication (steel)

Source: Li K.F. *et al.* 2015, Table 1. Reproduced with permission of ASCE.
RC: reinforced concrete; PC: prestressed concrete.

and wave data retained for project design are from Macau observation station. The design water levels are 2.74 m (high) and −1.27 m (low) for a return period of 10 years, and 3.47 m (high) and −1.51 m (low) for a return period of 100 years.

The main technical challenges of the whole project include the immersed tube tunnel, the artificial island construction in a harsh marine environment, and the deep pile foundations for the large-span navigable spans. One of the primary design targets is to ensure the service life of 120 years for the concrete structures in the HZM project in a rather aggressive marine environment.

Table 9.2 Environmental classification for RC elements in HZM project

Structure	Environment	Agents	Exposure condition	Element (RC/PC)
Bridge	Atmospheric	CO_2	Indoor, sheltered	Box girder (internal)
	Marine	Chloride ions	Immersed in seawater	Pile, bearing platform (part)
			Salt fog, 15 m above sea level	Deck, pylon (part), pier (part)
			Splashing, tidal actions	Bearing platform (part), pier (part), box girder
	Chemical	SO_4^{2-}, Mg^{2+}, CO_2 in seawater	SO_4^{2-}:1000–4000 mg/L Mg^{2+}:1000–3000 mg/L CO_2:30–60 mg/L	Bearing platform (underwater part)
Tunnel	Atmospheric	CO_2	Indoor, sheltered	Tube (internal)
	Marine	Chloride ions	Salt fog	Tube (internal)
			Seawater/air	Tube (external)
			Splashing/tidal actions	Anti-shock pier
	Chemical	SO_4^{2-}, Mg^{2+}, CO_2 in seawater	SO_4^{2-}:1000–4000 mg/L Mg^{2+}:1000–3000 mg/L CO_2:30–60 mg/L	Tube (external)
Artificial island	Marine	Chloride ions	Splashing/tidal actions (plain concrete)	Retaining wall, breakwater
	Chemical	SO_4^{2-}, Mg^{2+}, CO_2 in seawater	SO_4^{2-}:1000–4000 mg/L Mg^{2+}:1000–3000 mg/L CO_2:30–60 mg/L	Retaining wall, breakwater

Source: adapted from Li K.F. *et al.* 2015, Table 3. Reproduced with permission of ASCE.

9.1.2 Durability Design: The Philosophy

On the basis of the hydrology and meteorology data, the exposure conditions for the concrete elements are given in Table 9.2. The possible deterioration mechanisms of concrete elements include: (1) the carbonation-induced corrosion of reinforcement steel in the atmospheric environment, (2) chloride-induced corrosion of reinforcement steel by marine chlorides, and (3) physical and chemical attack of salts in seawater and seabed soil. In addition, the internal expansion reactions in structural concrete, like alkali–aggregate reaction and sulfate reaction, have to be taken into account given the long service life of the project and the crucial function of concrete elements.

The general philosophy of durability design with respect to these deterioration processes is to formulate the requirements on both a material level and a structural level, combining the material design with the structural design to achieve a working life of 120 years. Table 9.3 recapitulates all the requirements divided into material and structural levels. The risks of salt attack and internal expansion reactions, alkali–aggregate reaction and delayed ettringite formation, are controlled by limitations on the chemical composition of raw materials (e.g., cement, mineral admixture, and aggregates) and will not be subject to further model-based design for durability. Accordingly, the carbonation-induced corrosion and chloride-induced corrosion remain as the most critical processes, and durability models are used to help the choice of design parameters with the expected level of safety margin.

Table 9.3 Durability requirements for RC elements at material and structural levels

Requirements		Carbonation-induced corrosion	Chloride-induced corrosion	Salt attack		Internal expansion reaction	
				Physical	Chemical	AAR	DEF
Material level	w/b ratio	✓	✓	✓	✓	✓	✓
	Binder type		✓		✓	✓	✓
	Binder content	✓	✓	✓	✓	✓	✓
	Cl⁻ content		✓				
	C₃A content				✓		✓
	SO₃ content				✓		✓
	Alkali content					✓	
	Alkali reactivity						✓
Structural level	Cover quality	✓	✓	✓	✓		
	Cover thickness	✓	✓	✓	✓		
	Crack control	✓	✓	✓	✓		

Source: adapted from Li K.F. *et al.* 2015, Table 4. Reproduced with permission of ASCE.
AAR: alkali–aggregate reaction; DEF: delayed ettringite formation.

The design working life of the whole project is 120 years, and the design lives should be specified to RC/PC elements as the starting point of the model-based design process. The principle of this specification is that the principal elements adopt the same working life as the whole project (i.e., 120 years), and those of secondary or replaceable elements can be shorter. For carbonation-induced and chloride-induced corrosion processes, two DLSs can be defined: (a) corrosion initiation state; (b) corrosion to an acceptable extent.[1] In the HZM project, PC elements, principal RC elements, and RC elements with high maintenance difficulty should adopt (a), while secondary RC elements can adopt (b). Then the safety margin of durability design should be specified. Taking annual failure probability as 0.13% for irreversible processes in the SLS and corresponding reliability index $\beta_1 = 3.0$ (*fib*, 1999), the failure probability is estimated as 14.5% for $t_{SL} = 120$ years, corresponding to $\beta_{SL} = 1.06$. The reliability index corresponding to the DLS (a) is taken as $\beta_{SL} = 1.3$, and the corresponding failure probability $p_f = 9.68\%$. Further discussion of this value is given in Li K.F. *et al.* (2015) and Li Q.W. *et al.* (2015). Table 9.4 provides the principal PC/RC elements with the design service life, environmental actions (for model-based design), and the corresponding DLS.

9.1.3 Model-based Design for Chloride Ingress

On the basis of the specification in Table 9.4, the model-based design is performed for both carbonation-induced corrosion for atmospheric CO_2 exposure and chloride-induced corrosion for marine exposure. The results show that the chloride-induced corrosion dominates the

---

[1] These two DLSs correspond respectively to (C1a, C2a) and (C1b, C2b) in Table 7.2.

Table 9.4 Design lives for PC/RC elements and their durability limit states (DLSs)

Structure	Element	Design life (years)	Environmental actions	DLS	Crack control (mm)
Cable-stayed bridges (navigable spans)	Integral pylon–pier	120	Splashing	(a)	0.15
	Steel box girder	120	—	—	—
	Concrete pavement	120	Marine air	(a)	0.20
	Pier (auxiliary)	120	Splashing/tidal	(a)	0.15
	Bearing platform	120	Splashing/tidal	(a)	0.15
	Pile	120	Immersion	(a)	0.20
Box-girder bridges (non-navigable spans)	RC Deck	120	Marine air	(a)	0.20
	PC box girder (external)	120	Marine air	(a)	0.20
	PC box girder (external)	120	Splashing/tidal	(a)	0.15
	PC box girder (internal)	120	Atmospheric/ marine air	(a)	0.20
	Pier	120	Splashing/tidal	(a)	0.15
	Bearing platform	120	Splashing/tidal/ immersion	(a)	0.15
	Pile	120	Immersion	(a)	0.20
Immersed tube tunnel	Tube (immersed, external)	120	Splashing	(a)	0.15
	Tube (immersed, internal)	120	Atmospheric, marine air	(a)	0.20
	Tube (approach, internal/external)	120	Splashing/tidal	(a)	0.15
	Anti-shock pier	120	Splashing	(a)	0.15
Artificial island	Breakwater (plain concrete)	120	Splashing/tidal	—	—
	Retaining wall	120	Splashing/tidal	—	—

Source: adapted from Li K.F. *et al.* 2015, Table 5. Reproduced with permission of ASCE.

design results; the full design results can be found in Li K.F. *et al.* (2015), and the design procedure for chloride ingress is provided hereafter. The design model for chloride-induced corrosion is adapted from the widely used analytical model of Fick's second law (DuraCrete, 1998; *fib*, 2006). With the DLS specified as the corrosion initiation state (a), the design equation can be written as

$$G = C_{crit} - C_S \left[1 - \mathrm{erf} \left(\frac{x_d}{2\sqrt{D_{Cl}^{app} t_{SL}}} \right) \right] \geq 0 \qquad (9.1)$$

with C_{crit} and C_S (mass% binders) the threshold chloride content for steel corrosion and the concrete surface chloride content respectively, x_d (m) is the concrete cover thickness, D_{Cl}^{app} (m²/s) is the apparent chloride diffusion coefficient of concrete, t_{SL} is the design service life of structural elements, and erf is the mathematical error function. The chloride diffusion coefficient D_{Cl}^{app} adopts an ageing law to account for the decrease of this coefficient with concrete exposure age t:

$$D_{Cl}^{app}(t) = D_{Cl}^0 \left(\frac{t_0}{t}\right)^{n_{Cl}} = D_{Cl}^0 \eta(t_0, t, n_{Cl}) \tag{9.2}$$

with n_{Cl} the ageing exponent for the D_{Cl}^{app} decrease law and D_{Cl}^0 the diffusion coefficient for concrete age t_0. The term $\eta(t_0, t, n)$ is the ageing factor of the diffusion coefficient. The main mechanisms accounting for this decrease are detailed in Chapter 2. Since it is not rational to assume D_{Cl}^{app} decreases infinitely with exposure age, this decrease law is truncated at $t = 30$ years for durability design in HZM project:

$$\eta(t_0, t, n_{Cl})\big|_{t > t_D} = \eta(t_0, t_D, n_{Cl}) \quad \text{with} \quad t_D = 30 \text{ years} \tag{9.3}$$

In other words, the coefficient no longer decreases with exposure age after 30 years' exposure. Thus, this models contains five parameters in total: C_{crit}, C_S, x_d, D_{Cl}^0, and n_{Cl} (or η), for a give target service life t_{SL}. These parameters have important dispersions for a given exposure condition and their statistical nature must be taken into account to guarantee sufficient safety margins for durability design. The model-based design can be performed either by a full probabilistic format using the statistical properties of parameters directly or through a partial factor format with characteristic values of parameters and assigned partial factors (ISO, 1998). The two formats can be equivalent for a specified target reliability level provided that all statistical properties of parameters are known. Here, the design is illustrated through the partial factor format using partial factors calibrated from the statistical properties of these parameters and the target reliability index $\beta_{SL} = 1.3$. The design equation in partial factor format is written as:

$$G_1 = \frac{C_{crit}}{\gamma_C} - \gamma_S C_S \left\{ 1 - \text{erf}\left[\frac{x_d^{nom} - \Delta x_d}{2\sqrt{(\gamma_D D_{Cl}^0)(\gamma_\eta \eta) t_{SL}}}\right] \right\} \geq 0 \tag{9.4}$$

One crucial feature of partial factor design is that these factors should be calibrated from the local statistics of parameters; that is, the statistical properties adapted to the HZM environments. To this purpose, the data for the exposure station and from structural investigations in the past 30 years were collected in the same region as the HZM project site. The statistical properties are presented in Figure 9.2 for the parameters for different exposure zones. The statistics of surface chloride content C_S were performed on 351 concrete specimens in Zhanjiang Exposure Station (China) during 12 years, showing C_S obeys a normal distribution with the following mean values and standard deviations: 2.0% and 0.31% (atmospheric zone), 5.4% and 0.82% (splash zone), 3.8% and 0.58% (tidal zone), and 4.5% and 0.68% (immersed zone). The statistical properties of threshold content C_{crit} are regressed from 68 specimens at the same exposure station. A lognormal distribution, with 0.85% and 0.13% as mean value and standard deviation, is used to describe C_{crit} in the atmospheric zone, and a beta-distribution

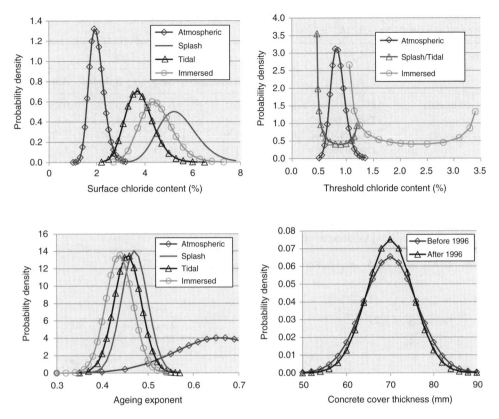

Figure 9.2 Statistical properties for model parameters of chloride ingress: surface chloride content, threshold chloride content, ageing exponent and cover thickness. *Source:* Li K.F. *et al.* 2015, Fig. 1. Reproduced with permission of ASCE.

is adopted for the immersed, tidal, and splash zones. The statistical properties of diffusion coefficient D_{Cl}^0 and the corresponding ageing exponent n_{Cl} are obtained from 395 concrete specimens exposed since 2002. The diffusion coefficient D_{Cl}^0 adopts a lognormal distribution with the variation coefficient of 0.2, and the ageing exponent n_{Cl} follows a normal distribution with a mean and standard deviation of 0.53 and 0.08 (atmospheric zone), 0.47 and 0.028 (splash zone), 0.46 and 0.029 (tidal zone), and 0.44 and 0.029 (immersed zone). The distribution of cover thickness x_d is regressed on 1904 measurements from in-place investigation on marine structures during last 30 years. For cover thickness over 50 mm, x_d observes a normal distribution, with a standard deviation of 6.1 mm for the concrete elements constructed before 1996 and 5.3 mm for elements after 1996.

On the basis of the statistical properties obtained, the partial factors and characteristic values for model parameters are calibrated through a full probabilistic scheme for a target reliability index of $\beta_{SL} = 1.3$, using the model in Equations 9.1–9.3. The detailed calibration process can be found in Li Q.W. *et al.* (2015). Following the custom of structural design, the concrete cover is not calibrated as characteristic value and partial factors, but specified as its

Table 9.5 Calibrated characteristic values and partial factors for chloride ingress design for service life of 50 years and 120 years ($\beta_{SL} = 1.3$)

Parameter		Zone			
		Immersed	Tidal	Splash	Atmospheric
C_{crit} (%binder)	Characteristic value	2.0	0.75	0.75	0.85
	Partial factor γ_c	2.0	1.7	1.7	1.1
C_S (%binder)	Characteristic value	4.50	3.82	5.44	1.98
	Partial factor γ_s	1.05	1.05	1.05	1.1
D_{Cl}^0 (10^{-12} m²/s)	Characteristic value	3.0	3.0	3.0	3.0
	Partial factor γ_D	1.05	1.1	1.1	1.1
Ageing factor η	Characteristic value	0.074	0.067	0.061	0.047
	Partial factor γ_η	1.05	1.05	1.05	1.35
Cover x_d (mm)	Characteristic value x_d^{nom}	60	80	80	50
	Operation error Δx_d	10	10	10	10

Source: Li K.F. *et al.* 2015, Table 6. Reproduced with permission of ASCE.

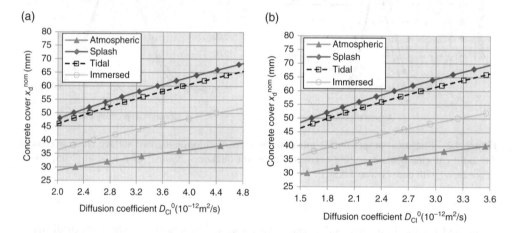

Figure 9.3 Concrete cover thickness and quality for RC elements in different exposure zones in terms of (a) 28-day chloride diffusion coefficient and (b) 56-day chloride diffusion coefficient ($\beta_{SL} = 1.3$, $t_{SL} = 50$ years). *Source:* Li K.F. *et al.* 2015, Fig. 2. Reproduced with permission of ASCE.

minimum requirement x_d^{min} and its nominal value x_d^{nom}, with the difference Δx_d as the operation error. Taking the average values in Figure 9.2 for x_d^{nom} and the 95% guarantee value as x_d^{min}, the operation error Δx_d is estimated as 10 mm. The calibrated partial factors are given in Table 9.5.

The design process takes the characteristic value of chloride diffusion coefficient D_{Cl}^0 and nominal value of concrete cover x_d^{nom} as design parameters. Using Equation 9.4, the characteristic values of these two parameters are calculated for the concrete elements in the HZM project exposed to immersed, tidal, splash, and atmospheric zones. For quality control purposes, the values of the diffusion coefficients are calculated both for 28 days and 56 days. The results are presented in Figures 9.3 and 9.4 for service lives of 50 years and 120 years respectively.

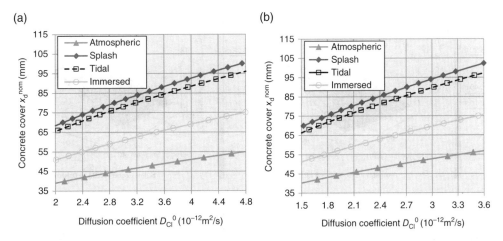

Figure 9.4 Concrete cover thickness and quality of RC elements in different exposure zones in terms of (a) 28-day chloride diffusion coefficient and (b) 56-day chloride diffusion coefficient ($\beta_{SL} = 1.3$, $t_{SL} = 120$ years). *Source:* Li K.F. *et al.* 2015, Fig. 3. Reproduced with permission of ASCE.

Taking $\beta_{SL} = 1.3$ as the target reliability index, Equation 9.4 actually provides a ternary relationship among the expected service life t_{SL}, the specified concrete cover thickness x_d^{nom} and the corresponding required chloride diffusivity. This ternary relation is given in Figure 9.5 with the chloride diffusivity specified at 28 days, and the adopted design parameters for concrete structures in the HZM project are summarized in Table 9.6. In the table, the design results for chloride diffusivity are noted as D_{NSSD} (i.e., non-steady-state diffusion coefficients) and the values of D_{NSSM} (non-steady-state migration coefficients) are for quality control by laboratory accelerated tests. These values will be discussed later.

9.1.4 Quality Control for Design

The design values for chloride diffusivity in Equation 9.4, D_{Cl}^0, are actually the non-steady-state diffusion coefficient of chloride in concrete D_{Cl}^0, corresponding to the apparent chloride diffusion coefficient at the exposed project site; see Figure 2.8. This apparent diffusivity provides an average diffusion rate, up to age t_0, for external chloride ingress and envelopes multiple other processes other than chloride diffusion (e.g., chloride disposition process on the concrete surface and pore saturation effect). By contrast, the rapid chloride migration (RCM) test (Nordtest, 1999) measures the instantaneous rate of chloride diffusion into concrete at a given age driven by an electrical potential; that is, D_{Cl}^{nssm}. Qualitatively, D_{Cl}^{nssd} should be smaller than the D_{Cl}^{nssm} value since the average rate of diffusion should be less than the instantaneous diffusion rate. This is why the design value D_{Cl}^{nssd} cannot be taken directly as the quality control value for the RCM test. Some detailed analysis on the relation between the diffusion and migration coefficients of chloride can be found in the literature (Tang, 1999; Castellote, 2001).

Without attempting to establish such a theoretical relation for the structural concretes in the HZM project, the correlation between the two chloride diffusivity values, D_{Cl}^{nssd} and D_{Cl}^{nssm}, was determined experimentally. Again, the long-term exposure data collected in the Zhanjiang

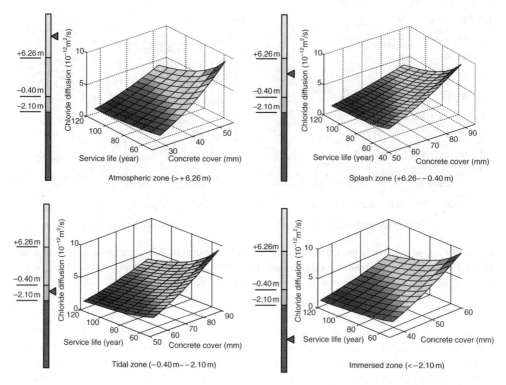

Figure 9.5 Ternary design chart for concrete structures in the HZM project for target reliability index $\beta_{SL} = 1.3$.

Exposure Station (see Figure 2.3) were used to determine the D_{Cl}^{nssd} values for various concrete specimens, and the concretes made from the same formulations were reconstituted in the laboratory and subject to RCM tests at ages of 28 days and 56 days to obtain the values of D_{Cl}^{nssm}. Then, the correlation between the D_{Cl}^{nssd} and D_{Cl}^{nssm} values at the same age is investigated, and the correlation for the age of 28 days is given in Figure 9.6. For the quality control of structural concrete in the HZM project, the D_{Cl}^{nssm} is taken conservatively as two times D_{Cl}^{nssd} in Table 9.6.

9.2 High-Integrity Container for 300 Years

9.2.1 High-Integrity Container and Near-Surface Disposal

The disposal of radioactive wastes (radwaste), generated from civil and military sectors, has been highlighted as an important environmental issue in recent years (Maringer *et al.*, 2013). In particular, the safe disposal of radwaste from nuclear power plants (NPPs) has become a public concern and a technical precondition for NPP design and operation (IAEA, 1995). For the low-level and intermediate-level radwaste (LILW), near-surface disposal is regarded as one economic option as long as the disposal site can be operated with enough safety for both the operator and the public. In near-surface disposal, the radwastes are first disposed or

Table 9.6 Durability design results for RC elements in the HZM project ($\beta_{SL} = 1.3$)

Element	Exposure	Design life (years)	DLS	Cover x_d^{nom} (mm)	Design value D_{Cl}^{nssd} (10^{-12} m²/s) 28 days	56 days	Control value D_{Cl}^{nssm} (10^{-12} m²/s) 28 days	56 days
Box girder	Salt fog	120	(a)	45	3.0	2.0	6.0	4.0
(exterior)	Splashing	120	(a)	80	3.0	2.0	6.0	4.0
Box girder	CO_2	120	(a)	35	–	–	–	–
(interior)	Salt fog	120	(a)	45	3.0	2.0	6.0	4.0
Pier, Pylon	Salt fog	120	(a)	50	3.5	2.2	7.0	4.5
(exterior)	Splashing	120	(a)	85	3.5	2.2	7.0	4.5
Pier (interior)	CO_2	120	(a)	35	–	–	–	–
	Salt fog	120	(a)	50	3.5	2.2	7.0	4.5
Bearing platforms	Splashing	120	(a)	85	3.5	2.2	7.0	4.5
	Immersed	120	(a)	65	3.5	2.2	7.0	4.5
Bored hole pile	Immersed	120	(a)	65	3.5	2.2	7.0	4.5
RC facilities	Splashing	50	(a)	60	3.5	2.2	7.0	4.5
(artificial islands)	CO_2	50	(a)	25	–	–	–	–
Tunnel tube	Splashing	120	(a)	80	3.5	2.2	7.0	4.5
(exterior)								
Tunnel tube	Salt fog	120	(a)	50	3.5	2.2	7.0	4.5
(interior)	CO_2	120	(a)	40	–	–	–	–

Source: Li K.F. *et al.* 2015, Table 8. Reproduced with permission of ASCE.

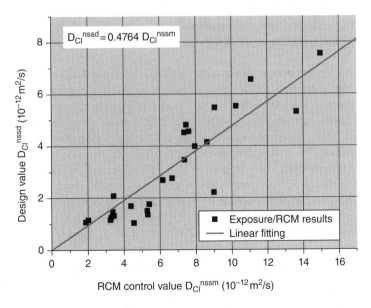

Figure 9.6 Correlation between D_{Cl}^{nssd} and D_{Cl}^{nssm} at age of 28 days for concrete specimens in the HZM project. *Source:* Li K.F. *et al.* 2015, Fig. 5. Reproduced with permission of ASCE.

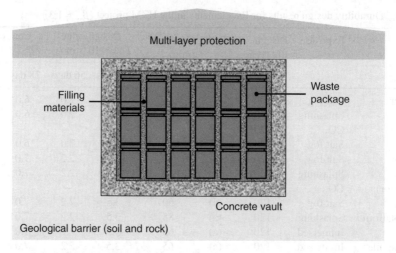

Figure 9.7 Schematic illustrations for near-surface disposal for LILW.

solidified in waste packages, the packages are stored in a concrete vault, and the concrete vault is buried with a multilayer protection on the ground level; see Figure 9.7. The near-surface disposal is designed as such to limit the possible release of radionuclides into the external environment by the multilayers of engineered barrier materials, including the waste package, the inter-package filling (buffer) materials, concrete containment structure (vault), and the geological barriers. Note that cement-based materials are used extensively as barrier materials in near-surface disposal facilities (IAEA, 2001). The life cycle of the disposal structure and facilities is expected to reach 300 years, equal to 10 times the half-life of decay for the radionuclide ^{137}Cs, which is one of the main radioactive sources in disposed LILW. After this duration, the remaining radioactivity of the radwastes is considered to be low enough to attain the exemption level.

The HIC is one design option for the waste package in Figure 9.7. The concept is to fabricate the container with high-performance materials so that the container itself has a long service life and can efficiently reduce the probability of internal radionuclide release (Matte *et al.*, 2004). The HIC can adopt steel, nodular cast iron, high-density polyethylene, or cement-based materials. This chapter is focused on the HIC using high-performance concrete as a constitutive material. The basic requirements for concrete HIC include very low permeability with respect to water ingress, very high durability with respect to the environmental actions, and very high mechanical resistance with respect to accidental shock and impact actions. Furthermore, these requirements should be observed for a service life more than 300 years. The first generation of concrete HICs has been available since the 1990s, and fiber-reinforced high-performance concrete was used to make the container; it was reported that these containers acquired a gas (nitrogen) permeability as low as 5×10^{-18} m^2 and extremely low diffusivity for radionuclides (1.0×10^{-3} cm^2/day for cesium and 1.5×10^{-3} cm^2/day for tritiated water) (Pech *et al.*, 1992). These containers are expected to achieve a service life of 300 years as a disposal unit (Matte *et al.*, 2004).

The container subject to durability design in this chapter belongs to the second generation of concrete containers for radwastes disposal developed at China Nuclear Power Engineering

(a) (b)

Figure 9.8 Concrete container of (a) the first generation according to EJ-914 and (b) the second generation (HIC) developed by CNPE. *Source:* courtesy of CNPE.

(CNPE) Co., Ltd. The first-generation concrete containers were fabricated during the 2000s according to Chinese Industrial Standard EJ-914 (CIS, 2000), and conventional concrete was used to fabricate containers with disposal volumes of 140–950 L and typical wall thicknesses of 150–400 mm; see Figure 9.8a. The second generation of concrete containers aims at the HIC level with disposal volumes of 260–600 L and typical wall thicknesses of 100–150 mm. The requirements of concrete properties are comparable to ANDRA requirements (Specification STE119.581S) for French HIC products (Pech *et al.*, 1992). Actually, the CNPE HIC is closing its research and development phase, the high-performance concrete formulation is fixed, and the sample HIC products have been fabricated, see Figure 9.8b.

Strictly speaking, the HIC is not a concrete structure but an industrial unit made from concrete materials. On the basis of the concrete formulation research and along with the mechanical resistance study of the container, the study in this chapter hereafter is on quantifying the concrete deterioration under environmental actions, and helping the choice of some key parameters in the design (e.g. wall thickness of container) with respect to the main function of the container to limit possible radioactivity release during a disposal life of 300 years.

9.2.2 Design Context

The life cycle of the HIC can be divided into the following phases: empty container storage in the NPP (~5 years), radwaste filling and storage in the NPP (5–10 years), preliminary disposal in the near-surface disposal unit (5 years), and permanent disposal in the near-surface disposal unit (~300 years). During these phases, the first two phases occur in the NPP, and the HIC is mainly subject to indoor environments; the third phase refers to the disposed HIC stored in not yet closed near-surface facilities, thus possibly subject to all actions from the atmospheric environment; the fourth phase corresponds to the permanent disposal period in which the

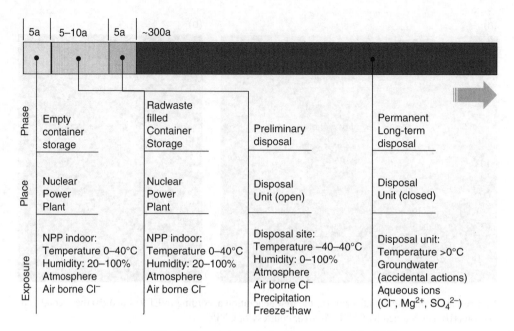

Figure 9.9   Different phases in the disposal life of the HIC.

near-surface facilities are closed and the HIC is subject to near-surface geological actions. These phases are illustrated in Figure 9.9. During these phases the most severe actions belong to the third phase, since both actions from the atmosphere and geology exist. But considering the relatively short duration of the first three phases compared with the permanent disposal phase, this study considers only the actions in the permanent disposal phase.

According to the Chinese technical conditions for near-surface disposal siting, the underground environmental conditions include: temperature above 0 °C, probable groundwater contact, sulfate concentration (SO_4^{2-}) between 100 and 800 mg/L, pH value of groundwater 5.0–9.2, aggressive CO_2 60 mg/L, chloride concentration (Cl^-) between 200 and 5000 mg/L. The HIC is supposed to maintain its main functionality under such environmental actions for a disposal life of 300 years. Under such underground conditions, the possible failure processes leading to deterioration of the containment functionality for the HIC concrete include: the migration of radionuclides from disposed wastes through the container wall to outside; the leaching of the external surface of the HIC by groundwater; the carbonation of the HIC wall and the corresponding corrosion effect on the steel bars/fibers; the ingress of chloride into the HIC wall and the corresponding effect on the corrosion of the steel bars/fibers; the sulfate attack on the HIC external wall; acid and microbes attack on the HIC external surface. These deterioration processes are classified into three categories according to their criticality during the permanent disposal life.

Control process: the occurrence of the process is determinant to the main functionality of the HIC and the near-surface disposal facilities. The process can have either high occurrence probability but slight consequence, or low occurrence probability but serious consequence. The release of radionuclides from the HIC to other engineered barriers, eventually into the environment, belongs to this category.

Table 9.7 Criticality analysis for failure (deterioration) processes for HIC used in near-surface disposal

Aggressive agent	Affected material	Deterioration process	Criticality
Radionuclides	HIC concrete	Radionuclide migration	Control
Groundwater	HIC concrete	HIC external surface leaching	Control
CO_2	HIC concrete/steel bar, fiber	Corrosion of steel bar/fiber	Probable (bar) Negligible (fibre)
Chloride	HIC concrete/steel bar, fiber	Corrosion of steel bar/fiber	Probable (bar) Negligible (fibre)
Salts	HIC concrete	Chemical attack	Probable
Acid	HIC concrete	Chemical attack	Probable
Microbes	HIC concrete	Microbe biological corrosion	Probable

Probable process: the disposal conditions inducing the process are not included in the standard technical conditions for near-surface disposal, but can exist in a particular near-surface siting. The process, once occurred, can have an important impact on the main functionality of the HIC and the disposal facilities. For example, the concentration of aggressive agents in groundwater surpassing the upper range specified in the standard technical conditions belongs to this category. The probable process should be considered in design once this situation is confirmed.

Negligible process: the occurrence of the process is low and the risk can be reduced to a satisfactory level, even screened out, through the raw material selection for HIC fabrication or standard technical requirements for near-surface siting. The sulfate salts attack can be regarded as an inactive process for permanent disposal once the concentration range in the standard technical conditions is met; that is, in the range 100–800 mg/L.

Table 9.7 summarizes the criticality analysis for the probable failure processes for the HIC used in near-surface disposal. Radionuclide migration through HIC wall and leaching by groundwater are retained as control processes for the HIC design. The two corrosion processes, by carbonation or by chloride ingress, are regarded as probable processes if steel bars are used in the HIC; otherwise, these two processes are considered to be negligible if only steel fibers are used in the HIC concrete. This judgement is made on the basis of the design option to adopt a highly corrosion-resistant steel fiber in the HIC concrete.

The design of the HIC includes two parallel but interactive aspects: the constitutive materials for the HIC and the structural design of the container. The former studies the formulation for a concrete with low permeability, high durability, and high mechanical resistance; the latter investigates the proper structural detailing of the container with respect to the possible environmental and mechanical actions during different phases of disposal life. Thus, a table for expected properties for HIC materials is first proposed as technical requirements for HIC materials study; and the structural design, on the basis of these requirements, determines the geometry parameters for the HIC container, including the dimension, the wall thickness, and the special detailing for transportation and placing. No attempt is made here to present all the aspects of the HIC design, and the following is focused on the design choice of the wall thickness for the HIC with specified material properties and selected control processes (i.e., the nuclide transport and the groundwater leaching). To this purpose, the properties requirements are given in Table 9.8. The requirements of the first CNPE container and ANDRA requirements

Table 9.8 Requirements of properties on HIC materials for CNPE container, CNPE HIC, and French HIC

Properties	CNPE container (CIS, 2000)	CNPE HIC	French HIC (Pech *et al.*, 1992)
Compressive strength (MPa)	≥55 (28 days)	≥60 (28 days)	≥50 (28 days)
Tensile strength (MPa)	≥5.0 (28 days)	≥5.5 (28 days)	≥4.5 (28 days)
Shrinkage (10^{-6})	≤600 (28 days)	≤300 (28 days)	≤300
Nitrogen permeability (10^{-18} m^2)	—	≤5.0	≤5.0
Strength decrease after γ-radiation (%)	—	≤20	≤20
Radionuclide diffusivity ^{137}Cs (10^{-3} cm^2/day)	—	≤1.0	≤1.0
Radionuclide diffusivity H$_3$O (10^{-3} cm^2/day)	—	≤1.5	≤1.5
Chloride diffusivity (10^{-12} m^2/s)	—	≤1.5	—
Electrical charge by ASTM C1202 (C)	≤1000	—	—
Porosity (%)		≤9	5–9
Strength decrease after thermal cycle (%)	≤20	—	≤20
Carbonation depth (mm)	≤1	≤0.1	—
Freeze–thaw resistance	—	F400	—
Sulfate resistance	—	KS150	—

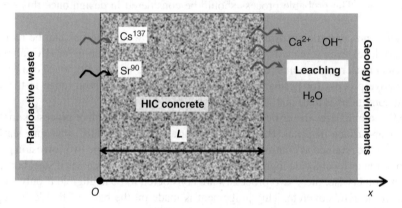

Figure 9.10   Illustration of the most unfavorable design scenario for radioactivity release from the HIC.

for the French HIC are also given in the table for comparison. The carbonation depth, freeze–thaw resistance, and sulfate resistance are measured by laboratory accelerated tests and are described in Section 8.2.3.

9.2.3   Design Models for Control Processes

The most unfavorable design scenario for radioactivity release is the radionuclide migration from the internal side to the external side of the HIC and, simultaneously, the leaching process from the external side to the internal side of the HIC. The leached concrete has enlarged porosity and much lower resistance to radionuclide migration; see Chapter 4. Figure 9.10 illustrates this design scenario: the migration of radionuclides in near-surface disposal considers the ions $^{137}Cs^+$ and $^{90}Sr^{2+}$. The performance criterion for the HIC containment, similar to the DLS in the structural durability design, is taken as the released radioactivity below an expected level at the end of 300 years of permanent disposal.

The radionuclide transport in the HIC wall is modeled as a single-ion transport process in unsaturated cement-based materials. The model from Bejaoui *et al.* (2007) and Itakura *et al.* (2010) is retained as

$$\left[\phi s_1 + \rho_{c,d} K_d^n\right]\frac{\partial c_n}{\partial t} = \nabla\left[\phi D_n(s_1)\frac{\partial c_n}{\partial x}\right] - \lambda_n\left(\phi s_1 + \rho_{c,d} K_d^n\right)c_n \tag{9.5}$$

In this above model, ϕ is the material porosity, s_1 is the volumetric water saturation in pores (0–100%), c_n (kg/m^3 or mol/m^3) is the nuclide concentration in the HIC concrete, D_n (m^2/s) is the nuclide diffusivity in the concrete pore solution, highly dependent on the pore saturation s_1, ρ_c (kg/m^3) is the dry density of the HIC concrete, K_d^n (m^3/kg) is the sorption distribution coefficient of the solid matrix of the HIC concrete with respect to the migrating radionuclides, and λ_n (s^{-1}) is the decay constant for the radionuclides. This model actually couples the transport of radionuclides, denoted by c_n, the decay of the radionuclide, scaled by the decay constant λ_n, and the transport of moisture, represented by the variable s_1. In this study, the moisture condition in the HIC concrete is represented by several constant saturation levels to make a limit value analysis, and no moisture transport is considered. For a specified saturation level s_1^d, Equation 9.5 can be further transformed as

$$\frac{\partial c_n}{\partial t} = D_n^a(s_1^d)\frac{\partial^2 c_n}{\partial x^2} - \lambda_n c_n, \quad D_n^a(s_1^d) = D_n(s_1^d)\left(s_1^d + \frac{\rho_{c,d} K_d^n}{\phi}\right)^{-1} \tag{9.6}$$

As the concrete pores are totally saturated (i.e., $s_1^d = 1.0$), this equation becomes

$$\frac{\partial c_n}{\partial t} = D_{n0}^a\frac{\partial^2 c_n}{\partial x^2} - \lambda c_n, \quad D_{n0}^a = D_{n0}\left(1 + \frac{\rho_{c,d} K_d^n}{\phi}\right)^{-1} \tag{9.7}$$

with D_{n0}^a and D_{n0} representing respectively the apparent diffusivity and effective diffusivity of radionuclide n in saturated concrete of the HIC. The dependence of radionuclide diffusivity on pore saturation follows the power law in Equation 2.3. The leaching process takes the L-2 model in Chapter 4, and the model is described comprehensively from Equations 4.17–4.20 with the parameters involved described in Equations 4.13–4.15.

9.2.4 Model-based Design for 300 Years

With the models described in Section 9.2.3, the problem illustrated in Figure 9.10 can now be solved. To help the choice of the HIC wall thickness, the radionuclide release is simulated for cases with and without external leaching for three different thicknesses (100, 80, and 70 mm) and three levels of radionuclide diffusivity. Two radionuclides are taken into account: $^{137}Cs^+$ and $^{90}Sr^{2+}$. The simulation parameters are provided in Table 9.9, and the diffusion coefficient of Sr is conservatively taken as 10 times the diffusion coefficient of Cs for the same diffusivity level for the HIC concrete.

Table 9.9 Simulation parameters for radioactive release of HIC for permanent disposal

Category	Parameter	Value
Radionuclide migration	Concrete dry density $\rho_{c,d}$ (kg/m³)	2400
	Concrete capillary porosity ϕ	0.09
	Concrete pore saturation s_l	1.0
	Saturated diffusivity of ^{137}Cs$^+$ in concrete D_{Cs0} (mm²/day)	0.01; 0.1; 1.0
	Distribution coefficient of ^{137}Cs$^+$ in concrete K_{d0}^{Cs} (m³/kg)	0.001
	Decay constant λ_{Cs} (s^{-1})	7.32×10^{-10}
	Saturated diffusivity of ^{90}Sr^{2+} in concrete D_{Sr0} (mm²/day)	0.1; 1.0; 10.0
	Distribution coefficient of ^{90}Sr^{2+} in concrete K_{d0}^{Sr} (m³/kg)	0.05
	Decay constant λ_{Sr} (s^{-1})	7.84×10^{-10}
Leaching	Water-to-binder ratio w/b	0.25
	Hydration degree α_c	0.80
	Initial calcium content in concrete $m_{Ca,0}^s$ (mol/L)	3.82
	Pore solution initial pH value	12.4
	Calcium concentration for C-S-H total dissolution c_{Ca}^l (mol/L)	0.0015
	Ambient temperature T (K)	293.15
	Volume fraction of coarse aggregates v_g	0.365
	Volume fraction of fine aggregates v_s	0.243
	Volume fraction of cement paste v_p	0.392
	Model parameters c_{agg}, d_{agg}	1.5, 0.86
	Ca^{2+} diffusivity in water D_{Ca}^0 at 293.15 K (m²/s)	1.37×10^{-9}
Design option	Thickness of HIC wall L (mm)	100; 80; 70

The HIC concrete is assumed to be saturated (i.e., the pore saturation $s_1 = 1.0$), intending to obtain conservative design results. Among the three levels of radionuclide diffusivity, the middle values (i.e., 0.1 mm²/day for Cs and 1.0 mm²/day for Sr) correspond to the HIC material requirements in Table 9.8. The lower and higher values are for sensitivity analysis for radioactivity release with respect to the diffusivity. For the simulation of radionuclide migration, the internal surface of the HIC is subjected to a decaying concentration of radionuclide:

$$c_n^b = c_n^0 \exp(-\lambda_n t) \quad \text{with} \quad n = {}^{137}\text{Cs}, {}^{90}\text{Sr} \tag{9.8}$$

In the following results, the radioactivity release is normalized through the ratio between the radioactive release flow at the external surface of the HIC and the radioactivity from the prescribed radionuclide concentration at the internal surface of the HIC; that is, c_n^0. Thus, the initial value of radionuclide concentration c_n^0 can adopt an arbitrary value in the analysis. For the leaching process, the concrete is assumed to contact with underground soft water, and the Ca^{2+} concentration at $x = L$ is set to zero. A finite-volume method is established to solve the combined migration–leaching process in Figure 9.10. The results show that the leaching depth at 300 years attains 6.5 mm and the accumulated Ca^{2+} flow sums up to 17 mol/m² at 300 years.

The simulation results for the normalized radioactive release for 100 mm thickness are presented in Figure 9.11. From the figure, one can see that the radioactive release of ^{90}Sr is very low compared with the release value from ^{137}Cs at the same diffusivity level, and this is due to its larger sorption distribution coefficient, slowing down substantially its migration in the HIC

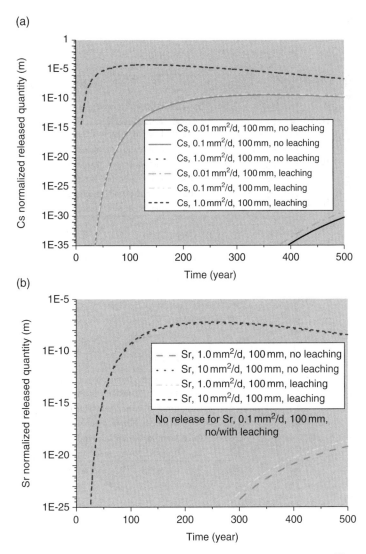

Figure 9.11 Normalized radioactive release with HIC thickness 100 mm for (a) ^{137}Cs and (b) ^{90}Sr.

concrete. The radionuclide diffusivity has a determinant impact on the radioactive release: for the ^{137}Cs case without leaching, the final release values of 300 years are 2.1×10^{-43} m, 3.3×10^{-10} m and 8.1×10^{-6} m for Cs diffusion coefficients of 0.01 mm^2/day, 0.1 mm^2/day, and 1.0 mm^2/day respectively. Note that the lowest diffusivity (e.g., 0.01 mm^2/day for ^{137}Cs$^+$) is not far from the diffusion results of real HIC material in the laboratory. The leaching process does have an impact on the radioactive release, increasing the radioactive release by around 20% for 0.01 mm^2/day and 0.1 mm^2/day cases. The results for ^{137}Cs migration are given for thicknesses of 80 mm and 70 mm in Figure 9.12. Similar observations can be made, and the influence of thickness on radioactive release is also very remarkable: for 80 mm cases without leaching, the Cs normalized radioactive release values are respectively 1.6×10^{-30} m, 8.9×10^{-9} m, and

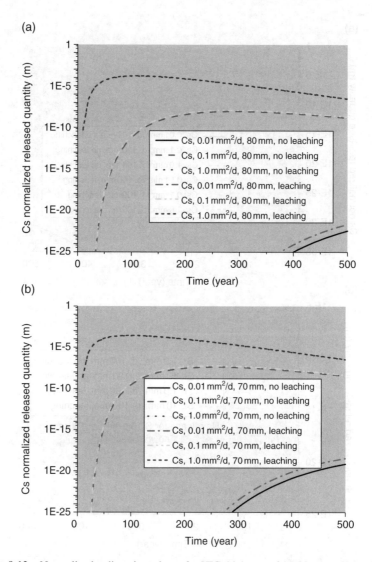

Figure 9.12   Normalized radioactive release for HIC thickness of (a) 80 mm and (b) 70 mm.

1.4×10^{-5} m for diffusivity levels of 0.01 mm²/day, 0.1 mm²/day, and 1.0 mm²/day at the end of 300 years, compared with 3.9×10^{-25} m, 3.8×10^{-8} m and 1.7×10^{-5} m for the 70 mm cases. To facilitate the reading, these results are reorganized in Figure 9.13 for the required values for radionuclide diffusivity in terms of different HIC wall thicknesses and leaching cases.

These data form the basis for the containment capacity of the HIC with respect to the internal radionuclides ($^{137}Cs^+$, $^{90}Sr^{2+}$) migration and external leaching. The safety assessment of near-surface disposal can use these results as basic data to evaluate the radioactive release of the whole disposal facility and the related environmental impact. These results also provide a sensitivity analysis of the radioactive release of the HIC unit with respect to the HIC thickness and external leaching. Both options with 100 mm and 80 mm thickness show a very low release level, compared with the results reported in the literature for an HIC concrete container

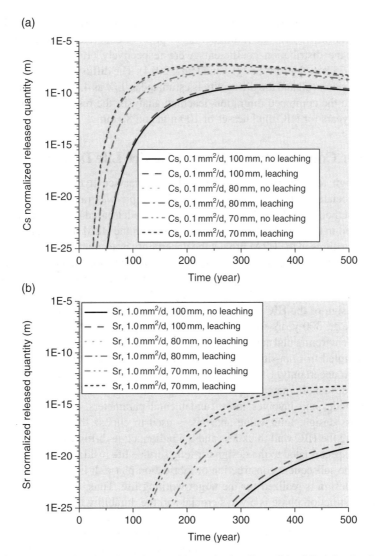

Figure 9.13 Normalized radioactive release with required radionuclide diffusivity for (a) ^{137}Cs and (b) [90]Sr.

(Bejaoui *et al.*, 2007). The mechanical study in parallel favored 100 mm thickness to limit the damage extent of the HIC under shock actions, which was confirmed by the damage patterns recorded on the sample HIC after a dropping test in the prefabrication factory. Accordingly, the thickness of 100 mm is finally chosen for the design of the CNPE HIC in cylinder form.

Further Analysis

The absolute value of radionuclide release from the HIC is also estimated by prescribed internal radionuclide concentrations and measured properties on the high-performance fiber-reinforced concrete developed for the CNPE HIC. The initial concentrations of radionuclides are taken as 10^{-8} mol/L for $^{137}Cs^+$ and 10^{-10} mol/L for $^{90}Sr^{2+}$ following the analysis in the NAGRA report (Bradbury and Sarott, 1994). The radionuclide sorption of concrete was measured for $^{137}Cs^+$ and $^{90}Sr^{2+}$, and the distribution coefficients were respectively $(1.40-1.46) \times 10^{-3}$ m^3/kg for $^{137}Cs^+$ and 70.7–75.8 m^3/kg for $^{90}Sr^{2+}$ (Li and Pang, 2014). The diffusivity $^{137}Cs^+$ was measured by a natural diffusion cell and the diffusivity is estimated as low as 4.7×10^{-3} mm^2/day. Putting these values into the combined migration–leaching analysis, the results show no radioactive release for 300 years for HIC thicknesses of 100 mm and 80 mm.

9.3   Further Considerations for Long Service Life Design

Durability design for concrete structures with long service lives first has to solve the uncertainty associated with the design process. This topic was raised in Section 7.2.3 for model-based methods for durability design. For the model-based design part, two distinct methods are used in this chapter for concrete structures in the HZM project and the HIC unit. For concrete structures in the HZM project the uncertainty is accounted for by considering the statistical properties of the design parameters, and the partial factors in the design equation are calibrated on the basis of these statistical properties. This design process provides the results with a specified safety margin, failure probability with respect to the specified DLS, and the model-based design of the HIC unit gives results with sensitivity information. Note that for very long life (e.g., 300 years for disposal), determining the statistical properties of design parameters and environmental actions can be a huge challenge in itself. Without the statistical properties calibrated from in-situ data, the probabilistic approach cannot help much in mastering the design uncertainty. Under this context, limit value analysis, which is deterministic in nature, provides an alternative approach to help in the design. This approach uses a more sophisticated model and involves more fundamental parameters. For the HIC migration–leaching process design, limit value analysis is used to choose the key parameters of the design, including the HIC wall thickness and the radionuclide diffusivity magnitudes.

The second aspect related to the design for long service life resides in the control of design parameters in the subsequent construction or fabrication phase. It is through this operational phase that the design is realized for the target working life. Thus, the quality control in the subsequent construction phase is always crucial and the durability design should extend into the construction phase to help the quality control of concretes. This control includes two issues: concrete workmanship and parameter-based requirements for concrete in construction. Good concrete workmanship is the basis and precondition for durability, since in the design the concrete is assumed to be homogeneous and contain no thermal or shrinkage cracks. The parameter-based requirements come from the results in the design phases; for example, the requirements on D_{NSSM} values using RCM tests derived from the D_{NSSD} values from the durability design. The transfer of parameter-based requirements should take into account the test method and the potential difference between the values specified from the design and the values measured in laboratory conditions (Tang, 1999; Castellote *et al.*, 2001). For concrete structures in the HZM project, this process is achieved through correlation analysis between the long-term exposure tests and the laboratory RCM tests. For the HIC the key parameter is the radionuclide diffusivity for a given HIC thickness. The radionuclide diffusion test is by no means a quality control one owing to its radioactive nature and long-duration constraints. Accordingly, quality control should adopt an indirect scheme: the concrete quality is to be checked by nondestructive methods on each HIC unit for a given age after fabrication, and a visual check on concrete quality is performed on a destroyed HIC unit (cut into half) at a ratio of $1:50$. The radionuclide diffusion test serves only as verification for the study of the concrete formulation.

The inspection and monitoring are the third aspect associated with long service life design. Through durability design and quality control, some uncertainty associated with the performance for structures with long service lives will still remain. To reduce the risk of unexpected failure of the structure or structural elements, preventive planning for inspection, monitoring, and related maintenance activities can be efficient in maintaining the durability performance during the service life. The maintenance design and the related LCC concepts were introduced and demonstrated in Section 7.3.3. Actually, the maintenance design in the detailed design phase can be a starting point of LCC planning and actions in the long service life. The maintenance planning can be optimized according to the inspection and monitoring results obtained from the structure during the service life, and the maintenance actions, in turn, can adjust the inspection and monitoring intensity. This concept applies to both concrete structures in the HZM project and near-surface disposal facilities using HIC units.

10

Codes for Durability Design

The requirements for durability in design codes, if properly specified with respect to the environmental actions, structural service conditions, and raw materials, can efficiently promote the durability of concrete structures through the design procedure. Nowadays, the design codes for concrete structures provide the design basis and the requirements for different environmental exposures. This chapter gives a review on the requirements of durability from several design codes in use, and particularly the design code of GB/T 50476 (CNS, 2008). This chapter starts with a state-of-the-art review on the durability requirements from different design codes, including ACI-318 (ACI, 2011), Eurocode (CEN, 2000, 2004), JSCE code (JSCE, 2010) and China codes (CCES, 2005; CNS, 2008). These codes specify the durability requirements from different aspects, and the specification from Eurocode is more systematic. The design code GB/T 50476, the first durability design code for concrete structures in China, is then reviewed comprehensively for its technical background and context, the design basis, design methods, and durability requirements, as well as the underlying reasoning behind these requirements.

10.1 Codes and Standards: State of the Art

10.1.1 Eurocode

Eurocodes are a set of technical specifications issue from the European Committee for Standardisation for the structural design of construction works in the European Union. Eurocodes prescribe the durability-related requirements for concrete both on the material level – in EN206-1, *Concrete-Part1: Specification, Performance, Production and Conformity* (CEN, 2000) – and on structural level – in EN 1992-1-1, *Eurocode 2: Design of concrete structures – part 1-1: General rules and rules for buildings* (CEN, 2004). In these codes, the

---

Durability Design of Concrete Structures: Phenomena, Modeling, and Practice, First Edition. Kefei Li.
© 2016 John Wiley & Sons Singapore Pte. Ltd. Published 2016 by John Wiley & Sons Singapore Pte. Ltd.

environmental actions are defined into different environmental classes and intensity degrees according to the respective deterioration mechanisms. The environmental classification is illustrated in Table 10.1 with the explanatory text for each environmental class and action intensity.

In EN206-1, the durability requirements for different exposure classes are formulated through the prescriptive approach, or the "deem-to-satisfy" approach, limiting the types of constituent materials, specifying the maximum water/cement ratio, minimum cement content, minimum concrete compressive strength class, and/or the minimum entrained air content; see CEN (2000: Article 5.3.2). The informative Annex F provides recommendations for the limiting values for concrete composition for an intended service life of 50 years, maximum aggregate size in the range of 20–32 mm, and cement strength class of 32.5. Alternatively, the requirements can also be formulated through performance-related methods, considering the type and form of structure, local environmental conditions, and level of execution, as well as the required working life, and the relevant principles are given in CEN (2000: Annex J).

The code EN1992-1-1 (CEN, 2004), by Article 4.3, specifies that the durability requirements for concrete structures should be included into structural conception, material selection, construction details, execution, quality control, inspection, verifications, and special measures. More particularly, the requirements for corrosion protection of steel bars include the density, quality, and thickness of concrete cover as well as the cracking width control of elements. Then the code establishes the structural classification for durability requirements on concrete cover thickness considering the exposure class and intensity, intended working life, concrete strength class, geometry of elements, and special quality control. Based on this structural classification, the minimum thickness of concrete cover for durability is prescribed for the conventional reinforcement and prestressing steel. In Annex E, the code recommends the indicative concrete strength classes for durability requirements.

Globally, Eurocodes provide a comprehensive framework for durability requirements considering the environmental actions and the intended working life of structures. The durability requirements from EN206-1 and EN 1992-1-1 are summarized in Table 10.1 for RC elements with an intended working life of 50 years, of which the maximum size of the aggregates is in the range 20–32 mm, and the cement strength class is 32.5. The specified values of minimum concrete cover correspond to the cases where the indicative minimum strength classes are used. Although the requirements in Table 10.1 follow strictly a prescriptive approach, performance-based design is always possible according to Article 5.3.3 of EN206-1.

10.1.2   ACI Code

The ACI 318 code (ACI, 2011) specifies the requirements for structural concretes in buildings. In its most recent version, ACI318-11, the durability requirements are given in Chapter 4 "Durability requirements." In this chapter, the environmental actions are classified into several categories:

- *Exposure F* refers to exterior concrete exposed to moisture and cycles of freezing and thawing, with or without deicing chemicals.
- *Exposure S* applies to concrete in contact with soil or groundwater containing deleterious amounts of water-soluble sulfate ions; that is, more than 0.1% for soils and more than 150 ppm for water.

Table 10.1 Environmental classification and durability requirements from EN206-1 and EN1992-1-1 (intended working life 50 years, maximum size of aggregates 20–32 mm, cement class 32.5)

Mechanism	Class	Exposure	Concrete composition						Minimum concrete cover (mm)	Cracking width (mm)
			Minimum strength class (cubic)	Maximum w/c	Minimum cement content (kg/m³)	Minimum air content (%)	Aggregate	Cement type		
—	X0	No corrosive	C15	—	—	—	—	—	10	0.4
Carbonation (XC)	XC1	Dry/immersed	C25	0.65	260	—	—	—	15	0.4
	XC2	Wet	C30	0.60	280	—	—	—	25	0.3
	XC3	Moderate humidity	C37	0.55	280	—	—	—	25	0.3
	XC4	Dry–wet cycle	C37	0.50	300	—	—	—	30	0.3
Chloride/deicing (XD)	XD1	Moderate humidity	C37	0.55	300	—	—	—	35	0.3
	XD2	Wet	C37	0.55	300	—	—	—	40	0.3
	XD3	Dry–wet cycle	C45	0.50	320	—	—	—	45	0.3
Chloride/seawater (XS)	XS1	Airborne salt	C37	0.50	300	—	—	—	35	0.3
	XS2	Submerged	C45	0.45	320	—	—	—	40	0.3
	XS3	Tidal, splashing	C45	0.45	340	—	—	—	45	0.3
Freeze-thaw (XF)	XF1	Moderate saturation, no salt	C37	0.55	300	—	—	—	45	0.3
	XF2	Moderate saturation, salt	C30	0.55	300	4.0	Freeze–thaw resistance (CEN, 2013)	—	—	—
	XF3	High saturation, no salt	C37	0.50	320	4.0		—	—	—
	XF4	High saturation, salt	C37	0.45	340	4.0		—	—	—
Chemical (XA)	XA1	Slightly aggressive	C37	0.55	300	—	—	Sulfate resisting	—	—
	XA2	Moderately aggressive	C37	0.50	320	—	—		—	—
	XA3	Highly aggressive	C45	0.45	360	—	—		—	—

- *Exposure P* pertains to concrete in contact with water requiring low permeability.
- *Exposure C* necessitates additional protection against corrosion of steel of reinforced or prestressed concrete.

For each environment several classes are specified according to the intensity of environmental actions. The durability requirements are formulated in terms of the concrete strength class and the maximum w/cm ratio (cm: cementitious materials). The additional requirements on the constituents of concrete are specified for each environment: limits on air content and cementitious materials for F exposure, cement type and calcium chloride admixture for S exposure, and maximum chloride content for C exposure. Except the mandatory requirements on the cementitious binders prescribed for sulfate environments, alternative cementitious binders are allowed as long as these binders give satisfactory results from the performance test of type ASTM C1012 (ASTM, 2013b).

The ACI-318 code prescribes the concrete protection for different reinforcement steel bars in cast-in-place and prefabricated elements in Section 7.7, but these requirements are not explicitly related to the exposure environments of elements. It is only stated in Article 7.7.6 that "In corrosive environments or other severe exposure conditions, the concrete cover shall be increased as deemed necessary and specified by the licensed design professional." The environmental classification and the related durability requirements from ACI-318 are summarized in Table 10.2. Note that no explicit link is made between the durability requirements and the intended working life of concrete structures. Thus, the sufficiency of the durability design using ACI-318 requirements should be subjected to quantitative methods to evaluate the working life of structures.

10.1.3   JSCE Code

The Japanese Specification for Concrete Structures – 2007 "Design" (JSCE Guideline for Concrete No.15) (JSCE, 2010) provides the general requirements and standard methods for design of concrete structures. In the "Design: General requirements" part, the code dedicates Chapter 8 to the "Verification of Durability," giving the durability requirements related to different exposure conditions. As the counterpart, in the "Design: Standard methods," "Part 3 Durability design" is implemented to provide the models and parameters to evaluate the criteria related to the durability requirements. Unlike Eurocode and ACI-318 code, the JSCE guideline adopts mainly a performance-based approach for durability design. The exposure conditions are classified into two categories: reinforcement corrosion and concrete deterioration. The requirements of durability are formulated with respect to these two main mechanisms.

For the reinforcement corrosion, the durability requirements are formulated into three criteria: (1) the crack width at the concrete surface is not greater than the specified value; (2) the chloride content at the steel surface remains less than the critical chloride content at the end of service life; (3) the carbonation depth does not reach the critical depth for steel corrosion at the end of service life. The environments of reinforcement corrosion are classified as follows in specifying the allowable crack width:

- *Normal environment*, referring to normal outdoor environment without airborne salts.
- *Corrosive environment*, referring to concrete exposed to cyclic drying and wetting, or in contact with groundwater containing aggressive agents, or concrete immersed in seawater.
- *Severely corrosive environment*, referring to reinforcement subjected to aggressive agents or marine structures subjected to tides and wave splashing.

Table 10.2 Environmental classification and durability requirements for structural concretes by ACI-318 code

Exposure	Class	Conditions	Minimum strength (psi)[a]	Maximum w/cm	Cementitious binder	Air content (%)	Maximum Cl⁻ content (%cm)	CaCl₂ admixture
F	F0	Not exposed to FT cycles	2500	—	—	—	—	—
	F1	FT cycles, occasional moisture exposure	4500	0.45	—	3.5–6.0	—	—
	F2	FT cycles, contact with moisture	4500	0.45	—	4.5–7.5	—	—
	F3	FT cycles, contact with moisture, with salts	4500	0.45	Limited	4.5–7.5	—	—
S	S0	<0.1% (soil), <150 ppm (water)	2500	—	—	—	—	—
	S1	0.1–0.2% (soil), 150–1500 ppm (water) or seawater	4500	0.50	MS[b]	—	—	—
	S2	0.2–2.0% (soil), 1500–10000 ppm (water)	4500	0.45	HS[b]	—	—	Not permitted
	S3	>2.0% (soil), >10000 ppm (water)	4500	0.45	HS + pozzolano/slag[b]	—	—	Not permitted
P	P0	Contact with water, low permeability not required	2500	—	—	—	—	—
	P1	Contact with water, low permeability required	4000	0.50	—	—	—	—
C	C0	Dry concrete, or protected from moisture	2500	—	—	—	1.00 (0.06)[c]	—
	C1	Exposed to moisture, without external chlorides	2500	—	—	—	0.30 (0.06)[c]	—
	C2	Exposed to moisture and external chlorides	5000	0.40	—	—	0.15 (0.06)[c]	—

FT: freeze–thaw.
[a]The unit psi is defined as lb/in², 1 MPa=145 psi.
[b]Notations according to ASTM C1157 (ASTM, 2011a).
[c]The values not in parentheses are for reinforced concrete, and the values in parentheses are for prestressed concrete.

Table 10.3 Durability requirements (partial) for reinforcement corrosion and concrete deterioration from JSCE guideline

Mechanism	Environment	Maximum w/cm	Minimum concrete cover (mm)	Performance requirements	
				Crack width[a]	Criterion
Reinforcement corrosion	Normal	$0.50^b/P^c$	$40^b/P$	$0.005c$	• Carbonation depth
	Corrosive	$0.50^b/P$	$40^b/P$	$0.004c$	• Chloride content
	Severely corrosive	P	P	$0.0035c$	
Freeze–thaw (FT)	Occasional contact with water	P	—	Relative dynamic elastic modulus 60% (70%)[d]	
	Contact with water, moderate FT cycles	P	—	Relative dynamic elastic modulus 60% (70%)	
	Contact with water, frequent FT cycles	P	—	Relative dynamic elastic modulus 70% (85%)	

[a] c is the thickness of concrete cover.
[b] Values for carbonation processes, beam elements and intended working life of 100 years.
[c] P designates the values determined by performance-based approach.
[d] Values not in parentheses are for general elements and values in parentheses are for thin elements.

In terms of these exposure conditions, the allowable crack width of concrete elements is specified; see Table 10.3. For the concrete deterioration, the durability requirements are formulated for both freeze–thaw actions and chemical attack. The criterion for freeze–thaw action states that the relative elastic modulus of concrete should not be less than a specified design value. The relative elastic modulus is obtained from laboratory tests, and the design values are specified for different freeze–thaw exposure conditions, taking into account the water saturation of concrete, the freeze–thaw intensity of the environment, and the section geometry of the elements. For chemical attack, only a conceptual design criterion is proposed: the expected degraded depth of concrete should not be greater than the specified value; for example, the reinforcement cover. No further environmental classification is implemented for chemical attack.

Except for these performance requirements, the JSCE guideline does provide a "deem-to-satisfy" design approach for requirements on reinforcement corrosion by concrete carbonation, giving the maximum w/c values and minimum concrete covers for different elements. According to the guideline, these specified values can be used as the durability requirements for carbonation actions instead of the performance criterion for carbonation. The prescription-based and performance-based requirements for durability are summarized in Table 10.3.

10.1.4 China Codes

Nowadays, in China, several design codes and technical standards exist that treat the durability requirements of concrete structures. The first systematic technical standard for the durability requirements of concrete structures is CCES01-2004 "Guide to the durability design and construction of concrete structures" issue by Chinese Civil Engineering Society (CCES) (CCES, 2005; Li *et al.*, 2008). In this guide, a general prescription framework is established for the requirements for concrete composition, crack control, and construction quality of concrete structures. Together with the requirements on raw materials of concrete, some performance-based indicators are proposed for different environmental actions. Based on this document, several codes and standards were later drafted, including code GB/T 50476 "Code for durability design of concrete structures" (CNS, 2008), and TB10005 "Code for durability design on concrete structure of railway" (CIS, 2010). This chapter is focused on the content of GB/T 50476.

Code GB/T 50476 was the first code for durability design of concrete structures in China at a national level, and its first version was finished and issued in 2008. The reason of being of this code originates from both the accumulated knowledge on durability of concrete structures and the urgent need to extend the service life of concrete structures in China. Since the 1980s China's concrete infrastructure construction has undergone a vast expansion, which is testified to by its cement consumption of 2.42 billion tons in 2013, approaching 60% of the global cement production (4.0 billion tons).[1] However, a report from the China Engineering Academy (Chen, 2004) revealed a critical situation in the service conditions of the concrete infrastructures at the beginning of 2000: for civil residence buildings, 50% needed retrofit and 10% necessitated urgent intervention; the deteriorated highway bridges amounted to 9597 by 2000; 18.8% of railway bridges manifested degradation of different extents; water infiltration failure cases were recorded to be as much as 54% for railway tunnels and 13.2% was recorded for tunnel lining cracking. The statistics of average service life were also alarming: 30–40 years for civil buildings, 20–30 years for industrial buildings, 10–20 years for marine ports before major interventions, 10–20 years for bridges exposed to deicing chemicals before major retrofits. Accordingly, for the concrete structures yet to be constructed, ameliorating their durability has both technical and economic significance.

The drafting of GB/T 50476 had a rather specific technical background; that is, such concrete practice standards as ACI-318 or EN206-1 were not yet available in China around 2000. As a consequence, GB/T 50476 was expected to provide the durability requirements for both concrete materials and concrete structures in a rather general scope. After the issue of GB/T 40476, the code served as the basis for durability technical standards in more specialized engineering sectors. Note that the link between CCES01-2004 and GB/T 50476-2008 is very tight since the two documents shared the same drafting team.

10.2 GB/T 50476: Design Basis

In the terminology, the code GB/T 50476 defines "durability" as "the ability of a concrete structure to maintain its service performance under environmental actions during its expected service life." Three basic concepts of durability design are involved herein: the "environmental action" describes the external solicitation, the expected "service life" defines the valid duration, and the "service performance" to be maintained is the reference limit states. The code GB/T 50476 follows basically a prescriptive approach; that is, the durability requirements are formulated following the environmental actions and the intended working

---

[1] http://www.cembureau.be/about-cement/key-facts-figures.

life of concrete structures. In addition to the prescribed requirements, the code proposes durability indicators related to the environmental actions and the intended working life. Further, the code offers the possibility to use a model-based method to verify the design results. Actually, the model-based method is required for design verification of important concrete structures exposed to chlorides, and the relevant principles are given in Annex A of CNS (2008). This annex provides all the necessary elements for the performance-based design, such as explicit durability limit state (DLS), safety margin for durability, and modeling issues. In this section, the design procedure is presented following the logic line of environmental classification, service life and DLS specification, and the design content and methods.

10.2.1 Environmental Classification

The environmental actions are classified following their respective deterioration mechanisms in concrete structures. In accordance with the state-of-the-art knowledge of concrete durability, the code treats the following processes: carbonation-induced and chloride-induced corrosion of reinforcement steel in concrete, concrete damage by freeze–thaw cycles, concrete scaling by salt crystallization, and concrete degradation by chemicals in groundwater, soil, and polluted atmosphere. The exposure is classified as follows.

- *Atmospheric exposure (Class I)* refers to conditions for corrosion of embedded steel bars in concrete induced by the carbonation of concrete cover.
- *Freeze-thaw exposure (Class II)* refers to the conditions for concrete damage by freeze–thaw cycles due to frost action in cold climates.
- *Marine exposure (Class III)* refers to the conditions for corrosion of embedded steel bars in concrete induced by the ingress of chlorides in a marine environment.
- *Deicing and other salts exposure (Class IV)* refers to the conditions for corrosion of embedded steel bars in concrete induced by the ingress of chlorides in environments other than a marine one.
- *Chemicals exposure (Class V)* refers to the conditions for concrete damage by aggressive agents in soils and groundwater, by polluted air, or by salt crystallization. The agents include SO_4^{2-}, Mg^{2+}, aggressive CO_2 in groundwater and soils, pH value of groundwater, and atmospheric pollutions like automobile exhaust and acid rain.

The subsequent step of environment classification is to define the intensity of each environmental action on concrete structures. The code grades the intensity of all the environmental actions from A to F, with A standing for the slightest action intensity and F the most severe. The environmental classification and the intensity grades under each class are presented in Table 10.4. The intensity grades under each environmental class are divided on the basis of the main influential factors for each deterioration process, which are to be detailed later. In the following, the environmental class refers to both the action type and its intensity. The durability requirements are then formulated on the basis of this classification. One should be cautious about the same intensity under different environment types; for example, I-C and III-C: the two classes just have similar deterioration intensity, but can have different environment-based requirements for durability. In addition to these deterioration processes, the code also treats the internal expansion reactions, such as delayed ettringite formation (DEF) and alkali–aggregate reactions (AARs), the long-term leaching of concrete by soft water, and concrete surface

Table 10.4 Environmental classification by GB/T 50476-2008

Class	Environment	Intensity	Exposure condition
I	Atmospheric	A	Indoor, dry; immersion in water
		B	Constant humid; outdoor without drying–wetting cycles
		C	Drying–wetting cycles
II	Freeze–thaw	C	Moderate frost climate, concrete of high saturation, no salt
			Cold region, concrete of moderate saturation, no salt
		D	Cold region, concrete of moderate saturation, with salt;
			Cold region, concrete of high saturation, no salt
			Moderate frost climate, concrete of high saturation, with salt
		E	Cold region, concrete of high saturation, with salt
III	Marine	C	Immersion in seawater
		D	Exposed to marine air and airborne salts
		E	Heavy salty air, tidal and splash zones in mild climate
		F	Tidal and splashing zones in hot marine climate
IV	Deicing and other salts	C	Slight deicing salt fog; immersion in chloride water; contact with water with low chloride content and drying–wetting cycles
		D	Deicing salt spray; water with medium chloride content and drying–wetting cycles
		E	Direct contact with deicing salts solution; heavy spray of deicing salts; water with high chloride content and drying–wetting cycles
V	Chemicals	C	Aggressive agents in low range
			Exposed to automobile exhaust gas
		D	Aggressive agents in medium range
			Exposed to acid rains with pH ≥ 4.5
		E	Aggressive agents in high range
			Exposed to acid rains with pH < 4.5

erosion and wear. These processes are not included in the environmental classification, and the relevant durability requirements are given either on concrete composition or from concrete cover specifications.

10.2.2 Design Lives and Durability Limit States

The DLS defines the acceptable deterioration extent of structural concrete subject to environmental actions; see Figure 7.1. The code combined the DLS, together with deformation, crack and fatigue control, into the serviceability limit state (SLS) of concrete structures. Note that, in some particular cases, the DLS can be more related to the ultimate limit state than the SLS; for example, the corrosion-induced failure of prestressing tendons or strands in high working stress. On the basis of Table 7.2, three DLSs were defined in the code:

1. *Corrosion initiation* refers to the initiation of the electrochemical process of steel corrosion, which corresponds to the carbonation front reaching concrete cover for carbonation-induced corrosion and the Cl⁻ accumulation reaching its critical content at the steel

surface for chloride-induced corrosion. This DLS applies to Classes I, III, and IV, recommended for elements with prestressed tendons, cold-working bars, rebars with diameter <6 mm.

2. *Corrosion to an acceptable extent* corresponds to a certain degree of steel corrosion where the corrosion depth of steel bars remains inferior to a nominal value (e.g. 0.1 mm) and no parallel cracking to steel bars is created. This DLS applies to Classes I, III, and IV, recommended for elements with conventional steel reinforcement and other embedded metals.

3. *Concrete damage to an acceptable extent* depicts concrete deterioration or scaling to such an extent not to impair the esthetical appearance and mechanical resistance of the structure. This DLS applies to Classes II and V, recommended for all structural concretes.

For model-based methods, the safety margin associated with these DLSs should be specified. The code prescribes a failure probability of 5–10% for these DLSs, corresponding to a reliability index $\beta = 1.3 - 1.6$. Strictly speaking, the prescribed failure probability should be formulated on the basis of annual failure probability (see Section 7.2.3), and the failure probability depends on the intended working life. Here, no such difference is made and the value range of failure probability is comparable to the values adopted by *fib* model code (*fib*, 2010).

The code defines the term "service life" as "the duration of time span for structures during which all intended functions are satisfied under expected loading and environmental actions and expected maintenance actions." So, the term "service life" in the code is equivalent to "design life" (ISO, 2008) and "design working life" (ISO, 1998). The code prescribes service life for three categories of civil structures:

1. *At least 100 years*, for symbolic, monumental buildings, large-scale public establishments, large-span or tall buildings, large-scale municipal bridges, viaducts, and large-scale municipal construction.

2. *At least 50 years*, for residential buildings, medium and small public establishments, medium and short-span bridges, normal municipal constructions, and large-scale industrial constructions.

3. *At least 30 years*, for temporary buildings and industrial constructions.

The term "major repair" is not used to define the duration of service life; that is, the service life is maintained as long as the maintenance, including major repairs, is expected. The building structures are better designed to have no major repairs during the service life, considering that their main function terminates as major repairs occur. However, the major repairs can be planned as maintenance activities during the service life for road and bridge structures because their main functions are not lost at major repairs. The detailed discussion can be found in Section 7.1.3.

10.2.3 Durability Prescriptions

For a given environmental class, expected service life and DLS, the code formulates the durability requirements on both material and structural levels. The material requirements include raw material constituents (content and composition) and curing conditions, as well as material performance indicators under specific environmental actions. Concrete raw material

---

Box 10.1

The cement type is prescribed according to the Chinese cement standard GB175 (CNS, 2007) as follows:

PO, ordinary Portland cement, 85–94% clinker, 6–15% mineral constituents
PI, Portland cement type I, 100% clinker
PII, Portland cement type II, 95–100% clinker, 0–5% admixture (slag, gypsum)
PS, blast-furnace slag Portland cement, 30–80% clinker, 20–70% slag
PF, fly ash–Portland cement, 60–80% clinker, 20–40% fly ash
PP, pozzolana–Portland cement, 50–80% clinker, 20–50% pozzolana
PC, composite–Portland cement, 50–85% clinker, 15–50% mineral constituents
SR, sulfate-resistant Portland cement, C_3S <55%, C_3A <5%, 95–100% clinker
HSR, high sulfate-resistant Portland cement, C_3S <50%, C_3A <3%, 95–100% clinker.

---

constituents and the curing conditions dictate the chemical stability, the microstructure, and the mechanical resistance of structural concretes. For raw material constituents, the code imposes limitations on cement type (see Box 10.1), binder (cementitious materials) content in concrete mixing, binder composition, and water-to-binder ratio. In addition, some environment-based parameters are given for raw materials, such as the chloride content in concrete mixing for corrosive environments III and IV. Owing to the extensive use of mineral admixture in concrete practice, the water-to-binder ratio is adopted instead of water-to-cement ratio. In the code, the reference mineral admixtures in binder include fly ash and slag with FA and SG signifying respectively the mass ratios of fly ash and slag in binder.

The prescription on curing condition is to assure concrete acquires sufficient maturity for the expected performance in construction. The requirements on curing are prescribed in terms of the wet curing period, the strength achieved at the end of curing, and the minimum exposure age of concrete to environments. The material performance indicators reflect the acquired properties in the hardened phase of concrete, including the conventional compressive strength at 28 days and the durability indicators based on laboratory tests for given deterioration processes. On a structural level, the requirements include the concrete cover to reinforcement, by its thickness correlated to the strength, and crack control adapted to different environments. Table 10.5 notes the requirements for the environments I–V, and the detailed requirements are given later. In addition, the code gives the limitation on soluble alkali content (equivalent Na_2O) for AARs and SO_3 content for internal sulfate reaction in structural concretes.

10.3   GB/T 50476: Requirements for Durability

Following the design philosophy and design basis in Section 10.2 in the code GB/T 50476, this section is dedicated to the detailed requirements of durability for different environmental classes and intended working lives. Moreover, the requirements for the prestressed tendons are addressed particularly.

Table 10.5 Durability prescription for requirements of GB/T 50476-2008

Durability prescription		Environmental class				
		I	II	III	IV	V
Material	Cement type	×	×	×	×	×
	Binder content in concrete	×	×	×	×	×
	Binder composition	×	×	×	×	×
	Water-to-binder ratio	×	×	×	×	×
	Environment-specific parameter		×	×	×	
	Curing condition	×	×	×	×	×
	Concrete strength at 28 days	×	×	×	×	×
	Durability indicator		×	×	×	
Structure	Concrete cover	×		×	×	
	Control of crack width	×	×	×	×	×

Source: Li *et al.* 2008, Table 3. Reproduced with permission of Springer.

10.3.1 Atmospheric Environment

In an atmospheric environment the durability design aims to protect RC elements against carbonation-induced corrosion of embedded steel. The corrosion process is composed of concrete carbonation and the subsequent steel corrosion. Apart from the chemical composition of cement hydrates, the concrete carbonation is substantially influenced by the internal humidity, or pore saturation. It is shown that there exists an optimal humidity, near 60%, for CO_2 transport in pores and the carbonation reaction between the dissolved CO_2 and the portlandite $Ca(OH)_2$ in the pore solution (Papadakis and Vayenas, 1991). For the subsequent corrosion process, low humidity makes concrete electrically very resistant, and extremely high humidity drives out the oxygen necessary for the electrochemical process, with the optimal humidity approaching 95% (Bohni, 2005). The detailed mechanisms can be found in Chapter 1. Accordingly the most favorable condition for the steel corrosion is exposure to drying–wetting cycles, under which both carbonation and corrosion processes can develop easily. The action intensity of the atmospheric environment is thus defined in terms of humidity condition, in particular the drying–wetting cycles. The action intensity is divided into I-A, I-B, and I-C, with the exposure conditions and the corresponding design cases defined in Table 10.6. For comparison, the equivalent classification of EN206-1 (CEN, 2000) and EN 1992-1 (CEN, 2004), if it exists, is also noted. The minimum requirements for durability design are then formulated on the basis of this classification.

Table 10.7 summarizes the requirements for reinforced concrete under atmospheric environment for a service life of 50 years, including the requirements on binder (content, cement type, and mineral admixture), maximum *w/b* ratio, strength class, minimum concrete cover, and allowable crack width. Owing to their adverse effect on concrete strength at 28 days and the supplementary curing period, the mineral admixtures, fly ash and slag, are not recommended in the concrete with *w/b* ratio larger than 0.55–0.60 and exposed to dry environments. And the type of cement is also limited to this purpose. Meanwhile, the mineral admixture content can increase in the concretes with lower *w/b* ratios and more favorable environments for curing; for example, I-A (immersion) and I-B (constantly humid). Note that the mineral admixture content, FA and SG,

Table 10.6 Action classification for atmospheric environment

Class	Exposure conditions	Design case	Eurocode class
I-A	Indoor, dry	Indoor members exposed constantly to relative humidity <60%	XC0, XC1
I-B	Immersion in water	Foundation immerged permanently in water	XC1
	Constantly humid	Members buried in humid soils	XC2
	Outdoor without drying–wetting cycles	Exposed but sheltered outdoor members	none
	Indoor, humid without drying–wetting cycles	Indoor members exposed to relative humidity >60%	XC3
I-C	Drying–wetting cycles	Unsheltered outdoor members exposed to rain	XC4

Source: Li *et al.* 2008, Table 4. Reproduced with permission of Springer.

in the table refers to both the minerals blended into cement during cement production and the part added into concrete at proportioning. For concrete curing, a minimum wet curing period of 3 days and an accomplished strength of 50% of 28-day strength are required for construction practice. The cement types in the table follow the Chinese cement standard (CNS, 2007).

It may be interesting to compare the durability requirements from different codes. Figure 10.1 illustrates the required concrete cover strength (cube specimen at 28 days) and thickness for the most severe exposure conditions of carbonation environment defined in GB/T 50476, Eurocode (CEN, 2000; CEN, 2004), and ACI-318 (ACI, 2011). Normally, it is difficult to compare directly the requirements from different codes even for the same deterioration process because different codes base the requirements on their own environment classification and concrete practice. Moreover, the concrete cover thicknesses prescribed by Eurocode and GB/T 50476 are durability requirements without construction tolerance, while the thickness from ACI-318 includes the tolerance. Assuming a construction tolerance of 5–10 mm, Figure 10.1 shows that the required strength and thickness of concrete cover are correlated almost linearly among different codes. The comparison illustrates the established concept that, under atmospheric environment, a higher quality of concrete cover needs less thickness to resist the carbonation process.

10.3.2   Freeze–Thaw Environment

In cold regions, the freeze–thaw cycles can induce phase changes and flow of pore water in concrete. As concrete is highly saturated, these processes can generate internal stress high enough to rupture the material (Coussy, 2005). Moreover, the concrete damage can be greatly enhanced with the presence of salts during the freeze–thaw cycles (Valenza and Scherer, 2006). On the basis of the mechanisms and influential factors present in Chapter 3, the action intensity of the freeze–thaw environment is divided into II-C, II-D, and II-E in Table 10.8 considering the following factors: the frost intensity (lowest temperature and daily temperature fluctuation), the pore saturation of concrete, and the presence of salts. Here, the cold region is defined as the climate of which the average temperature of the annual coldest month is lower than −3 °C, while the moderate frost climate pertains to the case of the lowest monthly

Table 10.7 Minimum requirements of reinforced concrete under atmospheric environment for service life of 50 years in GB/T 50476-2008

Class	Binder (cementitious materials)			Maximum w/b	Minimum strength (MPa)	Minimum cover (mm)	Crack width (mm)
	Content (kg/m³)	Cement type[a]	Mineral admixture				
I-A (dry)	260	PO, PI, PII	—	0.60	C25	20	0.40
	280	PO, PI, PII, PS, PF, PC	FA/0.2 + SG/0.3 ≤1	0.55	C30	15	0.40
I-A (immersion)	260	PO, PI, PII, PS, PF, PC	FA/0.5 + SG/0.7 ≤1	0.60	C25	20	0.40
	280	PO, PI, PII, PS, PF, PC	FA/0.5 + SG/0.7 ≤1	0.55	C30	15	0.40
I-B (constantly humid)	280	PO, PI, PII, PS, PF, PC	FA/0.5 + SG/0.7 ≤1	0.55	C30	20	0.30
	300	PO, PI, PII, PS, PF, PC	FA/0.5 + SG/0.7 ≤1	0.50	C35	15	0.30
I-B (humid, outdoor)	280	PO, PI, PII	—	0.55	C30	20	0.30
	300	PO, PI, PII, PS, PF, PC	FA/0.2 + SG/0.3 ≤1	0.50	C35	15	0.30
I-C	300	PO, PI, PII	FA/0.2 + SG/0.3 ≤1	0.50	C35	30	0.20
	320	PO, PI, PII	FA/0.25 + SG/0.4 ≤1	0.45	C40	25	0.20
	340	PO, PI, PII	FA/0.3 + SG/0.5 ≤1	0.40	C45	20	0.20

Source: Li *et al.* 2008, Table 5. Reproduced with permission of Springer.
[a]See Box 10.1 for the definition of cement types.

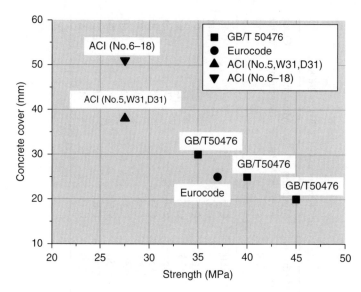

Figure 10.1 Required concrete strength and thickness of concrete cover for atmospheric environments from Eurocode, ACI-318, and GB/T 50476. *Source:* adapted from Li *et al.* 2008, Fig. 1. Reproduced with permission of Springer.

Table 10.8 Action classification for freeze-thaw environment

Class	Exposure condition	Design case	Eurocode class
II-C	Moderate frost climate, concrete of high saturation, no salt	Members exposed to transitional zone of ground water level	None
	Cold region, concrete of moderate saturation, no salt	Vertical surface exposed to rain	XF1
II-D	Cold region, concrete of moderate saturation, with salt	Member exposed to deicing salt spray	XF2
	Cold region, concrete of high saturation, no salt	Horizontal surface exposed to frequent precipitation	XF3
	Moderate frost climate, concrete of high saturation, with salt	Vertical surface exposed to transitional zone of groundwater level	None
II-E	Cold region, concrete of high saturation, with salt	Bridge deck exposed to deicing salts	XF4

Source: Li *et al.* 2008, Table 6. Reproduced with permission of Springer.

average temperature above −3 °C. The corresponding classification of Eurocode is also noted in the table, but this correspondence is not strictly equivalent because the frost intensity is not taken into account in Eurocode and the definition of cold region in GB/T 50476 is adapted to the climate conditions of China.

For material requirements, the content of entrained air is specific to the freeze–thaw environment, recommended as 4.5–6.0% depending on the aggregate size. For concrete practice it is required that at least 1 month should be left between the end of curing and the first exposure to frost actions. The concrete cube strength at 28 days and the durability factor (DF) based on laboratory rapid freeze–thaw test (ASTM, 2003) are chosen as the durability indicators. The pertinence of the indicator from accelerated tests to intended working life was debated in Section 8.2.3. Like the JSCE guideline, GB/T 50476 uses DF as a prescribed indicator in terms of frost intensity and intended working life to enhance durability requirements. The minimum requirements for service life of 50 years are summarized in Table 10.9. The control of crack width is for RC elements to avoid the penetration of aggressive agents into concrete in a freeze–thaw environment.

For concrete of high saturation and exposed to salts, the codes GB/T 50476, Eurocode, and ACI-318 all class it as the most severe condition in a freeze–thaw environment. Figure 10.2 compares the requirements in terms of concrete strength and entrained air content from these codes. For GB/T 50476 two classes are concerned: II-D (with salts in moderate frost climate) and II-E (with salts in cold region). It can be seen that GB/T 50476 and ACI-318 have the same air content requirement and the strength required by Eurocode is situated between the requirements of II-D and II-E from GB/T 50476.

10.3.3   Marine and Deicing Salts Environments

Protecting RC elements against chloride-induced corrosion is the objective of durability design in marine and deicing salt environments. The relevant mechanisms and influential factors have been analyzed in depth in Chapter 2. On this basis, the action intensity of a marine environment is divided into III-C, -D, -E, and -F, and the action intensity of deicing and other salts into IV-C, -D, and -E in Table 10.10. Again, drying–wetting is the most favorable condition for both chloride transport and the corrosion reaction. Table 10.10 provides the detailed exposure conditions and design cases, with the equivalent Eurocode classes also noted. Unlike Eurocode, GB/T 50476 attributes a higher class for airborne salt exposure than for a permanent immersion condition. This classification is based on the field investigations on steel corrosion in marine concrete structures in China, where steel corrosion was rarely observed for reinforcement in complete immersion cases. For Class IV, GB/T 50476 takes into account the groundwater containing chloride and its possible degradation effects on RC elements.

In addition to the conventional material requirements, the initial chloride content in concrete is specific to the marine and deicing salts environments. The chloride content is defined as the mass ratio between the chloride and the binder materials in concrete mixing. The code limits the initial chloride content to 0.1% for reinforced concrete and 0.06% for prestressed concrete. For concrete curing, GB/T 50476 imposes a minimum wet curing period of 7 days and a minimum achieved strength of 70% of standard 28-day strength at the end of curing. Moreover, it is recommended to retard the first exposure of concrete to salts. The durability indicators are chosen as the cube strength at 28 days and the chloride diffusion coefficient D_{RCM} from the rapid chloride migration (RCM) test (Nordtest, 1999). At the structural level, the concrete cover and crack control are specified. These requirements are summarized in Table 10.11 for a service life of 50 years. Figure 10.3 illustrates the requirements on concrete strength and thickness of concrete cover for the most severe chloride exposure conditions from GB/T 50476, Eurocode, and ACI-318. Note that the concrete cover thickness of ACI-318 includes

Table 10.9 Minimum requirements of reinforced concrete under freeze-thaw environment for service life of 50 years in GB/T 50476-2008

| Class | Binder (cementitious materials) | | | Air content (%) | DF (%) | Maximum w/b | Minimum strength (MPa) | Crack width (mm) |
	Content (kg/m³)	Cement type[a]	Mineral admixture					
II-C	320	PO, PI, PII	FA/0.2 + SG/0.2 ≤ 1	—	50	0.45	C40	0.20
	340	PO, PI, PII	FA/0.2 + SG/0.2 ≤ 1	—	50	0.40	C45	0.20
	280	PO, PI, PII	FA/0.3 + SG/0.4 ≤ 1	4.5	50	0.55	C30	0.20
II-D (no salts)	260	PO, PI, PII	FA/0.3 + SG/0.4 ≤ 1	6.0	70	0.50	C35	0.20
II-D (with salts)	280	PO, PI, PII	FA/0.3 + SG/0.4 ≤ 1	6.0	60	0.50	C35	0.20
II-E	280	PO, PI, PII	FA/0.3 + SG/0.4 ≤ 1	6.0	80	0.45	C40	0.15

Source: Li *et al*. 2008, Table 7. Reproduced with permission of Springer.
[a] See Box 10.1 for the definition of cement types.

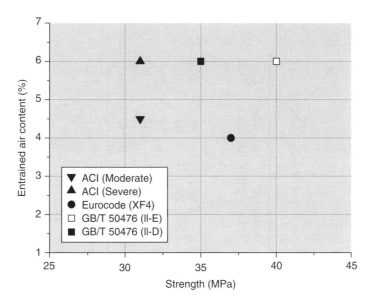

Figure 10.2 Required concrete strength and entrained air content for saturated concrete exposed to salts in freeze–thaw environments from Eurocode, ACI-318, and GB/T 50476. *Source:* adapted from Li *et al*. 2008, Fig. 2. Reproduced with permission of Springer.

Table 10.10 Action classification for marine and deicing salts environments

Class	Exposure condition	Design case	Eurocode class
III-C	Immersion in seawater	Bridge pier permanently in seawater, 1–1.5 m under lowest water level	XS2
III-D	Slight airborne salt	Members situated at 100–300 m from the coastline or 15 m above sea level	XS1
III-E	Heavy salty fog	Members situated under 15 m above sea level or <100 m from the coastline	XS1
	Tidal and splash zones in mild climate	Members in tidal and splash zones of seawater, or within 1.5 m under low water level in a mild climate	XS3
III-F	Tidal and splash zones in hot climate	Members in tidal and splash zones of seawater, or within 1.5 m under low water level in a hot climate	XS3
IV-C	Slight deicing fog	100 m away from traffic	XD1
	Immersion in chloride water	Foundation and basement under permanent groundwater level	XD2
	Low-chloride water with drying–wetting	Members exposed to transitional zone of ground water with Cl⁻ concentration between 100 and 500 mg/L	None
IV-D	Deicing salt spray	Viaduct retaining walls and piers within the reach of salt spray	XD3
	Medium-chloride water with drying–wetting	Members exposed to transitional zone of ground water with Cl⁻ concentration between 500 and 5000 mg/L, walls of seawater swimming pool	None
IV-E	Direct contact with deicing salt solution	Bridge decks	XD3
	Heavy spray of deicing salt	Bridge barriers, viaduct piers, members situated within 10 m from traffic	XD3
	High-chloride water with drying–wetting	Members exposed to transitional zone of groundwater with Cl⁻ concentration >5000 mg/L	XD3

Source: Li *et al*. 2008, Table 9. Reproduced with permission of Springer.

Table 10.11 Minimum requirements of reinforced concrete under marine and deicing salts environments for service life of 50 years in GB/T 50476-2008

Class	Binder (cementitious materials)			Cl⁻ content (%)	D_{RCM} (10^{-12} m²/s)	Maximum w/b	Minimum strength (MPa)	Minimum cover (mm)	Crack width (mm)
	Content (kg/m³)	Cement type[a]	Mineral admixture						
III-C, IV-C	300	PO, PI, PII	FA/0.5 + SG/0.8 ≤ 1	0.1 (RC),	—	0.50	C35	35	0.20
III-D, IV-D	320	PO, PI, PII	FA/0.5 + SG/0.8 ≤ 1	0.06 (PC)	≤10	0.45	C40	45	0.20
III-E, IV-E	340	PO, PI, PII	FA/0.5 + SG/0.8 ≤ 1		≤6	0.40	C45	50	0.15
	360	PO, PI, PII	FA/0.5 + SG/0.8 ≤ 1		≤6	0.36	C50	45	0.15
III-F	360	PO, PI, PII	FA/0.5 + SG/0.8 ≤ 1		≤6	0.36	C50	55	0.15
	380	PO, PI, PII	FA/0.5 + SG/0.8 ≤ 1		≤6	0.36	C50	50	0.15

Source: Li *et al.* 2008, Table 10. Reproduced with permission of Springer.

[a] See Box 10.1 for the definition of cement types.

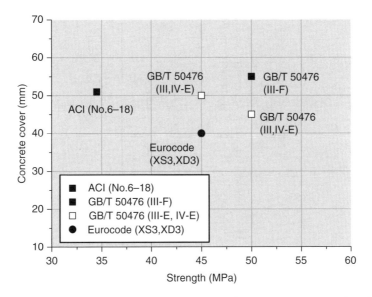

Figure 10.3 Required strength and cover thickness for reinforced concrete exposed to chloride environments from Eurocode, ACI-318, and GB/T 50476. *Source:* adapted from Li *et al.* 2008, Fig. 3. Reproduced with permission of Springer.

the construction tolerance (5–10 mm for cast-in-place concrete) and not for GB/T 50476 and Eurocode. Taking this aspect into account, Figure 10.3 shows that GB/T 50476 imposes thicker concrete cover and higher concrete strength than Eurocode and ACI-318.

10.3.4 Sulfate Environment

The code GB/T 50476 considers the aggressive agents for structural concretes present in natural environments in Class V. The identified aggressive agents include the ions SO_4^{2-}, H^+, and Mg^{2+}, aggressive CO_2 dissolved in groundwater and soils, as well as atmospheric pollution. The induced deterioration processes can be chemical, such as concrete dissolution by acid, or purely physical, such as salt crystallization in a dry climate. Admittedly, the knowledge related to concrete deterioration by these agents is far less advanced compared with other processes. The action intensity of aggressive agents in groundwater and soils is divided into V-C, -D, and -E according to their concentrations in Table 10.12. In the following, this section is focused on the durability requirements for the actions of sulfate salts.

Sulfate attack on structural concrete attracted the attention of engineers as early as 1945 (Lerch, 1945). However, the mechanisms are not yet clarified, and most of the knowledge on sulfate damage is rather laboratory based than experience based (Hobbs *et al.*, 1998; Neville, 2004). Today, it is believed that the detrimental effect of sulfate salts on concrete can be attributed to the chemical reactions between sulfates and cement hydrates, forming expansive products (gypsum and ettringite) and dissolving C-S-H in concrete, and to the physical crystallization of sulfate salts (Skalny *et al.*, 2002). Note that the crystallization of sulfate salts is detailed in Chapter 5 but the chemical reactions between the sulfate salts and cement hydrates

Table 10.12 Action classification for chemical agents in groundwater and soils

Class	SO_4^{2-} concentration (not dry or cold regions)[a]		SO_4^{2-} concentration (dry or cold regions)[a]		Mg^{2+} concentration in groundwater (mg/L)	Groundwater pH	Aggressive CO_2 in groundwater (mg/L)
	Groundwater (mg/L)	Soil (water soluble) (mg/kg)	Groundwater (mg/L)	Soil (water soluble) (mg/kg)			
V-C	200–1000	300–1500	200–500	300–750	300–1000	6.5–5.5	15–30
V-D	1000–4000	1500–6000	500–2000	750–3000	1000–3000	5.5–4.5	30–60
V-E	4000–10000	6000–15000	2000–5000	3000–7500	≥3000	≤4.5	60–100

[a] Dry region refers to the region with the drought factor greater than 2.0, and cold region refers to the region with altitude higher than 3000 m.

Table 10.13   Minimum requirements of reinforced concrete under sulfate attack for service life of 50 years in GB/T 50476-2008

Class	Binders (cementitious materials)			Maximum w/b	Minimum strength (MPa)
	Content (kg/m^3)	Cement type[a]	Mineral admixture		
V-C	320	PO, PI, PII, SR, HSR	FA/0.25 + SG/0.4 ≤ 1	0.45	C40
V-D	340	PO, PI, PII, SR, HSR	FA/0.30 + SG/0.5 ≤ 1	0.40	C45
	360	PO, PI, PII, SR, HSR	0.5 ≤ FA/0.5 + SG/0.8 ≤ 1	0.36	C50
V-E	360	PO, PI, PII, SR, HSR	0.5 ≤ FA/0.5 + SG/0.8 ≤ 1	0.36	C50

Source: Li *et al.* 2008, Table 12. Reproduced with permission of Springer.
[a] See Box 10.1 for the definition of cement types.

are not treated in this book. In Table 10.12, the intensity of sulfate attack is graded for both chemical and physical processes in terms of sulfate concentration, groundwater mobility, soil permeability, and environment temperature. It is intended to make a distinction between the chemical sulfate reaction and the salt crystallization process. According to the available field investigations, structural concrete deterioration by sulfate salt crystallization is more frequent, and much lower sulfate concentration was detected in crystallization damage, especially for concrete elements exposed partially to dry climate and partially to ground-water; see Figure 5.1.

For durability in a sulfate environment, it is of equal importance to achieve a compact concrete and to choose the appropriate binder materials. The detailed requirements are given in Table 10.13 for a service life of 50 years. The minimum requirements on the water-to-binder ratio w/b of concrete are illustrated in Figure 10.4 in terms of the sulfate concentration in groundwater from GB/T 50476, Eurocode, ACI-318, and CSA (CSA, 2004). Note that this figure is only for illustration purposes, and the difference among the requirements should be interpreted with caution because the concrete constituents are very region dependent and the notion of w/b ratio is not the same for the above codes. The codes GB/T 50476, ACI-318, and CSA count all the mineral constituents (slag and fly ash) in binder, while EN206-1 discounts the mineral constituents by a coefficient; see Article 5.2.5.2 in CEN (2000). Taking these into account, one can see that, roughly, GB/T 50476 imposes a w/b lower than ACI-318 and CSA, and GB/T 50476 and EN206-1 have nearly the same requirements for w/b ratios if following the same definition.

10.3.5   Post-tensioned Prestressed Structures

Prestressed concrete structures are generally more sensitive to environment-induced corrosion due to their high stress level in prestressed tendons. Among the prestress systems, the post-tensioned system has more components and relies on field operations to achieve the expected prestress on the structure, so it is more susceptible to construction quality failure and environment-induced deterioration processes. An early investigation on service conditions of prestressed concrete structures showed that, among all structural

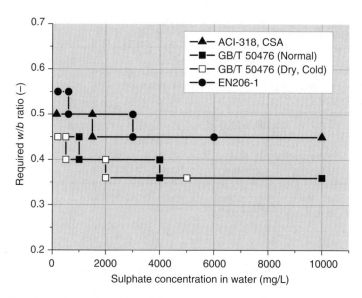

Figure 10.4 Required *w/b* ratio in sulfate environment in terms of SO_4^{2-} concentration in groundwater from EN206-1, ACI-318, GB/T 50476, and CSA. *Source:* adapted from Li *et al.* 2008, Fig. 4. Reproduced with permission of Springer.

prestress systems, the environment-induced service failure of post-tensioned systems occupied a large portion of total failure cases (Podolny, 1992). The durability problem of post-tensioned systems was highlighted by its ban in bridge structures by British authorities from 1992 to 1996 after some major structure failure cases (BDT, 1992). Recently, the practice standards worldwide have been updated and proposed more durable post-tensioned prestress systems for concrete structures (EOTA, 2002; BCS, 2013). In China, the service condition of post-tensioned concrete structures was alarming (Chen, 2004): according to the field investigations on post-tensioned railway bridges, the environment-induced failure cases amounted to nearly 10% of the total bridges, appealing for more protection for post-tensioned prestress systems.

To this purpose, GB/T 50476 dedicates one individual chapter to the protection of post-tensioned prestress systems, and the multi-barrier protection concept is applied. On the basis of available protection techniques, the multilayer protection concept (FDOT, 2004) is adopted and developed into an environment-based protection strategy for prestressed tendons and anchors. Figures 10.5 and 10.6 illustrate respectively the multi-barrier protections for the prestressed tendons (strands and monostrands) and the anchors (buried and exposed anchors) with the significance for each protection barrier. The barriers PS2a, PS3a, and PS4a are respectively the strengthened measures for PS2, PS3, and PS4. Among these protection barriers, all except PS3a concern internal tendons, and all except PS4(a) and PS5 pertain to the protection of external tendons. The available protection techniques for both buried and exposed anchors are summarized on five barriers from anchor component material till concrete external treatment. All barriers except PA3a concern buried anchorage, while all barriers except PA4(a) and PA5 pertain to exposed anchorage protection.

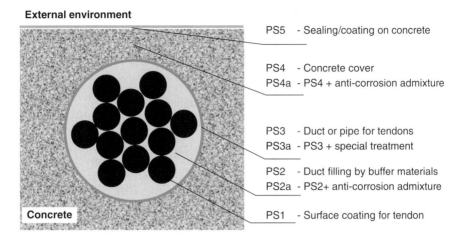

External environment

PS5   - Sealing/coating on concrete

PS4   - Concrete cover
PS4a - PS4 + anti-corrosion admixture

PS3   - Duct or pipe for tendons
PS3a - PS3 + special treatment

PS2   - Duct filling by buffer materials
PS2a - PS2+ anti-corrosion admixture

PS1   - Surface coating for tendon

Concrete

Figure 10.5   Multi-barrier protection illustrations for post-tensioned prestressed tendons.

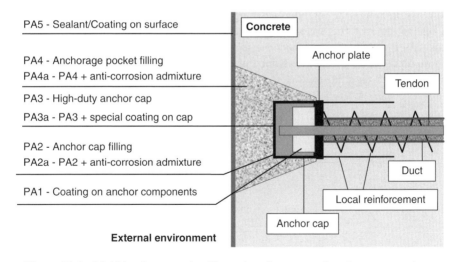

PA5 - Sealant/Coating on surface

PA4 - Anchorage pocket filling
PA4a - PA4 + anti-corrosion admixture

PA3 - High-duty anchor cap
PA3a - PA3 + special coating on cap

PA2 - Anchor cap filling
PA2a - PA2 + anti-corrosion admixture

PA1 - Coating on anchor components

External environment

Concrete

Anchor plate

Tendon

Duct

Local reinforcement

Anchor cap

Figure 10.6   Multi-barrier protection illustrations for post-tensioned prestress anchors.

Table 10.14 Multi-barrier requirements of protection for prestress systems from GB/T50476-2008

Class	Tendon (PS-)		Anchorage (PA-)	
	Internal	External	Buried	Exposed
I-A,B	PS2, PS4	PS2, PS3	PA4	PA2, PA3
I-C	PS2, PS3, PS4	PS2a, PS3	PA2, PA3, PA4	PA2a, PA3
II-C,D^a	PS2, PS3, PS4	PS2a, PS3	PA2, PA3, PA4	PA2a, PA3
II-D^b,E	PS2a, PS3, PS4	PS2a, PS3a	PA2a, PA3, PA4	PA2a, PA3a
III-C,D	PS2a, PS3, PS4	PS2a, PS3a	PA2a, PA3, PA4	PA2a, PA3a
III-E	PS2a, PS3, PS4, PS5	PS1, PS2a, PS3	PA2a, PA3, PA4, PA5	Not allowed
III-F	PS1, PS2a, PS3, PS4, PS5	PS1, PS2a, PS3a	PA1, PA2a, PA3, PA4, PA5	Not allowed
IV-C,D	PS2a, PS3, PS4	PS2a, PS3a	PA2a, PA3, PA4	PA2a, PA3a
IV-E	PS2a, PS3, PS4, PS5	PS1, PS2a, PS3	PA2a, PA3, PA4, PA5	Not allowed
V-C,D	PS2a, PS3, PS4	PS2a, PS3a	PA2a, PA3, PA4	PA2a, PA3
V-E	PS2a, PS3, PS4, PS5	PS1, PS2a, PS3	PA2a, PA3, PA4, PA5	Not allowed

Source: Li *et al.* 2008, Table 14. Reproduced with permission of Springer.
^{a} II-D exposure without salts.
^{b} II-D exposure with salts.

On the basis of environmental classes, GB/T 50476 recommends a multi-barrier and environment-based protection strategy in Table 10.14. Owing to the limited knowledge on deterioration and lack of experience, these requirements are related to the environmental class but not to the intended working life of the structure. These requirements are based on the state of the art of the protection techniques of post-tensioned prestress systems, and the strategy is susceptible to being updated by further technical advancement both on the level of post-tensioned prestressing kits and on the level of the protection materials and techniques.

References

AASHTO (2004) *LRFD Bridge Design Specifications*, third edn, American Association of State Highway and Transportation Officials, Washington DC, US.

Achintya, P.M.M. (2003) Behaviour of concrete in freeze-thaw environment of sea water. *Journal of the Institution of Engineers. India. Civil Engineering Division*, **84**, 96–101.

ACI (1988) *Cold Weather Concreting (306R-88)*, American Concrete Institute, Farmington Hills, PA.

ACI (2011) *Building Code Requirements for Structural Concrete (ACI-318-11)*, American Concrete Institute, Farmington Hills, US.

ACI (2013) *Requirements for Design & Construction of Concrete Structures for Containment of Refrigerated Liquefied Gases & Commentary (ACI 376-11)*, American Concrete Institute, Farmington Hills, MI.

AFGC (2007) *Concrete Design for a Given Structure Service Life-Durability Indicators*, Association Française de Génie Civil, Paris.

Ahmed, S., Azad, A.K. and Loughlin, K.F. (2012) Effect of the key mixture parameters on tortuosity and permeability of concrete. *Journal of Advanced Concrete Technology*, **10**, 86–94.

Allard B., Eliasson L., Andersson K. (1984) Sorption of Cs I and actinides in concrete system. SKB Technical Report 84-15, Swedish Nuclear Fuel and Waste Management Co., Stockholm.

Almeida I.R. (1991) Resistance of high strength concrete to sulphate attack: soaking and drying test. Concrete durability. In *ACI SP-100*. American Concrete Institute, pp. 1073–1092.

Al-Omari, A., Beck, K., Brunetaud, X. *et al.* (2015) Critical degree of saturation: a control factor of freeze–thaw damage of porous limestones at Castle of Chambord, France. *Engineering Geology*, **185**, 71–80.

Ammoura, L., Xueref-Remy, I., Fros, V. *et al.* (2014) Atmospheric measurements of ratios between CO_2 and co-emitted species from traffic: a tunnel study in the Paris megacity. *Atmospheric Chemistry and Physics*, **14**, 12871–12882.

Andrade C., Mancini G. (2011) *Modelling of Corroding Concrete Structures*. RILEM Bookseries, Vol. **5**, Springer, Dordrecht.

Andrade, C., Castellote, M. and Andrea, R. (2011) Measurement of ageing effect on chloride diffusion coefficients in cementitious materials. *Journal of Nuclear Materials*, **412** (1), 209–216.

Angst, U., Elsener, B., Larsen, C.K. and Vennesland, Ø. (2009) Critical chloride content in reinforced concrete – a review. *Cement and Concrete Research*, **39**, 1122–1138.

Archie, G.E. (1942) The electrical resistivity log as an aid in determining some reservoir characteristics. *Petroleum Transactions of AIME*, **146**, 54–62.

ASTM (2003) *Standard Test Method for Resistance of Concrete to Rapid Freezing and Thawing (ASTM C666-03)*, ASTM International, West Conshohocken, PA.

Durability Design of Concrete Structures: Phenomena, Modeling, and Practice, First Edition. Kefei Li.
© 2016 John Wiley & Sons Singapore Pte. Ltd. Published 2016 by John Wiley & Sons Singapore Pte. Ltd.

ASTM (2009) *Standard Test Method for Corrosion Potentials of Uncoated Reinforcing Steel in Concrete (ASTM C876-09)*, ASTM International, West Conshohocken, PA.

ASTM (2011a) *Standard Performance Specification for Hydraulic Cement (ASTM C1157-11)*, ASTM International, West Conshohocken, PA.

ASTM (2011b) *Standard Test Method for Determining the Apparent Chloride Diffusion Coefficient of Cementitious Mixtures by Bulk Diffusion (ASTM C1556-11a)*, ASTM International, West Conshohocken, PA.

ASTM (2012a) *Microscopical Determination of Parameters of the Air-void System in Hardened Concrete (ASTM C457-12)*, ASTM International, West Conshohocken, PA.

ASTM (2012b) *Standard Test Method for Scaling Resistance of Concrete Surfaces Exposed to Deicing Chemicals (ASTM C672-12)*, ASTM International, West Conshohocken, PA.

ASTM (2012c) *Standard Test Method for Flexural Performance of Fiber-Reinforced Concrete (Using Beam With Third-Point Loading) (ASTM C1609-12)*, ASTM International, West Conshohocken, PA.

ASTM (2012d) *Standard Test Method for Bulk Electrical Conductivity of Hardened Concrete (ASTM C1760-12)*, ASTM International, West Conshohocken, PA.

ASTM (2013a) *Standard Test Method for Density, Absorption, and Voids in Hardened Concrete (ASTM C642-13)*, ASTM International, West Conshohocken, PA.

ASTM (2013b) *Standard Test Method for Length Change of Hydraulic-Cement Mortars Exposed to a Sulfate Solution (ASTM C1012-13)*, ASTM International, West Conshohocken, PA.

ASTM (2013c) *Standard Test Method for Measurement of Rate of Absorption of Water by Hydraulic-Cement Concretes (ASTM C1585-13)*, ASTM International, West Conshohocken, PA.

ASTM (2014) *Standard Test Method for Compressive Strength of Cylindrical Concrete Specimens (ASTM C39-14a)*, ASTM International, West Conshohocken, PA.

Atkinson, A., Hearne, J.A. and Knights, C.F. (1989) Aqueous chemistry and thermodynamic modelling of CaO–SiO_2–H_2O gels. *Journal of the Chemical Society, Dalton Transactions*, **12**, 2371–2379.

Bamforth, P.B. (1999) The derivation of input data for modelling chloride ingress from eight-year UK coastal exposure trials. *Magazine of Concrete Research*, **51** (2), 87–96.

Banthia, N. and Bhargava, A. (2007) Permeability of stressed concrete and role of fiber reinforcement. *ACI Materials Journal*, **104**, 70–77.

Baroghel-Bouny, V. (2007) Water vapour sorption experiments on hardened cementitious materials. Part II: essential tool for assessment of transport properties and for durability prediction. *Cement and Concrete Research*, **37**, 438–454.

Baroghel-Bouny, V., Gawsewitch, J., Belin, P. *et al.* (2004) Aging of concrete in natural environments: an experiment for the 21st century, IV. Results on cores extracted from field-exposed test specimens of various sites as part of the first measurement sequence. *Bulletin des Laboratoires des Ponts et Chaussées*, **249**, 49–100.

Basheer, P., Andrews, R., Robinson, D. and Long, A.E. (2005) PERMIT ion migration test for measuring the chloride ion transport of concrete on site. *NDT & E International*, **38**, 219–229.

BCS (2002) *Durable Post-Tensioned Concrete Bridges*, 2nd edition. Technical Report No.47, British Concrete Society, UK.

BCS (2013) Durable post-tensioned concrete structures. Technical Report 72, British Concrete Society, Berkshire, UK.

BDT (1992) Standards for post-tensioned prestressed bridges to be reviewed. British Department of Transport, 25 September 1992, Press Notice No. 260, London.

Bejaoui, S., Sercombe, J., Mugler, C. and Peycelon, H. (2007) Modelling of radionuclide release from a concrete container. *Transport in Porous Media*, **69**, 89–107.

Bertolini, L., Elsener, B., Pedeferri, P. and Polder, R. (2004) *Corrosion of Steel in Concrete*, Wiley-VCH, London.

Bodin, J., Delay, F. and Marsily, G. (2003) Solute transport in a single fracture with negligible matrix permeability 1. Fundamental mechanisms. *Hydrogeology Journal*, **11**, 414–433.

Boel, V., Audenaert, K. and De Schutter, G. (2008) Gas permeability and capillary porosity of self-compacting concrete. *Materials and Structures*, **41**, 1283–1290.

Bohni, H. (2005) *Corrosion in Reinforced Concrete Structures*, Woodhead Publishing, Cambridge.

Bradbury M.H., Sarott F.-A. (1994) Sorption databases for the cementitious near-field of a UILW repository for performance assessment. Technical report 93-08, NAGRA, Wettingen, Switzerland.

BSI (2000) *Maritime structures – Part 1: Code of Practice for General Criteria (BS6349-1:2000)*, British Standard, UK.

BSI (2003) *Steel, Concrete and Composite Bridges (BS5400-1:1998)*, British Standard, UK.

BSI (2009) *Testing Hardened Concrete, Part 8: Depth of Penetration of Water under Pressure (BS EN 12390-8:2009)*, British Standard, UK.

Cady, P.D. (1983) Inflation and highway economic analysis. *Journal of Transportation Engineering, ASCE*, **109** (5), 631–639.

Castel, A., François, R. and Arliguie, G. (1999) Effect of loading on carbonation penetration in reinforced concrete elements. *Cement and Concrete Research*, **29**, 561–565.

Castel, A., Coronelli, D., François, R. and Cleland, D. (2011) Modelling the stiffness reduction for corroded reinforced concrete beams after cracking, in *Modelling of Corroding Concrete Structures* (eds C. Andrade and G. Mancini), Springer, Dordrecht, pp. 219–230.

Castellote, M., Andrade, C. and Alonso, C. (2001) Non-steady chloride diffusion coefficients obtained from migration and natural diffusion tests. Part II: different experimental conditions. Joint relations. *Materials and Structures*, **34**, 323–331.

CCES (2005) *Guide for Durability Design and Construction of Concrete Structures (CCES01-2004)*, China Civil Engineering Society, Beijing.

CEB (1990) *CEB-FIP Model Code 1990*, Comité Euro-International du Béton, Lausanne.

CECS (2007) *Standard for Durability Assessment of Concrete Structures (CECS220-2007)*, China Association of Engineering Construction Standardization, Beijing.

CEN (2000) *Concrete – Part 1: Specification, Performance, Production and Conformity (EN 206-1)*, European Committee for Standardization, Brussels.

CEN (2002) *Eurocode – Basis of Structural Design (EN1990:2002)*, European Committee for Standardization, Brussels.

CEN (2004) *Eurocode 2: Design of Concrete Structures – Part 1-1: General Rules and Rules for Buildings (EN1992-1-1:2004)*, European Committee for Standardization, Brussels.

CEN (2013) *Aggregates for Concrete (EN 12620:2013)*, European Committee for Standardization, Brussels.

Chaussadent T. (1999) *State-of-the-Art and Reflection on the Carbonation of Reinforced Concrete*. Etudes et Recherches des Laboratoires des Ponts et Chaussées, Série Ouvrages d'Art, No. OA 29. LCPC, Paris (in French).

Chen, J.J., Kwan, A.K.H. and Jiang, Y. (2014) Adding limestone fines as cement paste replacement to reduce water permeability and sorptivity of concrete. *Construction and Building Materials*, **56**, 87–93.

Chen, Z.Y. (2004) Safety and durability: state-of-the-art, problems and preventions, in *Report on Safety and Durability of Engineering Structures*, China Academy of Engineering, China Building Industry Publishing, Beijing, pp. 225–266.

Chindaprasirt, P., Homwuttiwong, S. and Jaturapitakkul, C. (2007) Strength and water permeability of concrete containing palm oil fuel ash and rice husk–bark ash. *Construction and Building Materials*, **21**, 1492–1499.

Chlortest (2005) Resistance of concrete to chloride ingress – from laboratory tests to in-field performance. Project GRD1-2002-71808, Swedish National Testing and Research Institute, Borås, Sweden.

CIS (2000) *Concrete Container for Low-and Intermediate-Level Radioactive Solid Wastes (EJ914-2000)*, China Industry Standard, Beijing.

CIS (2010) *Code for Durability Design on Concrete Structures of Railway (TB10005)*, China Industry Standard, Beijing.

CNS (2007) *Common Portland cement (GB175-2007)*, China National Standard, Beijing.

CNS (2008) *Code for Durability Design of Concrete Structures (GB/T50476-2008)*. China National Standard, Beijing (in Chinese).

CNS (2009) *Standard for Test Methods of Long-term Performance and Durability of Ordinary Concrete (GB/T50082-2009)*. China National Standard, Beijing (in Chinese).

Constantinides, G. and Ulm, F.-J. (2004) The effect of two types of C-S-H on the elasticity of cement-based materials: results from nanoindentation and micromechanical modelling. *Cement and Concrete Research*, **34**, 67–80.

Coussy, O. (2005) Poromechanics of freezing materials. *Journal of Mechanics and Physics of Solids*, **53**, 1689–1718.

Coussy, O. (2006a) Deformation and stress from in-pore drying-induced crystallization of salt. *Journal of the Mechanics and Physics of Solids*, **54**, 1517–1547.

Coussy, O. (2006b) *Poromechanics*, John Wiley & Sons, Ltd, Chichester, UK.

Coussy, O. (2010) *Mechanics and Physics of Porous Solids*, John Wiley & Sons, Ltd, Chichester.

Coussy, O. and Fen-Chong, T. (2005) Crystallization, pore relaxation and micro-cryosuction in cohesive porous materials. *Comptes Rendus Mécanique*, **333**, 507–512.

CSA (2000) *Canadian Highway Bridge Design Code (CAN-CSA-S6-00)*, CSA International, Toronto.

CSA (2004) *Concrete Materials and Methods of Concrete Construction (A23.1-04)*, Canadian Standard Association, Ontario.

DARTS (2004) Durable and reliable tunnel structures: data. Report 2.3, Contract G1RD-CT-2000-00476, Project GrD1-25633, European Commission.

Desmettre, C. and Charron, J.P. (2013) Water permeability of reinforced concrete subjected to cyclic tensile loading. *ACI Materials Journal*, **110**, 67–78.

Dong, Z.P., Niu, D.T., Liu, X.F. and Wang, Q.L. (2006) Calculation method for corrosion initiation time of rebars in atmospheric environment. *Jounal of Xi'an University of Architecture and Technology (Natural Science Edition)*, **38** (2), 204–209.

Duan, A., Jin, W.L. and Qian, J.R. (2011) Effect of freeze–thaw cycles on the stress–strain curves of unconfined and confined concrete. *Materials and Structures*, **44**, 1309–1324.

Dullien, F.A.L. (1992) *Porous Media: Fluid Transport and Pore Structure*, Academic Press, San Diego, CA.

Dunster, A.M. (1989) An investigation of the carbonation of cement paste using trimethylsilylation. *Advances in Cement Research*, **2** (7), 99–106.

DuraCrete (1998) Probabilistic performance based durability design of concrete structures: modelling of degradation. Document BE95-1347/R4-5, Netherlands.

DuraCrete (1999) Compliance testing for probabilistic design purposes: probabilistic performance based durability design of concrete structures. Document BE95-1347/R8, Netherlands.

DuraCrete (2000) Probabilistic performance based durability design of concrete structures. Final Technical Report, Document BE95-1347/R17, Netherlands.

Edvardsen, C. (1999) Water permeability and autogenous sealing of cracks in concrete. *ACI Materials Journal*, **96** (4), 448–454.

El-Dieb, A.S. and Hooton, R.D. (1995) Water-permeability measurement of high performance concrete using a high-pressure triaxial cell. *Cement and Concrete Research*, **25**, 1199–1208.

EOTA (2002) *Guideline of European Technical Approval of Post-tensioning Kits for Prestressing Structures (ETAG 013)*, European Organization for Technical Approvals, Brussels.

Estoup J.M. (1987) Study of the internal carbonation of concrete and its applications to the concrete manufactory industry. PhD thesis, University of Paris 6, Paris (in French).

Fabbri, A. and Fen-Chong, T. (2013) Indirect measurement of the ice content curve of partially frozen cement based materials. *Cold Region Science and Technology*, **90–91**, 14–21.

Fabbri, A., Coussy, O., Fen-Chong, T. and Monteiro, P.J.M. (2008) Are deicing salts necessary to promote scaling in concrete? *Journal of Engineering Mechanics. ASCE*, **134** (7), 589–598.

Fagerlund, G. (1976) Determination of pore-size distribution from freezing point depression. *Materials and Structures*, **6** (33), 215–225.

Fagerlund G. (1993) On the service life of concrete exposed to frost action. Report TVBM-7054, Lund Institute of Technology, Lund, Sweden.

Fagerlund G. (2002) Mechanical damage and fatigue effects associated with freeze-thaw of materials. In *Proceedings of 2nd International RILEM Workshop on Frost Resistance of Concrete*, eds. Setzer M.J., Auberg R., Keck H.-J. RILEM, Essen, pp. 117–132.

FDOT (2004) *New Directions for Florida Post-tensioned Bridges*, vol. **1**, Florida Department of Transportation, Tallahassee, FL.

Fernández, A.M., Baeyens, B., Bradbury, M. and Rivas, P. (2004) Analysis of the porewater chemical composition of a Spanish compacted bentonite used in an engineered barrier. *Physics and Chemistry of the Earth*, **29**, 105–118.

FHWA (2009) *Technical Manual for Design and Construction of Road Tunnels – Civil Elements (FHWA-NHI-10-034)*, Federal Highway Agency, USA.

fib (1999) *Structural concrete, Textbook on Behaviour, Design and Performance, Bulletin 1, volume 1–2*, Sprint-Druck, Stuttgart.

fib (2006) *Model Code for Service Life Design. fib* Bulletin 34. Federation Internationale des Bétons, Lausanne.

fib (2010) Model code 2010, first complete draft, volume 1. *fib* Bulletin 55, Fédération Internationale des Bétons, Lausanne.

Flatt, R.J. (2002) Salt damage in porous materials: how high supersaturations are generated. *Journal of Crystal Growth*, **242**, 435–454.

Fletcher, N.H. (1971) Structural aspects of the ice–water system. *Reports in Progress in Physics*, **34**, 913–994.

François, R. and Arliguie, G. (1999) Effect of microcracking and cracking on the development of corrosion in reinforced concrete members. *Magazine of Concrete Research*, **51**, 143–150.

Garboczi, E.J. and Bentz, D.P. (1992) Computer simulation of the diffusivity of cementitious materials. *Journal of Materials Science*, **27**, 2083–2092.

Garde, C., François, R. and Torrenti, J.M. (1996) Leaching of both calcium hydroxide and C-S-H from cement paste: modelling the mechanical behaviour. *Cement and Concrete Research*, **26**, 1257–1268.

Gastaldini, A.L.G., Isaia, G.C., Hoppe, T.F. *et al.* (2009) Influence of the use of rice husk ash on the electrical resistivity of concrete: a technical and economic feasibility study. *Construction and Building Materials*, **23**, 3411–3419.

Gauri, K.L., Chowdhury, A.N., Kulshreshtha, N.P. and Punuru, A.R. (1990) Geologic feature and durability of limestones at the Sphinx. *Environmental Geology and Water Sciences*, **15** (3), 217–223.

Gebhart, B. (1993) *Heat Conduction and Mass Diffusion*, McGraw-Hill, New York.

Gérard, B., Le Bellego, C. and Bernard, O. (2002) Simplified modelling of calcium leaching of concrete in various environments. *Materials and Structures*, **35**, 632–640.

Girodet, C., Bosc, J.L., Chambannet, M. and Pera, J. (1997) Influence of sand on the freeze-thaw resistance of the mortar phase of concrete, in *Frost Resistance of Concrete, RILEM Proceedings 34* (eds M.J. Setzer and R. Auberg), E&FN Spon, London, pp. 53–60.

Gouda, V.K. (1970) Corrosion and corrosion inhibition of reinforcing steel. I. Immersed in alkaline solutions. *British Corrosion Journal*, **5**, 198–203.

Guiglia, M. and Taliano, M. (2013) Comparison of carbonation depths measured on in-field exposed existing RC structures with predictions made using fib-model code 2010. *Cement & Concrete Composites*, **38**, 92–108.

Haga, K., Sutou, S., Hironaga, M. *et al.* (2005) Effects of porosity on leaching of Ca from hardened ordinary Portland cement paste. *Cement and Concrete Research*, **35** (9), 1764–1775.

Halamickova, P., Detwiler, R.J., Bentz, D.P. and Garboczi, E.J. (1995) Water permeability and chloride ion diffusion in Portland cement mortars: relationship to sand content and critical pore diameter. *Cement and Concrete Research*, **25**, 790–802.

Hall, C. (1989) Water sorptivity of mortars and concretes. *Magazine of Concrete Research*, **41** (147), 51–61.

Hartell, J.A., Boyd, A.J. and Ferraro, C.C. (2011) Sulfate attack on concrete: effect of partial immersion. *Journal of Materials in Civil Engineering, ASCE*, **23** (5), 572–579.

Haynes, H., O'Nell, R., Neff, M. and Mehta, P.K. (2008) Salt weathering distress on concrete exposed to sodium sulfate environment. *ACI Materials Journal*, **105** (1), 35–43.

Heukamp, F.H., Ulm, F.-J. and Germaine, J.T. (2001) Mechanical properties of calcium-leached cement pastes: triaxial stress states and the influence of the pore pressures. *Cement and Concrete Research*, **31**, 767–774.

Hobbs, D.W., Matthews, J.D. and Marsh, B.K. (1998) *Minimum Requirements of Durable Concrete: Carbonation- and Chloride-Induced Corrosion, Freeze–Thaw Attack and Chemical Attack*, British Cement Association, Crowthorne, UK.

Hodgson, D. and McIntosh, R. (1960) The freezing of water and benzene in porous Vycor glass. *Canadian Journal of Chemistry*, **38**, 958–971.

Horne, A.T., Richardson, I.G. and Brydson, R.M.D. (2007) Quantitative analysis of the microstructure of interfaces in steel reinforced concrete. *Cement and Concrete Research*, **37**, 1613–1623.

Hussain, S.E., Rasheeduzzafar, A.-M.A. and Al-Gahtani, A.S. (1995) Factors affecting threshold chloride for reinforcement corrosion in concrete. *Cement and Concrete Research*, **25** (7), 1543–1555.

HWA (1999) *Design of Road Tunnels, Design Manual for Roads and Bridges* (BD 78/99), Part 9. The Highway Agency.

HZMBA (2010) *Bidding Documents for General Contracting of Design and Construction of HZM Project Main Works in Artificial Islands and Tunnel*. Hong Kong–Zhuhai–Macau Bridge Administration, Zhuhai, China (in Chinese).

IAEA (1995) *The Principles of Radioactive Waste Management*, Safety Series No. 111-F. International Atomic Energy Agency, Vienna.

IAEA (1999) *Safety Assessment for near Surface Disposal of Radioactive Waste*, IAEA Safety Standards Series No.WS-G-1.1. International Atomic Energy Agency, Vienna.

IAEA (2001) Performance of engineered barrier materials in near surface disposal facilities for radioactive waste. IAEA TEC-DOC-1225. International Atomic Energy Agency, Vienna.

IAEA (2006) *Geological Disposal of Radioactive Waste – Safety Requirements*, IAEA Safety Standard No. WS-R-4. International Atomic Energy Agency.

ICOLD (1984) *Deterioration of Dams and Reservoirs: Examples and their Analysis*. International Committee on Large Dam, Balkema Publishers, Boorkfield, VT.

IPCC (2007) *Climate Change 2007 –The Fourth Assessment Report*, Cambridge University Press, Cambridge.

ISO (1998) *General Principles on Reliability for Structures (ISO2394:1998(E))*, International Organization for Standardization, Geneva, Switzerland.

ISO (2008) *General Principles on the Design of Structure for Durability (ISO13823:2008(E))*, International Organization for Standardization, Geneva, Switzerland.

Itakura, T., Airey, D.W., Chin, J.L. *et al.* (2010) Laboratory studies of the diffusive transport of ^{137}Cs and ^{60}Co through potential waste repository soils. *Journal of Environmental Radioactivity*, **101**, 723–729.

Jacobs, F. (1998) Permeability to gas of partially saturated concrete. *Magazine of Concrete Research*, **50**, 115–121.

Jain, J., Olek, J., Janusz, A. and Jozwiak-Niedzwiedzka, D. (2012) Effects of deicing salt solutions on physical properties of pavement concretes. *Transportation Research Record: Journal of the Transportation Research Board*, **2290**, 69–75.

JSCE (2010) Standard specification for concrete structures – 2007 Design, in *JSCE Guidelines for Concrete No.15*, Japan Society of Civil Engineering, Tokyo, Japan.

Kamali, S., Gérard, B. and Moranville, M. (2003) Modelling the leaching kinetics of cement-based materials – influence of materials and environment. *Cement & Concrete Composites*, **25**, 451–458.

Katz, A.J. and Thompson, A.H. (1987) Prediction of rock electrical-conductivity from mercury injection measurements. *Journal of Geophysical Research*, **22**, 599–607.

Klinkenberg, L.J. (1941) *The permeability of porous media to liquid and gases, In Drilling and Production Practice*, American Petroleum Institute, New York, pp. 200–213.

Kollek, J.J. (1989) The determination of the permeability of concrete to oxygen by the CemBureau method – a recommendation. *Materials and Structures*, **22**, 225–230.

Kośmider J. (2011) Fe–H_2O Pourbaix diagram and corrosion protection. http://commons.wikimedia.org/wiki/File:Fe-H2O_korozja.svg (accessed May 25, 2016).

Koubaa, A. and Synder, M.B. (2001) Assessing frost resistance of concrete aggregates in Minnesota. *Journal of Cold Regions Engineering, ASCE*, **15** (4), 187–210.

Kulik, D.A. and Kersten, M. (2001) Aqueous solubility diagrams for cementitious waste stabilization systems: II, end-member stoichiometries of ideal calcium silicate hydrate solid solutions. *Journal of the American Ceramic Society*, **84** (12), 3017–3026.

Kumar, R. and Bhattacharjee, B. (2003) Porosity, pore size distribution and in situ strength of concrete. *Cement and Concrete Research*, **33**, 155–164.

Lam, L., Wong, Y.L. and Poon, C.S. (2000) Degree of hydration and gel/space ratio of high-volume fly ash/cement systems. *Cement and Concrete Research*, **30**, 747–756.

Lea, F.M. (1970) *The Chemistry of Cement and Concrete*, Edward Arnold, London.

Leech, C., Lockington, D. and Dux, P. (2003) Unsaturated diffusivity functions for concrete derived from NMR images. *Materials and Structures*, **36** (6), 413–418.

Lerch W. (1945) *Effect of SO_3 Content on Durability of Concrete*, PCA Pamphlet, No.0285. Portland Cement Association, USA.

Li C.Q., Yang S.T., Saafi M. (2014) Numerical simulation of behaviour of reinforced concrete structures considering corrosion effects on bonding. *Journal of Structural Engineering, ASCE*, **140**, 04014092, 1–10.

Li F.X., Wang H., Wang Y.J. (2015) Corrosion evaluation of the groundwater and soils in Tianjin area. *Soil Engineering and Foundation*, **29**(2), 100–104 (in Chinese).

Li, K.F. and Li, C.Q. (2013) Modeling hydroionic transport in cement-based porous materials under drying-wetting actions. *Journal of Applied Mechanics, ASME*, **020904**, 1–9.

Li, K.F. and Pang, X.Y. (2014) Sorption of radionuclides by cement-based barrier materials. *Cement and Concrete Research*, **65**, 52–57.

Li, K.F. and Zeng, Q. (2009) Influence of freezing rate on cryo-damage of cementitious material. *Journal of Zhejiang University Science A*, **10** (1), 17–21.

Li, K.F., Chen, Z.Y. and Lian, H.Z. (2008) Concepts and requirements of durability design for concrete structures: an extensive review of CCES01. *Materials and Structures*, **41**, 717–731.

Li, K.F., Li, C.Q. and Chen, Z.Y. (2009) Influential depth of moisture transport in concrete subject to drying–wetting cycles. *Cement & Concrete Composites*, **31**, 693–698.

Li, K.F., Ma, M.J. and Wang, X.M. (2011) Experimental study of water flow behaviour in narrow fractures of cementitious materials. *Cement & Concrete Composites*, **33**, 1009–1013.

Li, K.F., Yang, L.H., Li, Q.W. and Wu, P. (2014) Maintenance design and optimization of long service life port structures considering crack control levels. *Advances in Structural Engineering*, **17** (4), 471–480.

Li, K.F., Li, Q.W., Zhou, X.G. and Fan, Z.H. (2015) Durability design of Hong Kong–Zhuhai–Macau sea-link project: principle and procedure. *Journal of Bridge Engineering, ASCE*, **04015001**, 1–11.

Li, H.M., Wu, J., Song, Y.J. and Wang, Z. (2014) Effect of external loads on chloride diffusion coefficient of concrete with fly ash and blast furnace slag. *Journal of Materials in Civil Engineering, ASCE*, **04014053**, 1–6.

Li, Q.W., Li, K.F., Zhou, X.G. *et al.* (2015) Model-based durability design of concrete structures in Hong Kong–Zhuhai–Macau sea link project. *Structural Safety*, **53**, 1–12.

Li, W.T., Pour-Ghaz, M., Castro, J. and Weiss, J. (2013) Water absorption and critical degree of saturation relating to freeze-thaw damage in concrete pavement joints. *Journal of Materials in Civil Engineering*, **24** (3), 299–307.

Lim, C.C., Gowripalan, N. and Sirivivatnanon, V. (2000) Microcracking and chloride permeability of concrete under uniaxial compression. *Cement & Concrete Composites*, **22**, 353–360.

Litvan, G.G. (1973) Frost action in cement paste. *Materials and Structures*, **6** (4), 293–298.

Liu, Y.P. and Weyers, R.E. (1998) Modeling the time-to-corrosion cracking in chloride contaminated reinforced concrete structures. *ACI Materials Journal*, **95** (6), 675–681.

Liu, X.M., Chia, K.S. and Zhang, M.H. (2011) Water absorption, permeability, and resistance to chloride-ion penetration of lightweight aggregate concrete. *Construction and building Materials*, **25**, 335–343.

Lockington, D., Parlange, J. and Dux, P. (1999) Sorptivity and the estimation of water penetration into unsaturated concrete. *Materials and Structures*, **32** (5), 342–347.

Lübeck, A., Gastaldini, A.L.G., Barin, D.S. and Siqueira, H.C. (2012) Compressive strength and electrical properties of concrete with white Portland cement and blast-furnace slag. *Cement & Concrete Composites*, **34**, 392–399.

Mahmoud, A.M., Ab, B.-N., Am, B.-N. and Alajmi, R. (2012) Conjugate conduction convection and radiation heat transfer through hollow autoclaved aerated concrete blocks. *Journal of Building Performance Simulation*, **5** (4), 248–262.

Mainguy M. (1999) Non-linear diffusion models in porous media. Application to leaching and to drying of cementitious materials. PhD Thesis, Ecole des Ponts ParisTech, Paris.

Mainguy, M. and Coussy, O. (2000) Propagation fronts during calcium leaching and chloride penetration. *Journal of Engineering Mechanics, ASCE*, **126** (3), 250–257.

Mainguy, M., Coussy, O. and Baroghel-Bouny, V. (2001) Role of air pressure in drying of weakly permeable materials. *Journal of Engineering Mechanics, ASCE*, **127** (6), 582–592.

Maringer, F.J., Suran, J., Kovar, P. *et al.* (2013) Radioactive waste management: review on clearance and acceptance criteria legislation, requirements and standards. *Applied Radiation and Isotopes*, **81**, 255–260.

Marinoni, N., Birelli, M.P., Rostagno, C. and Pavese, A. (2003) The effects of atmospheric multipollutants on modern concrete. *Atmospheric Environment*, **37**, 4701–4712.

Martys, N.S. and Ferraris, C.F. (1997) Capillary transport in mortars and concrete. *Cement and Concrete Research*, **27** (5), 747–760.

Matte V., Pernelle F., Marlet L. (2004) High performance fibre reinforced concrete containers: a way to ensure containment and durability over 300 years. In *Proceedings of CSNI/RILEM Workshop on Use and Performance of Concrete in NPP Fuel Cycle Facilities*. CSNI-RILEM, Madrid, pp. 147–156.

McGee, R. (1999) Modelling of durability performance of Tasmanian bridges, in *ICASP8 Applications of Statistics and Probability in Civil Engineering*, vol. **1** (eds R.E. Melchers and M.G. Stewart), Balkema, Amsterdam, pp. 297–306.

McPolin, O., Basheer, P.A.M. and Long, A.E. (2009) Carbonation and pH in mortars manufactured with supplementary cementitious materials. *Journal of Materials in Civil Engineering*, **21** (5), 217–225.

Mehta, P.K. (2000) Sulfate attack on concrete: separating myths from reality. *Concrete International*, **8**, 57–61.

Mehta, P.K. and Monteiro, P.J.M. (2006) *Concrete: Microstructure, Properties and Materials*, third edn, McGraw Hill, New York.

Meziani, H. and Skoczylas, F. (1999) An experimental study of the mechanical behaviour of a mortar and of its permeability under deviatoric loading. *Materials and Structures*, **32**, 403–409.

Millington, R.J. (1959) Gas diffusion in porous media. *Science*, **130**, 100–102.

Moon, H.Y., Kim, H.S. and Choi, D.S. (2006) Relationship between average pore diameter and chloride diffusivity in various concretes. *Construction and Building Materials*, **20**, 725–732.

Moreno, M., Morris, W., Alvarez, M.G. and Duffo, G.S. (2004) Corrosion of reinforcing steel in simulated concrete pore solutions effect of carbonation and chloride content. *Corrosion Science*, **46**, 2681–2699.

Nasser, A., Clément, A., Laurens, A. and Castel, A. (2010) Influence of steel–concrete interface condition on galvanic corrosion currents in carbonated concrete. *Corrosion Science*, **52**, 2878–2890.

NEA (2000) *Status Report on Nuclear Power Plant Life Management*, Nuclear Energy Agency, OCDE, Paris.

Nehdi, M.L. and Bassuoni, M.T. (2008) Durability of self-consolidating concrete to combined effects of sulphate attack and frost action. *Materials and Structures*, **41**, 1657–1679.

Neretnieks, I. (2014) Development of a simple model for the simultaneous degradation of concrete and clay in contact. *Applied Geochemistry*, **43**, 101–113.

Neville, A.M. (1995a) Chloride attack of reinforced concrete: an overview'. *Materials and Structures*, **28**, 63–70.

Neville, A.M. (1995b) *Properties of Concrete*, 4th edn, Pearson Education, London, UK.

Neville, A.M. (2004) The confused world of sulfate attack on concrete. *Cement and Concrete Research*, **34**, 1275–1296.

Newlands, M.G., Jones, M.R., Kandasami, S. and Harrison, T.A. (2008) Sensitivity of electrode contact solutions and contact pressure in assessing electrical resistivity of concrete. *Materials and Structures*, **41**, 621–632.

Ngala, V.T. and Page, C.L. (1997) Effects of carbonation on pore structure and diffusional properties of hydrated cement pastes. *Cement and Concrete Research*, **27** (7), 995–1007.

Nganga, G., Alexander, M. and Beushausen, H. (2013) Practical implementation of the durability index performance-based design approach. *Construction and Building Materials*, **45**, 251–261.

Nguyen T.Q. (2007) Physical–chemical modelling of the penetration of chloride ions in the cement-based materials. PhD thesis, Ecole des Ponts ParisTech, Paris (in French).

Nguyen, V.H., Colina, H., Torrenti, J.M. *et al.* (2007) Chemo-mechanical coupling behaviour of leached concrete. Part I: experimental results. *Nuclear Engineering and Design*, **237**, 2083–2089.

Nguyen, T.S., Lorente, S. and Carcasses, M. (2009) Effect of the environment temperature on the chloride diffusion through CEM-I and CEM-V mortars: an experimental study. *Construction and Building Materials*, **23**, 795–803.

Nordtest (1995) Concrete hardened: accelerated chloride penetration (NT Build 443). Nordtest Method. Nordtest, Espoo, Finland.

Nordtest (1999) Concrete, mortar and cement-based repair materials: chloride migration coefficient from non-steady migration experiments (NT Build 492), Nordtest Method. Nordtest, Espoo, Finland.

Novak, G.A. and Colville, A.A. (1989) Efflorescent mineral assemblages associated with cracked and degraded residential foundations in southern California. *Cement and Concrete Research*, **19**, 1–6.

Oh, B.H. and Jang, S.Y. (2004) Prediction of diffusivity of concrete based on simple analytic equations. *Cement and Concrete Research*, **34**, 463–480.

Otieno, M., Beushausen, H. and Alexander, M. (2012) Towards incorporating the influence of cover cracking on steel corrosion in RC design codes: the concept of performance-based crack width limits. *Materials and Structures*, **45**, 1805–1816.

Page, C.L. (1975) Mechanism of corrosion protection in reinforced concrete marine structures. *Nature*, **258**, 514–515.

Page, C.L., Lambert, P. and Vassie, P.R.W. (1991a) Investigations of reinforcement corrosion. 1. The pore electrolyte phase in chloride-contaminated concrete. *Materials and Structures*, **24**, 243–252.

Page, C.L., Lambert, P. and Vassie, P.R.W. (1991b) Investigations of reinforcement corrosion. 2. Electrochemical monitoring of steel in chloride-contaminated concrete. *Materials and Structures*, **24**, 351–358.

Pang, X.Y., Li, K.F. and Li, C.Q. (2013) Moisture transport properties of cement-based materials for engineered barriers in radioactive waste disposal, in *Cement-Based Materials for Nuclear Waste Storage* (eds F. Bart, C. Cau-dit-Coumes, F. Frizon and S. Lorente), Springer, New York, pp. 125–134.

Papadakis, V.G. and Vayenas, C.G. (1991) Experimental investigation and mathematical modelling the concrete carbonation problem. *Chemical Engineering Science*, **46** (5–6), 1333–1338.

Papadakis, V.G., Vayennas, C.G. and Fardis, M.N. (1991a) Physical and chemical characteristics affecting the durability of concrete. *ACI Materials Journal*, **88** (2), 186–196.

Papadakis, V.G., Vayennas, C.G. and Fardis, M.N. (1991b) Fundamental modelling and experimental investigation of concrete carbonation. *ACI Materials Journal*, **88** (4), 363–373.

Pech R., Schacher B., Verdier A. (1992) Sogefibre, fibre reinforced concrete containers: from concept to fabrication. In *Proceedings of 4th RILEM International Symposium on Fibre Reinforced Cement and Concrete*. RILEM, Sheffield, pp. 4–12.

Penttala, V. (1998) Freezing-induced strains and pressures in wet porous materials and especially in concrete mortars. *Advanced Cement Based Materials*, **7**, 8–19.

Philip, J.R. and De Vries, D.A. (1957) Moisture movement in porous materials under temperature gradients. *Transactions of American Geophysical Union*, **38** (2), 222–232.

PIANC (1998) *Life Cycle Management of Port Structures – General Principles (PTC-II WG31), Supplement to Bulletin no. 99*, International Navigation Association, Brussels.

Pigeon, M., Prévost, J. and Simad, J.M. (1985) Freeze–thaw durability versus freezing rate. *ACI Journal*, **9–10**, 684–692.

Podolny, W., Jr (1992) Corrosion of prestressing steel and its mitigation. *PCI Journal*, **5**, 34–55.

Powers, T.C. (1949) The air requirement of frost resistant concrete. *Highway Research Board Bulletin*, **29**, 184–211.

Powers T.C. (1960) Physical properties of cement paste. In *Proceedings of the Fourth International Conference on the Chemistry of Cement*. US National Bureau of Standards Monograph, 43, vol. **2**. US Department of Commerce, Washington DC, pp. 577–613.

Powers, T.C. and Helmuth, R.A. (1953) Theory of volume changes in hardened Portland cement paste during freezing. *Highway Research Board Bulletin*, **32**, 285–297.

Qin, Y.H. and Hiller, J.E. (2011) Modelling the temperature and stress distributions in rigid pavements: Impact of solar radiation absorption and heat history development. *KSCE Journal of Civil Engineering*, **15** (8), 1361–1371.

Ramachandran, A. and Sharama, M.M. (1969) Absorption with fast reaction in slurry containing sparingly soluble fine particles. *Chemical Engineering Science*, **24**, 1681–1686.

Raupach, M. (2006) Models for the propagation phase of reinforcement corrosion – an overview. *Materials and Corrosion*, **57** (8), 605–613.

Reinhardt H.W. (1997) *Penetration and Permeability of Concrete: Barriers to Organic and Contaminating Liquids*, RILEM Report 16. E & FN Spon, London.

RILEM (1996) CDF test – test method for the freeze-thaw resistance of concrete tests with sodium chloride solution (RILEM-TC 117-FDC). *Materials and Structures*, **29**, 523–528.

RILEM (1999) Permeability of concrete as a criterion of its durability (RILEM TC 116-PCD). *Materials and Structures*, **32**, 174–179.

RILEM (2001) Slab tests – freeze/thaw resistance of concrete – internal deterioration (RILEM TC 176-IDC). *Materials and Structures*, **34**, 526–531.

RILEM (2013) *Performance of Cement-based Materials in Aggressive Aqueous Environments*, Springer, New York.

Rodriguez-Navarro, C., Doehne, E. and Sebastian, E. (2000) How does sodium sulfate crystallize? Implications for the decay and testing of building materials. *Cement and Concrete Research*, **30**, 1527–1534.

Romer, M. (2005) Effect of moisture and concrete composition on the Torrent permeability measurement. *Materials and Structures*, **38**, 541–547.

Rostam S. (1989) *Durable Concrete Structures – CEB Design Guide*, 2nd edn, Bulletin d'Information No.182. Comité Euro-International du Beton (CEB), Lausanne.

Saetta, A.V. and Scotta, R.V. (1993) Analysis of chloride diffusion into partially saturated concrete. *ACI Materials Journal*, **5**, 441–451.

Saito, M., Ohta, M. and Ishimoro, H. (1994) Chloride permeability of concrete subject to freeze–thaw damage. *Cement & Concrete Composites*, **16**, 233–239.

Sandberg P. (1998) Recurrent studies of chloride ingress in uncracked marine concrete at various exposure times and elevations. Report TVBM-3080, Lund Institute of Technology, Lund, Sweden.

Sarja, A. (2002) *Integrated Life Cycle Design of Structures*, Spon, London.

Scherer, G.W. (1999) Crystallization in pores. *Cement and Concrete Research*, **29**, 1347–1358.

Scherer, G.W. (2004) Stress from crystallization of salt. *Cement and Concrete Research*, **34**, 1613–1624.

Segura, I., Molero, M., Aparicio, S. *et al.* (2013) Decalcification of cement mortars: characterisation and modelling. *Cement & Concrete Composites*, **35**, 136–150.

Sellier, A., Buffo-Lacarriere, L., El Gonnouni, M. and Bourbon, X. (2011) Behaviour of HPC nuclear waste disposal structures in leaching environment. *Nuclear Engineering and Design*, **241** (1), 402–414.

Sengul, O. and Gjorv, O.E. (2008) Electrical resistivity measurements for quality control during concrete construction. *ACI Materials Journal*, **105**, 541–548.

Sercombe, J., Vidal, R., Gallé, C. and Adenot, F. (2007) Experimental study of gas diffusion in cement paste. *Cement and Concrete Research*, **37**, 579–588.

Setzer, M.J. (2001) Micro-ice-lens formation in porous solid. *Journal of Colloid and Interface Science*, **243**, 193–201.

Shackelford, C.D. (1991) Laboratory diffusion testing for waste disposal – a review. *Journal of Contaminant Hydrology*, **7**, 177–217.

Shafiro, B. and Kachanov, M. (2000) Anisotropic effective conductivity of materials with nonrandomly oriented inclusions of diverse ellipsoidal shapes. *Journal of Applied Physics*, **87**, 8561–8569.

Skalny, J., Marchand, J. and Odler, I. (2002) *Sulfate Attack on Concrete*, E&FN Spon, London.

Sugiyama, T., Bremner, T.W. and Holm, T.A. (1996) Effect of stress on gas permeability in concrete. *ACI Materials Journal*, **93**, 443–450.

Suleiman, A.R., Soliman, A.M. and Nehdi, M.L. (2014) Effect of surface treatment on durability of concrete exposed to physical sulfate attack. *Construction and Building Materials*, **73**, 674–681.

Sun, Z.H. and Scherer, G.W. (2010) Effect of air voids on salt scaling and internal freezing. *Cement and Concrete Research*, **40**, 260–270.

Suryavanshi, A.K. and Scantlebury, J.D. (1996) Mechanism of Friedel's salt formation in cements rich in tri-calcium aluminate. *Cement and Concrete Research*, **26** (5), 717–727.

Tang, L.P. (1999) Concentration dependence of diffusion and migration of chloride ions. Part 1. Theoretical considerations. *Cement and Concrete Research*, **29**, 1463–1468.

Tang, L.P. and Nilsson, L.O. (1993) Chloride binding capacity and binding isotherms of OPC pastes and mortars. *Cement and Concrete Research*, **23**, 247–253.

Taylor, R., Richardson, I.G. and Brydson, R.M.D. (2010) Composition and microstructure of 20-year-old ordinary Portland cement–ground granulated blast-furnace slag blends containing 0 to 100% slag. *Cement and Concrete Research*, **40**, 971–983.

Thiery M. (2005) Modelling of the atmospheric carbonation of cement-based materials. Taking into account the kinetics effects and the microstructural and moisture modifications. PhD thesis, Ecole des Ponts ParisTech, Paris (in French).

Thiery, M., Dangla, P., Belin, P. *et al.* (2013) Carbonation kinetics of a bed of recycled concrete aggregates: a laboratory study on model materials. *Cement and Concrete Research*, **46**, 50–65.

Thomas, J.J., Rothstein, D., Jennings, H.M. and Christensen, B.J. (2003) Effect of hydration temperature on the solubility behavior of Ca-, S-, Al-, and Si-bearing solid phases in Portland cement pastes. *Cement and Concrete Research*, **33**, 2037–2047.

Thomas M.D.A., Bentz E.C. (2000) Life-365, computer program for predicting the service life and life-cycle cost of reinforced concrete exposed to chlorides. Concrete Corrosion Inhibitors Association, Potomac, MD.

Torrent, R. (1992) Torrent-a two chamber vacuum cell for measuring the coefficient of permeability to air of the concrete cover on site. *Materials and Structures*, **25**, 358–365.

Torrent, R.J., Armaghani, J. and Yan, T.B. (2013) Evaluation of port of Miami tunnel segments. *Concrete International*, **5**, 39–46.

TRB (2004) Concrete bridge deck performance, a synthesis of highway practice. NCHRP Synthesis 333, Transportation Research Board, Washington, DC.

Trethewey, K.R. and Roberge, P.R. (1995) Corrosion management in the twenty-first century. *British Corrosion Journal*, **30** (3), 192–197.

Tsivilis, S., Chaniotakis, E., Batis, G. *et al.* (1999) The effect of clinker and limestone quality on the gas permeability, water absorption and pore structure of limestone cement concrete. *Cement and Concrete Research*, **21**, 139–146.

Tsivilis, S., Tsantilas, J., Kakali, G. *et al.* (2003) The permeability of Portland limestone cement concrete. *Cement and Concrete Research*, **33**, 1465–1471.

Tsukamoto, M. and Wörmer, J.D. (1991) Permeability of cracked fiber-reinforced concrete, Darmstad concrete. *Annals of Concrete and Concrete Structures*, **6**, 123–135.

Tuutti, K. (1982) *Corrosion of Steel in Concrete*, Swedish Cement and Concrete Research Institute, Stockholm.

Ulm, F.-J. and Coussy, O. (2001) What is a "massive" concrete structure at early-ages: some dimensional arguments. *Journal of Engineering Mechanics, ASCE*, **129**, 512–522.

Ulm, F.-J., Lemarchand, E. and Heukamp, F.H. (2003) Elements of chemomechanics of calcium leaching of cement-based materials at different scales. *Engineering Fracture Mechanics*, **70**, 871–889.

Ushiyama H., Goto S. (1974) Diffusion of various ions in hardened Portland cement pastes. In *VI International Congress on the Chemistry of Cement*, Moscow, Vol. **II-1**, pp. 331–337.

Valenza, J.J., II and Scherer, G.W. (2006) Mechanism for salt scaling. *Journal of American Ceramic Society*, **89** (4), 1161–1179.

Valipour, M., Pargar, F., Shekarchi, M. *et al.* (2013) In situ study of chloride ingress in concretes containing natural zeolite, metakaolin and silica fume exposed to various exposure conditions in a harsh marine environment. *Construction and Building Materials*, **46**, 63–70.

Van Genuchten, M.T. (1980) A closed-form equation for predicting the hydraulic conductivity of unsaturated soils. *Soil Science Society American Journal*, **44** (5), 892–898.

Vargaftik, N.B., Volkov, B.N. and Voljak, L.D. (1983) International tables of the surface tension of water. *Journal of Physical and Chemical Reference Data*, **12** (3), 817–820.

Verbeck, G.J. and Klieger, P. (1957) Studies of "salt" scaling of concrete. *Highway Research Board Bulletins*, **150**, 1–17.

Villain, G., Baroghel-Bouny, V., Kounkou, C. and Hua, C. (2001) Mesure de la perméabilité aux gaz en fonction du taux de saturation des bétons. *European Journal of Environmental and Civil Engineering*, **5**, 251–268.

Wang, L.C. and Ueda, T. (2014) Mesoscale modelling of chloride penetration in unsaturated concrete damaged by freeze–thaw cycling. *Journal of Materials in Civil Engineering, ASCE*, **26** (5), 955–965.

Wang, X.M. and Li, K.F. (2011a) Adsorption behaviour of ions on fracture surfaces of cement-based materials. *Journal of the Chinese Ceramic Society*, **39** (1), 1–6.

Wang, X.M. and Li, K.F. (2011b) Leaching behaviour of fracture surfaces of cement-based materials. *Journal of the Chinese Ceramic Society*, **39** (3), 525–530.

Wei Q.J., Li B.Y., Li X.N. (2014) Analysis on the characters of soil salinization in the Turpan Basin. *Journal of Arid Land Resources and Environment*, **4**, 163–167 (in Chinese).

Wittmann F.H., Zhao T.J., Zhang P., Jiang F.X. (2010) Service life of reinforced concrete structures under combined mechanical and environmental loads. In *Proceedings of 2nd International Symposium Service Life Design for Infrastructure*, eds. van Breugel K., Ye G., Yuan Y. RILEM, Delft, Netherlands, pp. 91–98.

Woodward, D.G. (1997) Life cycle costing – theory, information acquisition and application. *International Journal of Project Management*, **15** (6), 335–344.

Yang, C.C. (2006) On the relationship between pore structure and chloride diffusivity from accelerated chloride migration test in cement-based materials. *Cement and Concrete Research*, **36**, 1304–1311.

Yang, K., Basheer, P.A.M., Bai, Y. *et al.* (2014) Development of a new in situ test method to measure the air permeability of high performance concretes. *NDT & E International*, **64**, 30–40.

Yang, L.H., Li, K.F. and Pang, X.Y. (2013) Design and optimization of maintenance strategies for a long life-span port project. *Materials and Structures*, **46**, 161–172.

Yang Z.F., Brown H., Cheney A. (2006) Influence of moisture conditions on freeze and thaw durability of Portland cement pervious concrete. In *Proceedings of 2006 NRMCA Concrete Technology Forum – Focus on Pervious Concrete*, Nashville, TN.

Yao, Y., Wang, L. and Wittmann, F.H. (2013) *Publications on Durability of Reinforced Concrete Structures under Combined Mechanical Loads and Environmental Actions: An Annotated Bibliography*, Aedificatio Publishers, Freiburg.

Yildirim, S.T., Meyer, C. and Herfellner, S. (2015) Effects of internal curing on the strength, drying shrinkage and freeze–thaw resistance of concrete containing recycled concrete aggregates. *Construction and Building Materials*, **91**, 288–296.

Yokozeki, K., Watanabe, K., Sakata, N. and Otsuki, N. (2004) Modelling of leaching from cementitious materials used in underground environment. *Applied Clay Science*, **26**, 293–308.

Yoshida, N., Matsunami, Y., Nagayama, M. and Sakai, E. (2010) Salt weathering in residential concrete foundations exposed to sulfate-bearing ground. *Journal of Advanced Concrete Technology*, **8** (2), 121–134.

Yssorche, M.P., Bigas, J.P. and Ollivier, J.P. (1995) Mesure de la perméabilité à l'air des bétons au moyen d'un perméamètre à charge variable. *Materials and Structures*, **28**, 401–405.

Yu, B., Yang, L.F., Wu, M. and Li, B. (2014) Practical model for predicting corrosion rate of steel reinforcement in concrete structures. *Construction and Building Materials*, **54**, 385–401.

Zaharieva, R., Buyle-Bodin, F. and Wirquin, E. (2004) Frost resistance of recycles aggregate concrete. *Cement and Concrete Research*, **34**, 1927–1932.

Zeng, Q., Fen-Chong, T. and Li, K.F. (2014) Freezing behaviour of cement pastes saturated with NaCl solution. *Construction and Building Materials*, **59**, 99–110.

Zeng, Q., Li, K.F. and Fen-Chong, T. (2015) Heterogeneous nucleation of ice from supercool NaCl solution confined in porous cement paste. *Journal of Crystal Growth*, **409**, 1–9.

Zhang J., Wang L., Sun M., Liu Q. (2004) Effect of coarse/fine aggregate ratio and cement matrix strength on fracture parameters of concrete. *Engineering Mechanics*, **21**(1), 136–142 (in Chinese).

Zhang, J., Leung, C.K.Y. and Xu, S.L. (2010) Evaluation of fracture parameters of concrete from bending test using inverse analysis approach. *Materials and Structures*, **43**, 857–874.

Zhang, T., Wu, C., Liao, H.S. and Hu, Y.H. (2005) 3D numerical simulation on water and air two-phase flows of the steps and flaring gate pier. *Journal of Hydrodynamics Series B*, **17** (3), 338–343.

Zheng, Q.S. and Du, D.X. (2001) An explicit and universally applicable estimate for the effective properties of multiphase composites which accounts for inclusion distribution. *Journal of the Mechanics and Physics of Solids*, **49**, 2765–2788.

Zhou, C.S., Li, K.F. and Pang, X.Y. (2011) Effect of crack density and connectivity on the permeability of microcracked solids. *Mechanics of Materials*, **43**, 969–978.

Zhou, C.S., Li, K.F. and Han, J.G. (2012a) Characterizing the effect of compressive damage on transport properties of cracked concretes. *Materials and Structures*, **45**, 381–392.

Zhou, C.S., Li, K.F. and Pang, X.Y. (2012b) Geometry of crack network and its impact on transport properties of concrete. *Cement and Concrete Research*, **42**, 1261–1272.

Index

Note: page numbers followed by "t" indicate tables.

AAR *see* alkali–aggregate reaction
absorption
  rates, 201
  surface, 34, 35, 191
  water, 118, 146, 149, 150
ACC *see* carbonation, accelerated
accelerated leaching test, 84, 204
accelerated migration test, 200
accelerated test under combined actions, 204
action intensity, 235, 241, 245, 246,
    249, 253
additional protection measures, 55, 165
adsorption
  capacity, 127, 128
  capacity of fracture surface, 127
  final/equilibrium, quantity, 126
  ions, 118, 123
  kinetics of ions, 126
  layer, 128
  layer thickness, 128
adsorption rate, 126, 127
  coefficient, 126
  initial, 126
advection–dispersion equation, 123
AEA *see* air entrainment, agents
aggregates, 59, 64, 68, 69
  natural, 69
  recycled, 69
  maximum, size, 187, 235

air entrainment, 57, 60, 64, 67, 72, 77, 111
  agent, 67
air voids, 61, 65, 67, 71, 185, 186
  average spacing factor, 67, 77, 186
alkali–aggregate reaction, 213, 214, 241, 244
alternative current method, 201
aluminates hydrates, 9
ammonium nitrate solution, 84
anchorage
  buried, 256
  exposed, 256
anode reaction, 18, 48
Archie's law, 186, 194, 196, 197, 201
artificial islands, 210, 212
aspect ratio, 128, 129
asphalt pavement, 162
atmospheric pressure, 61, 72, 74

barrier materials, 222
bentonite, 80
binder content, 244, 245
binding capacity of chlorides, 35, 44
Biot coefficient, 63, 116
Biot's tangent modulus, 116
boundary
  condition(s), 86, 90, 92, 133, 134, 148
  layer, 90, 92, 94
  layer theory, 92, 135
  layer thickness, 92

Durability Design of Concrete Structures: Phenomena, Modeling, and Practice, First Edition. Kefei Li.
© 2016 John Wiley & Sons Singapore Pte. Ltd. Published 2016 by John Wiley & Sons Singapore Pte. Ltd.

boundary value, 15
bridge(s), 163, 210, 212, 215
bridge decks, 57
bridge structures, 5
building, 161, 163
    industrial, 4
    residential, 4
bulk modulus, 63, 116

C_2S *see* dicalcium silicates
C_3A *see* tricalcium aluminates
C_3S *see* tricalcium silicates
C_4AF *see* tetracalcium aluminoferrites
calcium
    chloride admixture, 237
    dissolution, 127
    equilibrium, 184, 196, 198
    equilibrium diagram, 81–83, 87
    initial concentration, 88
    initial content, 87, 88
    leaching, 78, 79, 87, 90, 92
    in pore solution, 81
    solid content, 184
calcium silicate hydrates, 3, 8–10, 13, 15, 16, 34–37,
    80–83, 85, 87, 88, 90, 184, 186, 194
    content, 95, 183, 184, 196, 198, 199, 207, 208
    dissolution, 81, 82, 88, 94, 184
    solubility, 8
capillary
    free water in, pores, 135
    porosity, 86–88, 90 (*see also* porosity)
    pressure, 34, 137, 140, 144, 146
    suction of salt solution, 102
carbonation
    accelerated, 12, 15, 17, 18, 203
    accelerated tests, 203, 208
    C-S-H, 8
    inverse resistance, 18
    front, 13–15, 22
    mechanism, 13, 14
    natural, 12, 17, 18, 203
    reaction, 4, 9, 13, 15
    resistance, 6, 7, 12, 15, 18
carbonation depth, 6–9, 11–16, 18–20, 22, 23, 185,
    202, 203, 209, 226, 237
    initial, 14, 15
    residual, 20, 22, 23
carbonation-induced corrosion, 19, 27, 28, 48,
    51, 183, 207, 213, 214, 245 *see also* corrosion,
    general
carbon dioxide
    aggressive, 224
    binding capacity, 17
    concentration, 6, 11–13, 15, 18
    diffusivity, 9, 10, 14

effective diffusion coefficient, 17
    infiltration, 9, 11
    local concentration, 11
    mass conservation, 14
Ca/Si ratio, 8, 15, 83
cathode reaction, 20, 48
    steel corrosion, 188
CemBureau method, 189, 199
cement
    composition, 15, 34, 35
    hydrates, 3, 13, 15, 34, 36
    hydration, 132
    minimum content, 235, 236
    strength class, 235
    type, 236t, 237, 244–246, 247t, 250t, 252t, 255t
cement-based materials, 222, 227
cement-based porous materials, 62
cement paste, 8, 9, 15, 84, 86–88, 90
    hardened, 59, 62, 64, 69, 184
CH *see* portlandite
characteristic curve
    phase change, 186
    pore crystallization, 102
    pore freezing, 66
characteristic diameter, 197, 198
characteristic length, 135
characteristic values, 216–218
chemical attack, 239
chloride(s)
    adsorption, 127
    air-borne, 30
    bound, 35
    conductivity, 209
    deposition, 38
    ingress, 29–32, 34, 36, 42, 43, 45, 46, 49–51, 54,
    158, 161, 169, 178, 183, 191, 204, 205, 210,
    214, 215, 218, 219, 225
    marine, 30
    molar mass, 50
    profiling method, 38, 41, 45
    sodium, 97, 103
    source, 30
chloride binding, 32, 34–36, 42, 46
    capacity, 36, 37, 43, 44, 49, 50, 82
    chemical, 34
    linearized capacity, 43
    physical, 34
chloride concentration, 30, 32, 35, 38, 47, 48, 50,
    53, 224
    critical ratio, 49
    free, 51
    initial, 44
    pore solution, 34, 42, 48, 49, 51
    seawater, 30 (*see also* seawater)
    surface, 36, 41, 44

chloride content, 178, 237, 242, 244
    absolute, 50
    critical, 50, 51, 178, 185, 198, 199, 207, 237
    initial, 45, 46, 249
    surface, 38, 45, 46, 178, 215, 216
    threshold, 215
    total, 43, 50, 51
chloride diffusion, 219
    apparent, coefficient, 215, 219
    coefficient, 35, 178, 206, 216, 218 (*see also* chloride
        diffusivity)
    effective, coefficient, 43
chloride diffusivity, 33–36, 38, 41, 43, 191, 198–200,
        205, 208, 210, 219, 226
    ageing exponent, 43, 45, 46, 178, 216, 217
    ageing factor, 216, 218
    ageing law, 43, 45, 216
    apparent, 33, 41, 45
    effective, 43, 44
    laboratory characterization, 41
    thermal dependence, 43
chloride migration coefficient, 41, 45, 46
chloride-induced corrosion, 29, 48, 54, 183, 213–215,
        241, 243, 249 *see also* corrosion, pitting
chloroaluminate calcium hydrates, 34
$C_{20}H_{14}O_4$ *see* phenolphthalein solution
Clapeyron equation, 148
CO_2 *see* carbon dioxide
coastal areas, 30
code
    ACI-318, 234, 235, 237, 240, 246, 249, 253, 255
    CCES01, 239, 240
    EN206-1, 234, 235, 240, 245, 255
    EN1992-1, 234, 235, 245
    Eurocodes, 234, 235
    GB/T50476, 234, 240, 244, 246, 248, 249,
        253, 255, 256
cold region, 56, 57, 242, 246, 248
cold weather concreting, 77
cold-working bars, 243
combined migration-leaching, 228, 232
compressive loading, 118
concrete
    carbonation, 3–7, 9, 11, 12, 14, 18
    carbonation phenomena, 3
    chemomechanical softening, 94
    compactness, 6, 18, 21, 27, 28, 54, 99
    composition, 235, 240, 242
    density, 50
    electrical pole, 97
    gallery, 79
    materials, 3, 4, 9, 27, 35, 46, 54
    microstructure, 34, 35
    pavement, 70
    piers, 57
    prestressed, 237, 249
    reinforced, 4–6, 22, 23, 25, 27, 28, 237, 245, 249
    road slab, 57
    roofs, 4
    splitting tensile strength, 25 (*see also*, strength,
        tensile)
    strength class, 235, 237
concrete cover, 4, 6, 20–25, 27, 28, 51, 55, 159, 161,
        165, 167, 175, 178, 179, 237, 241,
    242, 244, 246, 249, 253
    minimum, 235, 239, 245
    operation error, 218
    thickness, 22, 25, 28, 215, 219, 235, 246, 249
concrete deterioration, 223, 237, 239, 243,
        253, 255
    freeze–thaw actions, 159
    leaching, 159
    salts crystallization, 159
concrete–environment interaction, 45, 106
concrete structure(s), 4, 6, 11, 38, 157–159, 161–164,
        166, 167, 170, 171, 174, 177, 183, 195, 210, 212,
        219, 232–235, 237, 239–242, 249, 256
    prestressed, 159, 255
    reinforced, 29–31
concrete surface, 30, 34, 38–40, 45, 53, 55, 135, 144,
        146, 147, 151, 153
    emissivity, 135, 136
concrete workmanship, 232
construction
    phase, 164, 165, 171, 172
    quality, 240, 255
    tolerance, 246, 253
contact
    angle, 60, 67
    moisture, 238t
    water, 238t
container wharf, 177
control process, 210, 224–226
convection depth, 38, 45
convection zone, 45
conventional carbon steel, 21
correlation analysis, 232
corrosion
    to an acceptable extent, 243
    crevice, 159
    current, 49–51, 53–55
    current density, 50, 53
    general, 20, 29, 48, 50, 51, 53
    macro-cell, 187
    pitting, 48–51, 53
    product, 25
    reinforcement, 237, 239
    steel, 144
corrosion initiation, 8, 19, 20, 22, 48–51, 178, 184,
        187, 242

corrosion mode, 21
   cathode-reaction-controlled, 20–22
   oxygen diffusion, 49
   resistance-controlled, 21, 22, 49
corrosion rate, 20–22, 24, 25, 29, 54
cost
   construction, 171, 172
   demolition, 172
   indirect, 172, 173
   life-cycle, 55, 172, 173
   maintenance, 171, 172, 174, 176, 178, 180, 182
      (*see also* maintenance)
   user, 172, 175, 178, 179, 182
crack
   connectivity, 129, 130
   control, 240, 244, 249
   density, 128–130
   initiation stress, 74
   isotropic distribution, 128, 129
   length distribution, 130
   opening, 121, 123, 126, 128–130
   patterns, 118, 130
   percolation, 129
   self-healing, 123
   single, 118, 128
crack network, 118
   geometry, 130
crack width, 55, 237, 245, 249
   allowable, 237, 239, 245
   surface, 26
creep, 117
criticality analysis, 225
cryo-storage facilities, 56
crystallization
   characteristic curve, 196, 198
   damage, 97, 99, 104–106, 110, 111
   gypsum, 97
   laboratory tests, 105
   pressure, 100, 104, 109, 110
   stress model, 72
C-S-H *see* calcium silicate hydrates
curing
   conditions, 243–245
   factor, 17
   wet, period, 244, 246, 249
cyclic loading scheme, 118

damage
   to an acceptable extent, 243
   concrete, 241, 246
   cracking, 115
   D-cracking, 68
   extent, 118
   factor, 118
Darcy flow, 35, 43

Darcy's law, 33, 138, 187, 188
DEF *see* delayed ettringite formation
deicing chemicals, 235, 240
delayed ettringite formation, 213, 214, 241
demolition, 170–172 *see also* cost, demolition
design
   alternatives, 171, 173
   codes, 234, 239
   conceptual, 164, 165
   detailed, 164, 165, 168, 173
   model-based, 210, 213, 214, 216, 227, 232
   preliminary, 164, 165, 173
   procedure, 159, 162, 164, 167, 168
   scenario, 226
   uncertainty, 210, 232 (*see also* uncertainty)
   water levels, 212
   working life, 161, 162, 211, 215
design approach
   'deem-to-satisfy', 166, 235
   performance-based, 157, 166–168, 237, 239
   prescriptive, 166, 167, 235, 240
design format
   deterministic, 168, 169
   probabilistic, 168, 170, 174, 216
deterioration
   initiation, 160
   level for maintenance, 174–176, 178, 179, 182
   pattern, 158, 159
   processes, 158–161, 168, 174, 183, 187, 192,
      195, 201, 207, 210, 213, 224
   RC structures, 30
   reinforced concrete, 29
   specified extent, 160
dew point, 97, 104
DF *see* durability, factor
DI *see* durability indicator
dicalcium silicates, 9, 13, 15
differential scanning calorimetry, 66
diffusion, 32, 34–38, 41–43, 45, 49
diffusivity, 104
   ageing, 38
   chloride, 32, 33, 43 (*see also* chloride diffusivity)
   effective, calcium, 86–88
   gas, 189, 190, 199, 200, 207
   ion, 83, 84, 191, 197, 207
direct current method, 201
Dirichlet condition, 134
discount rate, 163, 172, 179
disposal
   deep geological, 79
   life, 210, 223–225
   near surface, 79, 153, 220, 222–225, 230, 233
   permanent, 223–225, 227, 228
   radwaste, 220 (*see also* radioactive waste)
   unit, 222, 223

dissolution, 118, 123, 127
  calcium, 78
  constant, 80, 81, 83, 88, 127
  equilibrium, 78, 80, 81, 83, 90, 127
  front, 85, 86, 88
  soluble phases, 123
dissolution rate, 84
  coefficient, 127
dissolution–diffusion, 85
dissolution–transport, 81, 84, 85
DLM *see* deterioration, level for maintenance
DLS *see* durability, limit state
drying
  condition, 147, 151
  extent, 104, 107, 109, 145
  front, 147–149
drying–wetting
  action, 38, 96, 109
  cycle, 38, 97, 101, 109, 144, 149, 150
  drying-dominated, 149
  equilibrium, 149
  equilibrium time ratio, 149, 150
  wetting-dominated, 150
durability
  design, 3, 27–29, 41, 46, 54, 76–78, 83, 94, 96, 97,
    99, 111, 157, 161, 163–171, 174, 205, 208, 210,
    213, 214, 216, 221, 222, 227, 232, 233
  factor, 203, 249
  limit state, 76, 111, 157, 159–162, 164, 165,
    167, 168, 170, 171, 174, 175, 178, 205, 214,
    215, 221, 227, 232, 241–243
  long-term, 3
  performance-based, design, 208
  performance tests, 71, 198, 201
  process, 3
  redesign, 210
  requirement, 55, 167, 210, 214, 234–240, 242, 243,
    246, 249, 253
  specifications, 208
durability indicator, 12, 34, 71, 94, 183, 195, 205–209,
    241, 244, 249
  groups, 206, 208
  probable set, 206
dynamic elastic modulus, 203
dynamic viscosity
  gas, 189
  water, 187

ecological impact, 162
economic cycle, 162
economic life, 162, 163
effective self-consistent scheme, 128
efflorescence, 102
electrical capacitance method, 66
electrical conductivity, 194, 196

electrical current, 20–22, 24
electrical resistance, 20, 21, 24, 27
  concrete, 50, 53
electrical resistivity, 6, 24, 27, 194, 196, 199,
    201, 207
  concrete, 6
  time dependence, 24
element(s)
  cast-in-place, 237
  concrete, 4–6, 11, 20, 96, 97, 105, 106, 111, 213,
    217, 218, 239, 255 (*see also* RC element)
  partially buried, 102
  permanent, 165, 170, 173
  prefabricated, 237
  prestressed concrete, 214
  principal, 160, 161, 164, 214
  replaceable, 160, 161, 165, 171, 173, 214
  secondary, 214
  sheltered, 164
  structural, 158, 159, 164–167, 171, 173
engineered barriers, 224
entrained air content, 71, 235, 249
environment(s)
  atmospheric, 245, 246, 247t (*see also* carbonation)
  freeze–thaw, 246, 248, 249, 250t (*see also*
    freeze–thaw)
  marine, 30, 38, 41, 55
  marine and deicing salts, 249, 251t, 252t (*see also*
    chloride)
  sulfate, 237, 253, 255 (*see also* sulfate)
environmental action(s), 115, 144, 234, 235, 237,
    240–243
  artificial, 183, 198
environmental classes, 235, 244, 258
environmental classification, 235, 237, 239,
    241, 242
environmental factor, 17, 18, 24
epoxy-coating, 111
equilibrium
  liquid–gas, 100
  liquid–vapor, 100
  soild–liquid, 78, 81
error
  function, 42, 149, 216
  single operator, 201
esthetical appearance, 243
evaporation–diffusion process, 147–148
exchange
  boundary condition, 134, 135
  coefficient, 90, 92, 135, 136
  ion, 34–36, 194
exothermic reaction, 132
exposure station, 31, 32
  La Rochelle, 31
  Maurienne, 12

Qeshm island, 32
Träslövsläge harbor, 32
Zhanjiang, 38, 39, 216

FA *see* fly ash
failure probability, 170, 174, 175, 178–180, 243
    annual, 170, 214, 243
    target, 170
Faraday's law, 24
Fick's law, 42, 189, 191
Fick's second law, 42, 215
finite-difference method, 90
finite-volume method, 228
flowing water, 78, 79, 84, 85, 92, 95
flow rate, 33, 38, 118, 123, 126
    coefficient, 121, 126
    initial, 123, 126
fly ash, 6, 8, 17
formation factor, 196–198
Fourier's law, 132, 133
fracture
    concrete cover, 25 (*see also* concrete cover)
    criterion, 109
    damage, 101(*see also* damage)
    resistance, 64, 111, 195, 196, 199, 201, 207
    surface, 121, 123, 126–128
    toughness, 201
freeze–thaw
    action, 56, 57, 70, 76
    cycle(s), 57, 58, 60, 65, 69, 76, 241, 249
    damage, 56, 60, 64, 65, 158, 183, 186, 196
    resistance, 226
    test, 57, 65, 69, 203
freezing
    characteristic curve, 196, 198 (*see also* pore,
        freezing)
    cryo-suction, 62
    point depression, 60, 62
    range, 62, 69, 71, 74
    rate, 69, 71
Freundlich law, 35
Friedel's salt, 33, 34, 194
frost
    action, 56, 57, 63, 64, 67, 69–72, 76
    damage, 57, 61, 64, 65, 67, 69–72, 74, 76, 77
    intensity, 246, 248, 249
FT *see* freeze–thaw
fusion entropy, 60, 74

GGBS *see* granulated blast furnace slag
Gibbs–Thomson equation, 60
glue–spall mechanism, 64
granulated blast furnace slag, 6, 8, 27
gravimetry method, 198, 200, 201
gravity quay walls, 163

Great Sphinx, 97
greenhouse emission, 11
groundwater, 29, 31, 57, 79, 96, 97, 105, 106, 224, 225,
        237, 241, 248, 249, 251, 253–255
    dissolved salts, 96 (*see also* salt)
    pH value, 224

halite, 97
HCP *see* cement paste, hardened
heat capacity, 132, 136, 142
    concrete, 141
heat flow, 132–134
heat transfer, 132–134
Heaviside function, 87
HIC *see* high-integrity container
high-driven piles, 177
high-integrity container, 210, 222–232
    concrete, 225, 227, 228, 230
    life cycle, 223
    wall thickness, 227, 230, 232
high-pile wharf, 30, 163
high-strength steel wires, 159
Hong Kong–Zhuhai–Macau project, 210–214, 216,
        218–221, 232, 233
humidifying condition, 151, 153
humidity, 96, 102, 104, 107
    annual average, 211
    environmental, 9–11, 13, 15, 27 (*see also* relative
        humidity)
    gradient, 149, 151
    optimal, 245
hydration
    degree of cement, 87, 88
    extent, 9, 10, 15
hydraulic pressure, 63, 69
hydraulic structure, 57, 79, 92, 94
hydrophobic treatment, 153
    theory, 57, 63, 69, 75
HZM *see* Hong Kong-Zhuhai-Macau project

ice crystal, 60–63, 67
    heterogeneous nucleation, 62
IDD *see* interaction direct derivative method; model,
        interaction direct derivative
ideal gas, 136
    constant, 99, 136, 148
    law, 136
immersed condition, 48
immersion test, 41
indicative protection duration, 175
influential depth, 149–151
    moisture, 153
    thermal transfer, 135
inland salt lakes, 97
inspection, 160, 165, 233

integrated life-cycle design, 171
interaction direct derivative method, 128
interface
  bond loss of concrete–steel, 25
  concrete–steel, 4, 22, 25, 49, 184, 186, 187, 196
  crystal–liquid, 99
  ice–liquid, 61
  ice–water, 59
  liquid–gas, 100, 101, 136, 144, 147, 148
  water–concrete, 84
interface energy, 186
  ice–water, 60, 74
interface transition zone, 64, 69, 187
ITZ *see* interface transition zone

Jennite, 83
JSCE guideline, 237, 239

Katz–Thompson relationship, 197, 198
Kelvin equation, 100
Klinkenberg coefficient, 189
KS index, 204

laboratory accelerated test(s), 57, 69, 71, 76, 219, 226
Langmuir law, 35
Laplace equation, 100
LCC *see* life cycle, costing
LCCA *see* life cycle, cost analysis
leaching, 183, 184, 196, 202, 204, 207, 210,
    224–230, 241
  depth, 78, 83, 85, 87, 90–92, 95, 228
  extent, 84
  rate, 83–85
legal constraints, 162
life cycle
  cost analysis, 157, 171, 173, 178
  costing, 171
  engineering, 157, 163, 171–173, 210
  management, 171
limit value analysis, 227, 232
liquid film, 99, 104
  thickness, 109
liquid flow
  cracks, 123
  nonsteady, 123
liquid layer, 60
liquid-tightness, 118
liquid viscosity, 121, 142
liquid water, 57–59, 61, 63, 65, 74, 75, 77
  contact, 144, 146, 149, 151, 153
  density, 136
  mass flux, 138
  supercooling, 62
  viscous flow, 62–63, 69
low-cycle fatigue, 64
low-pressure gradient, 123

maintenance
  action(s), 162, 165, 174, 178, 233
  design, 157, 173, 174, 233
  planning, 164, 165, 173–178, 233
  techniques, 174, 175, 178
maintenance level
  mandatory, 174, 175
  necessary, 174, 175, 182
  preventive, 174
maintenance scheme, 165, 173–175, 177, 179, 180, 183
  coating-synchronized, 179, 182
  synchronized, 179
  unsynchronized, 175, 179
major repair, 243
mechanical loading, 115–118, 128, 130
mechanical resistance, 222, 223, 225
mercury intrusion porosimetry, 67, 102, 198
micro-lens theory, 58
mineral admixture, 15
  content, 245
MIP *see* mercury intrusion porosimetry
mirabilite, 99, 102, 105, 106
model(s)
  critical saturation, 71 (*see also* pore saturation,
      critical)
  durability, 157, 168
  empirical, 3, 13, 15, 29, 42, 43, 168
  interaction direct derivative, 128 (*see also* interaction
      direct derivative method)
  macro–macro, 206
  mechanism-based, 3, 13, 29, 35, 42, 43, 51
  micro–macro, 205
model-based methods, 168
moisture
  characteristic curve, 137, 151, 191, 196, 198
      (*see also* characteristic curve)
  condition, 227
  diffusivity, 138–140, 142, 143, 151, 152, 191,
      196, 200
  equilibrium, 136, 146
  isotherm, 136–138, 140, 145, 150
  molar fraction, 99, 100, 106
  molar volume of crystal, 99
  monitoring, 165, 233
  Monte Carlo simulations, 178
  moving boundary problem, 86, 148
  multi-barrier protection, 256
  multi-field problems, 141

NAC *see* carbonation, natural
natural precipitation, 97, 106
negligible process, 225
Newmann condition, 134
nitrogen, 189, 200
NPP *see* nuclear power plant
nuclear power plant, 163, 220, 223

osmotoic theory, 58

parameter-based requirements, 232
partial factors, 169, 170, 216–218, 232
partial immersion test, 98, 105, 110
PC *see* concrete, prestressed
performance
    criterion, 227
    deterioration, 157–159, 162, 168, 174, 178
    model, 206 (*see also* model)
    structural, 157, 159, 161, 170, 171
performance-based indicators, 240
permeability, 104, 106, 111
    apparent water, 200
    concrete, 33–35
    equivalent, 128
    gas, 118, 188, 189, 199–201, 207–209, 222
    intrinsic, 33, 34, 138, 152, 187, 189, 197, 200
    liquid, 83
    oxygen index, 209
    relative, 35, 139
    relative liquid, 138
    relative water, 187
    saturated, 33, 187
    water, 187, 188, 196, 199–201, 207–209
permeation, 36, 38
phase change, 99 *see also* characteristic curve,
        phase change
phenolphthalein solution, 8, 9, 12, 15, 19, 20, 22
Poiseuille's law, 120
pore(s)
    capillary, 185–187, 198 (*see also* capillary, porosity)
    confined, 96, 106
    connectivity, 186
    dissolved salts, 97 (*see also* salt)
    freezing, 57, 59, 62, 65, 69, 72, 74, 75
    gel, 185
    ice content, 186
pore crystallization, 31, 63, 72, 96, 97, 100–102,
        104, 106, 107, 109, 111
    onset, 106
    theory, 58
pore pressure, 116
    capillary, 191
    mean, 63, 74, 101, 104, 108, 109
pore saturation, 11, 14, 17, 20, 21, 24, 33–36, 42–44, 46, 60,
        63–65, 71, 72, 76, 138, 140–142, 144, 147, 148, 153,
        186, 187, 189–191, 194, 200, 201, 203, 219, 227, 228
    critical, 65, 68, 71, 72, 76
    critical, theory, 58 (*see also* model, critical
        saturation)
pore size
    attainable, 102
    critical, 198
    distribution, 66, 67, 72, 74, 102, 104, 107, 109, 186, 207
    maximum, 102, 104

pore solution, 3, 5, 7–9, 11, 18, 32–37, 41–43, 47–53,
        78–85, 87–92, 94, 96, 97, 101, 106, 111
    alkaline, in concrete, 3
    alkaline ions, 49
    alkalinity, 3, 79
    oxygen content, 48
    pH value, 3, 8, 9, 18, 32, 36, 49, 88, 184
pore structure, 3, 9, 64, 66, 67, 71, 72, 74, 100, 102,
        136, 138, 145, 184, 186, 187, 189–191, 194–199
poroelastic medium, 116
poromechanics, 62, 63, 75
    theory, 58
porosity, 4, 8–10, 13–15, 33, 43, 44, 60, 64–66, 69, 74,
        77, 116, 152, 185–187, 189, 191, 196, 199, 207–209
    concrete, 10, 15, 50, 71, 72, 74, 185, 193, 198
    open, 118
    total, 102, 104, 109
porous building materials, 96
port, 163
portlandite
    carbonation, 8, 16
    content, 6, 7, 9, 10, 15, 28, 85, 87, 90, 94, 95, 184,
        187, 196, 198, 199, 207, 208
    content in concrete, 6, 15, 28
    dissolution, 85, 184
    dissolution depth, 94, 95
    dissolution rate, 127
    leaching depth, 87, 90–92, 95
    solubility, 83
post-tensioned system, 255, 256
potential
    electrode, 47, 48
    half-cell, 198, 199
    pitting corrosion, 47
Pourbaix diagram, 18, 20, 46, 184
pressure gradient, 121, 147
prestressed tendons, 243–256
    corrosion, 159
    external, 256
    internal, 256
    protection, 160
prestressing steel, 235
probability of driving rain, 18
probable process, 225
properties
    chemical, 183, 184
    durability, 183, 198 (*see also* durability)
    durability-related, 116
    electrochemical, 183, 184
    linear physical, 128
    material, 158, 168–170, 183, 196, 205, 208
    mechanical, 183, 194, 195, 200
    microstructure and related, 184, 185
    physical, 183, 195
    statistical, 168, 170, 178, 216, 217, 232
PSD *see* pore size, distribution

qualitative analysis, 168
quality control, 210, 218–220, 232, 233, 235
quantitative analysis, 168

radioactive release, 228–230, 232
  normalize, 228
radioactive waste (radwaste), 220, 222, 223
  containers, 163
  disposal, 79, 210
radioactivity release, 223, 226, 228 *see also*
  radionuclide, release
radionuclide(s), 222, 224, 226, 227, 230, 231
  apparent diffusivity, 227
  concentration, 228, 231
  decay constant, 227
  diffusivity, 227–230, 232
  effective diffusivity, 227
  half-life of, decay, 222
  release, 222, 227, 231
rapid chloride migration, 41, 200
rapid chloride migration test, 219, 220, 232, 249
raw material(s), 234, 244
  concrete, 240
  constituents, 243, 244
RC *see* concrete, reinforced
RC element(s), 4, 5, 22, 23, 25, 27, 29–32, 38, 42, 50,
  53–57, 70, 74, 76, 77, 79, 82, 83, 94, 177, 178,
  187, 213–215, 221, 235, 245, 249
  indoor, 31
  loading capacity, 27, 53
  rigidity, 53
  stiffness, 25
  thin-wall, 76
RCM *see* rapid chloride migration; rapid chloride
  migration test
reaction rate, 13
reinforcement steel
  conventional, 235
  corrosion resistance, 184
  electrochemical stability, 184
relative elastic modulus, 239
relative humidity, 6, 9–12, 14–21, 24–26, 36, 98, 100,
  136–138, 140, 148
  influence factor, 24
reliability
  index, 214, 243
  indices, 164
  level, 159, 164, 170
  target, index, 216, 217, 219
  theory, 170
repeatability, 183, 200, 203
representative elementary volume, 100
reproducibility, 183, 201, 203
residential foundations, 97
residual channeling, 123

return period, 162, 163
reuse, 171
REV *see* representative elementary volume
roughness
  fracture, 64 (*see also* fracture)
  fracture surface, 123
  surface, 121, 123, 126 (*see also* surface)

safety
  assessment, 79
  margin, 159, 160, 164, 169–171, 174, 213, 214,
    216, 232
  margin for durability, 241
salinity, 57, 70, 78, 79
salt(s)
  airborne, 96, 97, 106, 237, 242
  attack, 213, 214
  concentration, 99, 100, 104, 106, 107
  crystallization, 96–99, 102, 104, 111, 144, 183, 195,
    196, 207, 241, 253, 255 (*see also* crystallization)
  crystallization damage, 158
  deicing, 29–31, 38, 39, 45, 57, 70
  deposition, 102
  deposition rate, 106
  soil-borne, 96
  solubility, 104
  weathering, 96
saturation degree, 136 *see also* pore saturation
saturation depth, 74
scaling
  mass, 203
  salt, 57, 64, 65, 71 (*see also* salt)
  surface, 56, 57, 60, 64, 69, 76, 96, 97, 111
SCM *see* secondary cementitious materials
seawater, 29–32, 45, 57, 70, 236–238, 242
  ionic species, 31
  penetration, 144
  pH value, 211
  salinity, 211
secondary cementitious materials, 6–8, 15, 21, 28, 95
sensitivity analysis, 228, 230
serviceability, 158, 159
serviceability limit state, 25, 27, 158–160, 164, 170,
  214, 242
service condition(s), 158, 159, 234, 240, 255
service life, 157–167, 169, 170, 172–175, 177, 237,
  240, 241, 243, 245, 247, 249, 250, 252, 255
  design, 161–163
  intended, 235
service phase, 164, 165, 172
shortwave solar radiation, 134
shrinkage, 115, 118, 128
  autogeneous, 115
  drying, 115 (*see also* drying)
silane impregnation

SLS *see* serviceability limit state
smectite, 80
SO₃ *see* sulfur trioxide content
social constraints, 162
social discount rate, 172
soft water, 78, 79, 81, 83, 85, 91, 92
soils, 235, 238, 241, 246, 253–255
  saline, 31, 97, 106, 111
solid matrix
  average stress, 64, 66, 72, 74
  concrete, 32, 34, 81, 94, 106, 110
  stress, 106, 108, 109
soluble alkali content, 244
sorption distribution coefficient, 228
sorption isotherm
  chloride, 34, 38, 45, 194
  ion, 194
  linear, 194
  water vapor, 198, 200
sorptivity, 104, 106, 109
  chloride, 194, 199, 200
  relative water, 36
  water, 4, 34, 35, 65, 69, 118, 146, 191–193,
    196, 199–201, 207, 209
spalling, 60, 64, 74, 76
  concrete cover, 25, 27 (*see also* concrete cover)
splash zone, 38–40, 45
stainless steel, 49, 55
steel bars
  corrosion, 3, 5 (*see also* corrosion)
  corrosion depth, 26
  corrosion protection, 235
  diameter, 25
  electrochemical stability, 3
  reduced ductility, 27
Stefan–Boltzmann constant, 135, 136
stoichiometric number, 81, 83
strain
  residual, 118
  volumetric, 63
  volumetric residual, 118
strength
  compressive, 8, 11, 12, 18, 185, 191, 193–195,
    199, 204, 207, 208, 235, 244
  tensile, 201, 208
structural materials, 158, 159
subflorescence, 102
sulfate
  attack, 31, 253, 255
  concentration, 224, 255
  crystallization damage, 204
  magnesium, 97
  reaction, 213
  resistance, 226
  resistance test, 204

salts, 97
  sodium, 102, 103
sulfate reaction
  chemical, 255
  internal, 244
sulfur trioxide content, 244
supersaturation, 99, 101, 102, 104, 107, 109
  critical, 106, 111
  degree, 99, 100, 102, 107
  salt, 186
  solution, 99
surface
  energy, 99
  erosion and wear, 242
  fractal dimension, 123, 126
  tension, 142, 144
  topography, 123
  treatment, 111

technical working life, 162
temperature, 4–7, 9, 11, 12, 21, 24, 26, 27, 36–38, 43,
  46, 48–50, 53, 96, 99, 102, 104, 107, 109
  ambient, 97, 104
  annual average, 211
  atmospheric, 11
  fluctuation, 96
  influence factor, 24
tetracalcium aluminoferrites, 34–36
thenardite, 99, 102, 106
thermal-activation, 37, 46
thermal conduction, 132, 135
thermal conductivity, 132, 133, 135, 136, 142
  concrete, 141
thermal convection, 132
thermal expansion coefficient, 63
thermal–hydraulic effect, 104
thermal radiation, 132
thermo-hydro-couplings (TH couplings), 141, 142
thermo-hydro-field analysis, 144
three-point bending, 201
time of wetness, 18
tortuosity, 33
  factor of capillary porosity, 86
  function, 139, 150
  pore structure, 194 (*see also* pore structure)
transfer parameter, 45, 46
transport
  altered, properties, 118
  modes, 145
  moisture, 138, 141–144, 147, 149–153 (*see also*
    moisture)
  properties, 102, 104, 116–118, 130, 153, 184–187,
    194, 195, 205 (*see also* properties)
  radionuclide, 210, 227 (*see also* radionuclide)
  rate of salts, 104

tricalcium aluminates, 34–36
tricalcium silicates, 9, 13, 15
tunnel, 163
    immersed tube, 144, 210, 212
    Môquet, 5
    Traffic, 5

ULS *see* ultimate limit state
ultimate limit state, 158–160, 164, 242
ultrasonic pulse velocity, 118
UPV *see* ultrasonic pulse velocity
uncertainty
    artificial, 168, 170, 174
    inherent, 168, 174
urban viaduct, 31

vapor
    condensation of water, 146
    density, 142
    diffusivity, 142, 148 (*see also* diffusivity)
    diffusivity in air, 138, 150
    mass flux of water, 138
    molar mass, 136
    saturated pressure, 136, 148, 150

viscosity, 33, 38
    water, 142 (*see also* water)

waste package, 222
water
    deionized, 84
    head, 33, 187
    ingress, 222
    saturation profile, 153
    static, 84
water retention curve, 136
    hysteresis, 146
water to binder ratio, 185, 187, 197, 244
    maximum, 245, 247t, 250t, 252t, 255t
water to cement ratio, 9, 10, 15, 16, 19, 21, 87,
        88, 244
    maximum, 236t, 239
w/b *see* water to binder ratio
w/c *see* water to cement ratio
weather coefficient, 18
weathering actions, 96
wetting, 146, 147, 149, 151, 153
    condition, 151, 152
    time, 72